THE BOOK OF
Naturalists

The Book of Naturalists

An Anthology
of the Best Natural History

EDITED BY

WILLIAM BEEBE

PRINCETON UNIVERSITY PRESS
PRINCETON, NEW JERSEY

Published by Princeton University Press, 41 William Street,
Princeton, New Jersey 08540
In the United Kingdom: Princeton University Press, Guildford, Surrey

First Princeton Paperback printing, 1988

Reprinted by arrangement with Alfred A. Knopf, Inc.

Library of Congress Cataloging-in-Publication Data

The Book of naturalists.

Reprint. Originally published: New York: Knopf, 1944.
Bibliography: p.
1. Natural history.   I. Beebe, William, 1877–1962.
QH81.B715   1988   508   87–32727
ISBN 0–691–08466–1 (alk. paper)
ISBN 0–691–02408–1 (pbk.)

Printed in the United States of America by Princeton University Press,
Princeton, New Jersey

CONTENTS

Part I

Part II

v

CONTENTS

Contents

PREFACE

THE DEVELOPMENT and growth of natural history are reflected in the writings of many naturalists, from Aristotle to the present day, and the inspiration for these writings comes from interest and love of living animals and plants observed under natural conditions. Other fields of science, such as classification, anatomy, and economic biology, all have their distinguished exponents, but these are beyond the scope of this book.

The work falls rather naturally into two parts. The first deals with the very beginnings of natural history, its roots, its foundations as discerned in the early gropings of those primitive men who had distanced their ape-like predecessors, stood firmly erect, looked around, and been able to say, or at least think: "I am I." From Aristotle, the greatest naturalist of all early times, we pass in review the painfully slow development of natural history through the Middle Ages and beyond the Renaissance.

Part Two begins with Darwin, the second landmark of greatness, and extends from the date of the *Origin of Species* to the present day. Here we have the flowering, the fruition surmounting the roots and the trunk of the Tree of Knowledge. Among this exciting company of naturalists, some, with the greatest reputations and achievements, have provided the least quotable words and phrases; others, whose names will be almost or quite unknown to many readers, are here because, as Kipling puts it, "The magic of Literature lies in the words, and not in any man."

The cave paintings of the Dordogne provide us with the inception of natural history, and the middle of the fourteenth century initiates the beginning of inspired ability to write about it. Steadily, up to the present moment, there has been an ever continuous progression in increase of knowledge, an evolution worthy of comparison with that other evolution of animal life on our planet — the foundation of all our labors and our love of science. All this has taken place in spite of numberless plagues, famines, revolutions, and wars throughout human history. No corresponding menace or peril, however global, now or in the future, can ever put an end to this advance of the mind and the spirit of man.

I have asked and received help and valuable suggestions from a host of friends, but Miss Jocelyn Crane of my scientific staff has been of such pre-eminent assistance that her name should rather be on the title page as co-author than merely as recipient of my sincere thanks and appreciation.

The rereading of several hundred volumes as prelude to selection or rejection has resulted in a varied stirring up of emotions. Some old favorites proved to be impossible of acceptance. Others were subjectively absorbing, full of vital interest, but because of technical or beclouded diction quite inappropriate when judged for selective quotability. Still other authors, almost forgotten and of doubtful remembered value, proved more than adequate in scientific soundness and literary quality. Original spelling and syntax have of course been preserved.

These labors of an anthologist are and should be full of pleasure, but his reward is invariably lost in sarcastic if not acrimonious criticism both of omission and of commission. This, too, is as it should be, since the personal equation is so great. More than once I have been strongly tempted to base a decision on the toss of a coin, and I should have been wise to include several blank pages in each volume with an invitation to every reader to damn my selection and satisfy his personal convictions with his own choice.

My publishers allowed me one fifth of a million words as the maximum, so the size of the type should not afford discomfort to the reading eye. Before I knew it, my list of *musts* contained 248,000 words and I was not through. The subsequent cutting and slicing here and there was sheer mental agony. The list of authors as a whole could easily have been doubled and the book kept within limits, but this would mean unfairly short contributions instead of a more thorough presentation of subject, permitting a better appraisal of personality.

To make more real the acute difficulty, the cruel necessity, of choosing for an Anthology such as this I append a list of authors who were considered, any or all of whom might with propriety have been included. Of course this list in turn is far from complete: Albertus Magnus, Alcock, Aldrovandus, Allen, Banks, Bates, Belon, Boulenger, Buffon, Bullen, Bragg, Coues, Cuvier, Dewar, Eddington, Forel, Gibson, Gosse, Goethe, Gray, Haeckel, Hingston, Haldane, Jeans, Jefferies, Johnstone, Jordan, Kellogg, Lamarck, Lankester, Lewis and Clark, Lowe, Lubbock, Lydekker, Magellan, Marco Polo, Moseley, Moufet, Muller, Murphy, Peckham, Phillips, Platt, Poulton, Pycraft, Raleigh, Ray, Renshaw, Riley, Rondelet, Schmidt, Schomburgk, Sharp, Swammerdam, Teale, Thayer, D'Arcy Thompson, Vaughan Thompson, Wyville Thomson, Townsend, Turner, von Baer, Walton, Wells, Wilson, Wood.

The problem of sequence is easily solved. The reason for chronological and ordinal sequence depends not on the date of birth of the various naturalists, but on the date of publication of the quoted material. This represents, in the majority of cases, a high point in their creative output.

With so many equally excellent aspirants for inclusion, making choice perplexing, I gave weight to several considerations, such as wide distribution throughout time, nationality, geography, and subject matter. The result of this attempt may be judged from the following lists:

AUTHORS' PUBLICATION DATES BY CENTURIES

B.C. *Fourth:* Aristotle.

A.D. *First:* Pliny.

Eleventh: Theobaldus.

Thirteenth: Frederick II.

Sixteenth: Gesner.

Seventeenth: Leeuwenhoek.

Eighteenth: Réaumur, Linnæus, White, Bartram.

Nineteenth: Humboldt, Waterton, Audubon, Thoreau, Darwin, Wallace, Agassiz, Belt, T. H. Huxley, Hudson, Muir.

Twentieth: Maeterlinck, Fabre, Roosevelt, Thomson, Wheeler, Levick, Burroughs, Farrer, Stefansson, Akeley, Osborn, Digby, Seton, Roule, Eckstein, Heard, Ionides, J. Huxley, Chapman, Haskins, Peattie, Armstrong, Klingel, Carson.

NATIONALITIES OF AUTHORS

American: Bartram, Thoreau, Hudson, Muir, Roosevelt, Wheeler, Burroughs, Akeley, Osborn, Seton, Eckstein, Heard, Chapman, Haskins, Peattie, Klingel, Carson.

British: White, Waterton, Darwin, Wallace, T. H. Huxley, Belt, Thomson, Levick, Farrer, Digby, J. Huxley, Ionides, Armstrong.

French: Réaumur, Audubon, Maeterlinck, Fabre, Roule.

Swiss: Gesner, Agassiz.

Italian: Theobaldus, Frederick II.

Dutch: Leeuwenhoek.

German: Humboldt.

Greek: Aristotle.

Icelandic-Canadian: Stefansson.

Roman: Pliny.

Swedish: Linnæus.

SUBJECT MATTER

Plants: Himalayan flowers (Farrer) ; first bacteria (Peattie) ; seed behavior (Peattie) .

Protozoa: Leeuwenhoek.

Mollusks: oysters (Roule) ; pearls (Pliny) ; nautilus (Aristotle) ; octopuses (Klingel) .

Arthropods: scorpion (Fabre) ; winged ants (Réaumur) ; ants' social ties (Haskins) ; army ants (Belt) ; bee swarms (Maeterlinck) ; termites (Wheeler) .

Fish: eels (Aristotle) ; eel migration (Carson) ; torpedo and fishing frog (Aristotle) ; fierasfer (Roule) .

Amphibians: tadpoles (Thomson) .

Reptiles: tortoise (White) ; rattlesnake (Bartram) .

Birds: penguins (Levick) ; guillemots (Armstrong) ; turkeys (Audubon) ; cuckoos (Aristotle) ; arctic owl (Stefansson) ; water ouzel (Muir) ; various species in England, the United States, and Patagonia (White, Burroughs, and Hudson, respectively) .

Mammals: sloth (Waterton) ; monkeys (Wallace; Chapman) ; mice (Gesner) ; rats (Eckstein) ; bats (White) ; arctic foxes and polar bears (Stefansson) ; coatimundis (Chapman) ; sea otters (Seton) ; whale (Theobaldus) ; elephants (Pliny, Osborn, Akeley) ; mammoths (Digby) ; antelopes and zebras (Roosevelt) ; rhinoceros (Gesner) .

Evolution of Man and His Brain: early man (Heard) ; comparison between animals and men (J. Huxley) ; growth of ideas about the universe (Ionides) .

General Problems: evolution of species (Darwin) ; protective coloration and mimicry (Wallace) ; embryology and evolution (Agassiz) ; paleontology (T. H. Huxley) ; differences between the sexes (Aristotle) ; sleep (Aristotle) ; bird migration (Frederick II) ; aims, adventures, and tribulations of naturalists (Agassiz, Fabre, Hudson, Roosevelt, Akeley, Farrer) ; early classification (Linnæus) .

GEOGRAPHY OF SUBJECTS (WHERE DEFINITE)

United States: Eastern (Bartram, Audubon, Thoreau, Burroughs); Western (Muir) .

American Tropics: Panama (Chapman) ; Guiana (Waterton) ; Amazon (Humboldt) ; Brazil (Darwin) ; Galapagos (Darwin) .

Southern South America: Hudson.

Central Africa: Akeley, Roosevelt.

Europe: North Ireland (Armstrong) England (White; T. H. Huxley) ; France (Fabre, Réaumur) .

Asia: Himalayas (Farrer) .

Arctic Regions: general northland (Seton) ; far north (Stefansson) ; Siberia (Digby) .

Antarctic Regions: Levick.

Underwater: eel travels (Carson) ; oysters and their guests (Roule); diving and octopuses (Klingel) .

ACKNOWLEDGMENTS

THE EDITOR gratefully acknowledges the kindness of authors and publishers in giving permission to reproduce copyright material in *The Book of Naturalists* as follows:

Mrs. Carl Akeley: for selections from "Elephant Friends and Foes" from *In Brightest Africa* by Carl Akeley. Reprinted by permission.

D. Appleton-Century Company, Inc.: for the selection from "The Water-Ouzel" from *The Mountains of California* by John Muir; for "The Big Almendro" from *Life in an Air Castle* by Frank M. Chapman, copyright 1938. Reprinted by permission of D. Appleton-Century Company, Inc.

Edward Arnold & Co.: for the selections from *On the Eaves of the World* by John Farrer. Reprinted by permission of Edward Arnold & Co.

The Bobbs-Merrill Company: for the selections from *Stars and Men* by Stephen A. and Margaret L. Ionides, copyright 1933. Reprinted by permission of the Bobbs-Merrill Company.

Dodd, Mead & Company, Inc.: for the selection from *The Mason Bees,* by Jean Henri Fabre, translated by Alexander de Mattos, copyright 1914; for the selection from *Life of the Scorpion* by Jean Henri Fabre, translated by Alexander de Mattos, copyright 1923; for the selection from *Inagua* by Gilbert Klingel, copyright 1940; for the selection from *The Life of the Bee* by Maurice Maeterlinck, translated by Alfred Sutro, copyright 1901. Reprinted by permission of Dodd, Mead & Company.

E. P. Dutton & Co., Inc.: for "Seen and Lost" from *The Naturalist in La Plata;* for "The Plains of Patagonia" from *Idle Days in Patagonia* by W. H. Hudson. Reprinted by permission of E. P. Dutton & Co., Inc.

Harper & Brothers: for "The Uniqueness of Man" from *Man Stands Alone* by Julian Huxley, copyright 1939 by Julian Huxley; for "Two Lives" from *Lives* by Gustav Eckstein, copyright 1932 by Harper & Brothers. Reprinted by permission of Harper & Brothers.

Harcourt, Brace & Company, Inc.: for "The Emergence of the Half Man" from *The Emergence of Man* by Gerald Heard, copyright 1932. Reprinted by permission of Harcourt, Brace & Company, Inc.

Acknowledgments

Houghton Mifflin Company: for "Old Friends in New Places" from *Under the Apple Trees* by John Burroughs; for selections from *Walden* by Henry David Thoreau. Reprinted by permission of Houghton Mifflin Company.

The Macmillan Company: for selections from "Exploring the New Land" and "Marooned on the Ice" from *The Friendly Arctic* by Vilhjalmur Stefansson; for selections from *Diary of the Voyage of H.M.S. Beagle* by Charles Darwin, edited from the MS by Nora Barlow, 1933. Reprinted by permission of The Macmillan Company.

Robert McBride & Company: for selections from *Antarctic Penguins* by G. Murray Levick. Reprinted by permission of Robert McBride & Company.

Andrew Melrose, Ltd.: for "Spring" from *The Biology of the Seasons* by J. Arthur Thomson. Reprinted by permission of Andrew Melrose, Ltd.

W. W. Norton & Company, Inc.: for "The Tomb of Pearl" from *Fishes and Their Ways of Life* by Louis Roule, translated by Conrad Elphinstone. Reprinted by permission of W. W. Norton & Company, Inc.

Oxford University Press: for "Playboys of the Western World" from *Birds of the Grey Wind* by Edward A. Armstrong. Reprinted by permission of Oxford University Press.

Prentice-Hall, Inc.: for "The Ties that Bind" from *Of Ants and Men* by Caryl P. Haskins. Reprinted by permission of Prentice-Hall, Inc.

G. P. Putnam's Sons: for selections from *Flowering Earth* by Donald Culross Peattie, copyright 1939. Reprinted by permission of G. P. Putnam's Sons.

Ernest Thompson Seton: for selection from "The Sea Otter" from *Lives of Game Animals*. Reprinted by permission.

Charles Scribner's Sons: for selection from *A Book-Lover's Holidays in the Open* by Theodore Roosevelt; for selections from *African Game Trails* by Theodore Roosevelt. Reprinted by permission of Charles Scribner's Sons.

Simon & Schuster, Inc.: for "Journey to Sea" and "Return" from *Under the Sea-Wind* by R. E. L. Carson. Reprinted by permission of Simon & Schuster, Inc.

THE BOOK OF
Naturalists

To Laurance *and* Mary Rockefeller

PART I

INTRODUCTION

IN THE very heart of Borneo I waited for my interpreter to return from a three-mile paddle down stream in search of a missing dugout. From up the Rejang came two great war canoes filled with Dyaks, and soon several chiefs approached and began a friendly but utterly incomprehensible duet at me. Here we were, two types of human beings quite unable to communicate. I tore a page out of my journal and did my best to draw an argus pheasant. Before I had finished, heavy odorous breaths poured down my neck, and there arose a chorus of excited outcries. Vigorous noddings of feather-crowned heads indicated a unanimous burst of recognition: *"Ruoi burong! Ruoi burong!"*

I then delineated in uncertain outline three white eggs. Murmurs of negation and disagreement arose and grimy fingers sought to do something to my tremulous ovals. I crossed out one and the villagers' chorus arose affirmatively again. I erased another egg and the entire company expressed disapproval. All knew the argus pheasant and that it laid two white eggs.

This almost forgotten incident stands for me as typical of the first manifestation of communicated natural history on the earth. In the days of our cavemen ancestors, tens of thousands of years ago, in pre-Tegumai times, there were men typified by that "enterprising wight,"

> Unusually clever he,
> Unusually brave,
> And he drew delightful mammoths
> On the borders of his cave.

The first discovery of this prehistoric natural history was in itself interesting. A Spanish nobleman, searching a cave near his home in Santander, was bringing to light bones of extinct animals — horses, mammoths, and reindeer — when his attention was distracted by his little daughter who was sitting near. She interrupted his labors by her cries of *"Toros! Toros!"* and pointed to the roof of the cavern. There, painted in bright colors, was a whole herd, life-size, of deer, horses, and the ancient extinct bison of Europe. This was the first of many similar paintings found scattered through dozens of caves in southern France and Spain. Some of the animals were anatomically correct and artistically superb, and all were wholly unexpected products of these early, primitive peoples. This art perhaps antedated

3

any real language; certainly no written letter or word had been formed. It shows that our forebears had already passed beyond the stage when food and fear were the only bonds between them and the wild animal life which surrounded them. These admirable paintings implied the knowledge of a hunter, together with a certain leisure of life won by the beginning of human dominance of the earth. It was the inception of an artistic record of the history of nature — of natural history.

The only relics left by the earliest races of manlike beings are stone implements, and from age to age we find these increasing in complexity, adaptive symmetry, and cunning shape. Yet long before the end of the Stone Age, artists gave us these incredibly skillful drawings; mammoths, reindeer, bison, bears, horses, deer, wild boars and rhinoceroses, some grazing singly, others in combat or running at full speed. Sheltered within dry caverns, these pictographs have persisted unharmed, unchanged, their pigments still bright, although applied in days long before the last ice age, in an antiquity which makes the work of Egyptian artists seem ultramodern.

Think of the law of compensation of being a Stone Age youth, as compared with one of today: to him everything in nature fresh, a continual surprise, nothing named, the life of no creature printed in any book. Yet this point of view becomes real and reincarnated in any modern boy who is born a predestined naturalist. When he first sees, without knowing beforehand or why, the caterpillar changing into a chrysalis and this to a butterfly, or the tadpole-fish turning before his eyes into a land animal, nothing in later life can quite equal it.

To the primitive youngster there was added the exciting uncertainty of any possible rising of next day's sun, and the terror at the vanishing of leaves, insects, and birds at the beginning of what we call autumn and the onset of winter. Perhaps the boy had even the greater fear of a still more terrible end of all things when he was taken on a long trek to the south to avoid the onrush of glaciers — the whole world turning to ice.

Think of the thousands of lonely human beings in succeeding millenniums who must have been stirred by the beauty of wild creatures, tortured by the awakening desire in their dull minds to understand it all and then to express their thoughts in the slow evolution of picture writing and finally of words. With nothing but sticks and chipped flints for weapons, implements, and tools; with a bit of skin for clothing and a cave for a home; with dogs and horses and cows still wild and untamed; with wheat, corn, and fruit only scattered

4

weeds or forest trees; with wolves and other carnivores ever hungry, ever present dangers: to them the only classification of animal life was edible or inedible, dangerous or harmless. What honor we must pay to the patient founders of language and art, to the first, forever unknown naturalists, who did not know it themselves, but whose accumulated efforts slowly lighted the early history of what we now call China, Egypt, and Greece!

We have no record of the tribe or race or the time of the very first talk or writing. The earliest ape men such as the Heidelberg Man were without them, but little by little, sounds were made to stick to this and that; a picture was stripped of details and became a symbol. In that unscientific and wholly delightful tale of Kipling's, "How the Alphabet was Made," there may be more of truth than we shall ever know.

A short, slender man of forty years of age, showing already a hint of baldness, with fine features and head, and eyes bright with intelligence and interest, has fastened his tunic high up about his middle. He is wading on spindly legs in the beautiful, almost landlocked lagoon of Pyrrha on the island of Lesbos. Now and then he scrambles ashore and deposits piles of shells and other water creatures near his wife and little daughter. A group of pupils crowds about the small aquatic things, sorts them, and later listens absorbed in the words of their teacher.

If from this present year of 1944 we go back and back and back to the year 1600, when Shakespeare in his turn was about forty years old, we should find him, 344 years ago, at work on *A Midsummer Night's Dream*. If Aristotle, poking about in his Pyrrhic pool, could look forward 344 years, he would envisage the birth of Jesus Christ.

But now, all unconscious of the value of what he is doing and thinking, of the influence it will exert for two full millenniums, this middle-aged, vital Greek labors in the evening by the flickering light of a bronze lamp, perhaps one of the graceful little Grecian ones which have recently come into use. Patiently he scratches with his stylus upon a long roll of papyrus. How wonderful if we could have a picture of Aristotle at work, or fragments of the papyrus itself!

We wonder on what day and under what conditions he wrote:

"Out of the Lagoon of Pyrrha all the fishes swim in wintertime, except the sea-gudgeon; they swim out owing to the cold."

"Of shellfish, and fish that are finless, the scallop moves with greatest force and to the greatest distance, impelled along by some internal energy."

5

"Of molluscs the sepia is the most cunning, and is the only species that employs its dark liquid for the sake of concealment as well as from fear. The octopus and calamary make the discharge solely from fear." And so on, as we shall see later in this volume.

Thus he records facts and ideas, facts and theories all new, fresh, first-hand, dependent upon no book learning, for there are no nature books as yet. His loneliness, his pre-eminence as a pioneer in natural history, are evident from his own words: "I found no basis prepared; no models to copy. Mine is the first step, and therefore a small one, though worked out with much thought and hard labour. It must be looked at as a first step and judged with indulgence. You, my readers, or hearers of my lectures, if you think I have done as much as can fairly be required for an initiatory start, as compared with more advanced departments of theory, will acknowledge what I have achieved and pardon what I have left for others to accomplish."

Yet from the dim aisles of human history long before Aristotle there have come to us hints of early strivings toward expression. We know for certain that about 3500 B.C. writing was still sheer pictographs on stone. When clay began to be used, these complete drawings were often smudged and distorted, and there seemed to have evolved from these thumbed smears a sort of syllabic shorthand. The change from this to the separate characters of the Babylonians was probably very gradual. On the very same type of clay tablets, looking like fat little pillows, we find, in cuneiform, notes by pre-Grecian Assyrians concerning the positions and the movements of stars. So these first written accounts of nature were inscribed as long before the birth of Aristotle as our year 2000 will be after his death.

If the writings of Aristotle seem unvarnished by literary embellishments let us recall that many are isolated facts jotted down for use in lectures; others are notes taken by pupils as the master talked. Of his written dialogues, carefully composed, literary, lucid, only the merest fragments remain. Aristotle was fortunate in that he was appreciated in his day and country. Not only was he a successful lecturer in his own school or lyceum, but his one-time pupil Alexander the Great gave him a grant toward the production of his classic *Natural History of Animals and Plants* to the amount of 800 talents, or about $200,000. Natural history through all the ages, has owed a great deal to generous patrons, whether emperors, kings, queens, yachtsmen, or assorted gentlemen of affluence; whether by way of talents to Aristotle, precious jewels to Columbus, pounds and shillings to Captain Cook, rubles to Pallas, or, in our day, less romantic-sounding but equally effective dollars and cents.

After the death of Aristotle, and throughout the growth of the Roman Empire there was a shift from philosophy and natural history to politics and world conquest, and from the mental to the material. The last feeble flame of early interest in nature was the thirty-seven volumes of Pliny the Elder, but this was a pseudo-natural history, an uncreative compilation of mingled fact and myth. With the rise of Christendom creative observation and learning sank to an all-time low level, and mental activities during the medieval ages were confined chiefly to religious, metaphysical, and abstract disquisitions. If an argument arose as to the number of teeth possessed by a horse, there would ensue long and bitter oral and written debates and disputes, without its occurring to anyone to look into the mouth of the animal. This method of simian science, offspring of conceit, sheer laziness, and disregard of truth, is never wholly expunged from human mentality. Witness Plutarch's *Morals*, and our contemporaries who believe that the tongue of a serpent is its sting, and that a woodcock binds up its broken leg with clay and wooden splints. Believers in this latter tale should recall that eighteen centuries ago Pliny went them one better: When a hippopotamus had become over-fat from excess of eating, he sought out a sharp reed and, pressing against it, punctured his thigh. When sufficient blood-letting had ensued he covered up the wound with mud!

About sixteen hundred years after the death of Aristotle an adequate translation of his *Natural History* was made from Greek into Latin, and for two more centuries, up to about A.D. 1500, his every word was held to be absolute truth. The only notable exception to this, curiously enough, was Frederick II, the Holy Roman Emperor who caused the translation to be made. Frederick was too powerful to care about criticism, and as for the Inquisition, he helped to found it, so he could flout it. In his own admirable treatise on the *Art of Falconry* written about 1245, he included excellent chapters on the anatomy and habits of birds, and even disproved some of Aristotle's second-hand statements. The eleven chapters on migration, as we shall later see, could reasonably have been written today.

One of the first suggestions of the value of letters and knowledge came from an English Benedictine monk who, about the year 1500 wrote: "The comyn people . . . Whiche without lutterature and good informacyon Ben lyke to Brute beestes." The first "informacyon" which would concern the "comyn people" would be that of greatest usefulness; the necessary lore of the "hunter, trapper, woodsman, fisherman, herdsman, husbandman, gardener, herbalist, midwife and medicineman." Close in the wake of these when there came the least surcease from the worry about the next meal and dread of mental persecution, curiosity and wonder began to sprout and spread.

The middle of the sixteenth century or thereabouts was the zero hour for a reincarnation of natural history. From then on ignorance began to have a bad time, especially in the sense of Artemus Ward's definition; "Ignorance does not mean not knowing, but knowing so many things that ain't so." In spite of continued inquisitions and fanatical opposition the study of man and animals never ceased. With the invention of the telescope and microscope the scales began to fall from myopic, biased mental visions, and little by little freedom of observation, experimentation, and logical deductions assumed their deserved place. A desire developed even to clothe naked facts and ideas in some semblance of pleasing phraseology, together with apt and accurate simile.

A perfect spate of writing ensued — Turner, Belon, Rondelet, Aldrovandus, and Moufet. All were more or less under the influence of Aristotle, but all supplemented the study of his works with observations of their own. Some, like Aldrovandus, promptly got into trouble with the Inquisition in its last throes, but persisted in their studies. Aldrovandus, like Rondelet and Gesner, produced huge tomes concerned with plants, insects, fish, birds, and animals. The excellence of much of the type and the beauty of reproduction of illustration are all the more remarkable when we remember that the invention of printing was only one hundred years old. Thus we have the evolution of artistic representation, from the amazing paintings in the Dordogne caves, through the gold and silver illuminated vellum of patient medieval artists, to mechanical reproduction — the final phase serving the art of the naturalist.

Singer tells us that "There is a special reason why Gesner should be remembered by nature lovers. It is strange that neither in Antiquity nor in the Middle Ages nor at the Renaissance of Learning was there any real appreciation of mountain scenery. Mountains were regarded with dread or even disgust. Gesner was among the first to voice a feeling for mountains. In a letter to a friend he speaks of the wonders of mountain scenery, and declares his intention of climbing at least one each year, not only to collect plants, but also for air and exercise. He wrote a description of Mount Pilatus which is probably the earliest work on mountaineering."

Collecting soon began to go hand in hand with observing. Desultory botanical gardens and zoos came into being, and "cabinets" of natural history. I can clearly recall the polished cowries, the cards of mounted Syrian grasshopper wings, and the dusty white coral which my seafaring forebears had brought home and which reposed in a glass case in my grandmother's best parlor. Museums crystallized and large private collections and scientific societies were inaugurated. Yet natural-history writing in the modern sense was still impossible. Writers continued to face greater and greater difficulties in telling about plants and animals. Their trouble was the same as that of Adam; no creature had a real name. The more different kinds of organisms became known, the longer and more intricate became necessary descriptive labels.

An unconscious handicap lay in the apparent changelessness of all forms of life. Aristotle's birds and fishes and bees were the same in Gesner's day after two thousand years, and as probably they always had been since creation or when they walked down the gangway of the Ark.

Relief from the handicap of uncertain names came to all scientists and naturalists in the year 1758 with the publication of the tenth edition of Linnæus's *Systema Naturæ*. What the alphabet is to speech, and numerals to mathematics the universal adoption of two definite names for each species of plant and animal on the earth is to natural history. Only Plato's name of "featherless biped" could ever compete with *Homo sapiens*! The quills of writers were now freed to inscribe as they pleased the habits and courtships, the homes and the ways of life of living creatures.

For example, Thomas Moufet in his *Theatre of Insects* thus defines grasshoppers and locusts: "Some are green, some black, some blue. Some fly with one pair of wings, others with more; those that have no wings, they leap, those that cannot either fly or leap, they walk; some have longer shanks, some shorter. Some there are that sing, others are silent. And as there are many kinds of them in nature, so their names were almost infinite, which through the neglect of naturalists are grown out of use. Now all locusts are either winged or without wings. Of the winged some are more common and ordinary, some more rare; of the common sort, we have seen six kindes all green, and the lesser of many colors."

This translation was published in 1658. Exactly one hundred years later these one hundred and twelve words could have been summed up in one word, *Orthoptera*, which would have been clear even to Aristotle, as it was his native tongue; or even in a single family name, *Locustidæ*.

Linnæus seems to me to be as dim a figure historically as Buffon or Cuvier. But suddenly we come across names, quite or almost contemporaneous, such as Gilbert White and Charles Waterton, names of men which seem to have been friends whom I have missed knowing rather by the obstacle of geography than by the passing of time.

A few sentences will suffice to show the various fields of activity of a few of the better-known naturalists from 1500 to the middle of the nineteenth century. Leeuwenhoek, Swammerdam, Malpighi, and Hooke spent much of their lives at the eye-pieces of their primitive microscopes, and made wonderful discoveries and exquisite drawings of the new world of life thus made visible.

As extremes, Captain Cook traveled far and wide, and through his corps of naturalist passengers, headed by Sir Joseph Banks, added greatly to our knowledge of plants and animals in foreign lands, while Gilbert White and Isaak Walton were content with what they could see and describe close to their homes.

Ray, Pallas, and Linnæus traveled less extensively than Cook and used their eyes to good advantage, but Linnæus in addition contributed immeasurably to the orderliness of the nomenclature of science. Both Linnæus and Cuvier believed in the fixity of species, but Buffon and Erasmus Darwin sowed the first seeds of discontent, which Lamarck developed still farther, leaving Charles Darwin to bring to full flower the idea of the uninterrupted evolution of all forms of life on the earth.

BISON

painted on the wall of the Altamira Cave in Spain, by a cave-man many thousands of years ago

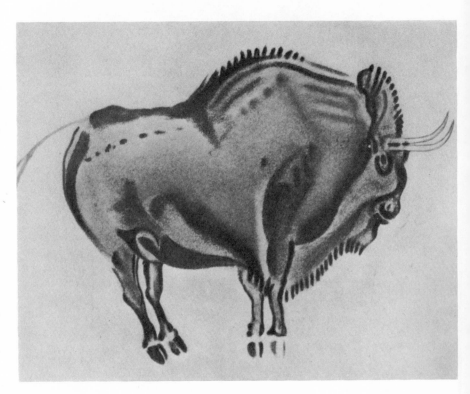

WILD HORSE

drawn by some early cave-man on the wall

of a Dordogne cave in France

WILD BOAR

from the Altamira Cave in Spain, showing an attempt at illustrating continuity of motion

ARTIST: AN UNKNOWN MAGDALENIAN
(*Circa* 25,000–20,000 B.C.)

Cave Paintings from the Stone Age

[IN THE Introduction (pages 3, 4) I have told briefly the story of the ancient cave paintings, three of which are here reproduced. I wish only to call attention to a new marvel concerning the natural history of this art of the dim dawn of man's evolution.

This discovery was imparted to me by my friend Andrey Avinoff of the Carnegie Museum. The subject is a well delineated wild boar which is furnished with eight legs, all anatomically correct when we interpret them rightly.

I quote from Dr. Avinoff's letter:

"The multiped boar with the eight legs was drawn by some caveman with, I assume, a deliberate intention to indicate motion. Another similar animal is depicted alongside without any suggestion of additional limbs. An Italian painter, Balla, was one of the first exponents of recording motion in painting by multiplying the legs of a running dog. This was thought to be an original attempt to overcome the static and momentary limitations of painting. Duration, change and motion have been the futile aspirations of pictorial art, but seem to become accessible only in terms of suggestion, anticipation or after-effect. The method of the Italian artist is repeated in the multiple photographic exposures, the stroboscopic photograph of a golf champion swinging his club which spreads in a fan-like fashion reminiscent of a parading peacock. These photographs are of the same order of illustration as the experiments of Balla, and his prodrome of twenty thousand years ago. The only predicament of labelling this movement as Futurism is the fact that it was initiated by these early antecedents in such a remote past."]

11

ARISTOTLE (384–322 B.C.)

Fishing-Frogs, Cuckoos, and Other Things

From *Historia Animalium*. CIRCA 344 B.C.

[ARISTOTLE, the Greek philosopher of the fourth century before Christ, was the founder of natural history and the greatest naturalist of all time. Reality and comprehensiveness are the two words by which his genius has been defined. "More truly than any thinker before him, or for twenty centuries to follow, did he measure the sphere within which thought is possible; more fully than any other did he embrace the content of that sphere." In his studies he neglected no field of nature; no living creature known to him was exempt from careful, intensive observation.

In the Introduction I have written briefly of him, but, to emphasize only one more aspect of his versatility, again and again in reading his works we find his mind reaching out far beyond his time and faintly adumbrating problems such as the psychology of the mind in sleep, hibernation, color change, the meaning of sexual dimorphism, division of labor among bees, the whys of migration, and even the effects of the activity of the endocrine glands. These are problems with which scientists of today are still concerned, if not, indeed, still mystified.

From the riches of his *Natural History of Animals* several extracts have been chosen, almost at random.]

I N MARINE CREATURES one may observe many ingenious devices adapted to the circumstances of their lives. For the accounts commonly given of the so-called fishing-frog are quite true; as are also those given of the torpedo. The fishing-frog has a set of filaments that project in front of its eyes; they are long and thin like hairs and are round at the tips; they lie on either side, and are used as baits. Accordingly, when the animal stirs up a place full of sand and mud and conceals itself therein, it raises the filaments, and when the little fish strike against them, it draws them in underneath into its mouth. The torpedo narcotizes the creatures that it wants to catch, overpowering them by the power of the shock that is resident in its body, and feeds upon them; it also hides in the sand and mud, and catches all the creatures that swim in its way and come under its narcotizing influence. This phenomenon has been actually observed in operation. The sting-ray also conceals itself, but not exactly in the same way. That the creatures get their living by this means is obvious from the fact that, whereas they are peculiarly inactive, they are often caught with mullets in their interior, the swiftest of fishes. Furthermore, the fishing-frog is unusually thin when he is caught after losing the tips of his filaments, and the torpedo is known to cause a numbness even in human beings. Again, the hake, the ray, the flat-fish, and the angel-fish burrow in the sand, and after concealing themselves angle with the filaments on their mouths, that the fishermen call their fishing-rods, and the little creatures on which they feed swim up to the filaments taking them for bits of sea-weed, such as they feed upon.

[These observations of Aristotle might have been written today in so far as their clarity and accuracy are concerned. Every word is true and shows careful and patient observation and reflection. The "shock" of the torpedo fish, which "numbs" and "narcotizes," was not to be identified with electricity until twenty-one centuries had passed.]

The nautilus (or argonaut) is a poulpe or octopus, but one peculiar both in its nature and its habits. It rises up from deep water and swims on the surface; it rises with its shell down-turned in order that it may rise the more easily and swim with it empty, but after reaching the surface it shifts the position of the shell. In between its feelers it has a certain amount of web-growth, resembling the substance between the toes of web-footed birds; only that with these latter the substance is thick, while with the nautilus it is thin and like a spider's web. It uses this structure, when a breeze is blowing, for a sail, and lets down some of its feelers alongside as rudder-oars. If it be frightened, it fills its shell with water and sinks. With regard to the mode of generation and the growth of the shell knowledge from observation is not yet satisfactory; the shell, however, does not appear to be there from the beginning, but to grow in their case as in that of other shell-fish; neither is it ascertained for certain whether the animal can live when stripped of its shell.

[As to the nautilus's righting its shell, spreading wide its sail, manning its arms as oars and rudder, and thus navigating the high seas, my reluctant rebuttal is somewhat enfeebled when I take from my shelves a conchology published in my own lifetime, more than two millenniums since Aristotle, and find repeated the selfsame tale with all details, in both text and illustration.]

Of the animals that are comparatively obscure and short-lived the characters or dispositions are not so obvious to recognition as are those of animals that are longer-lived. These latter animals appear to have a natural capacity corresponding to each of the passions: to cunning or simplicity, courage or timidity, to good temper or to bad, and to other similar dispositions of mind.

Some also are capable of giving or receiving instruction — of receiving it from one another or from man: those that have the faculty of hearing, for instance; and, not to limit the matter to audible sound, such as can differentiate the suggested meanings of word and gesture.

In all genera in which the distinction of male and female is found, Nature makes a similar differentiation in the mental characteristics of the two sexes. This differentiation is the most obvious in the case of human kind and in that of the larger animals and the viviparous quadrupeds. In the case of these latter the female is softer in character, is the sooner tamed, admits more readily of caressing, is more apt in the way of learning; as, for instance, in the Laconian breed of dogs the female is cleverer than the male.

In all cases, excepting those of the bear and leopard, the female is less spirited than the male; in regard to the two exceptional cases, the superiority in courage rests with the female. With all other animals the female is softer in disposition than the male, is more mischievous, less simple, more impulsive, and more attentive to the nurture of the young; the male, on the other hand, is more spirited than the female, more savage, more simple and less cunning. The traces of these differentiated characteristics are more or less visible everywhere, but they are especially visible where character is the more developed, and most of all in man.

The fact is, the nature of man is the most rounded off and complete, and consequently in man the qualities or capacities above referred to are found in their perfection. Hence woman is more compassionate than man, more easily moved to tears, at the same time is more jealous, more querulous, more apt to scold and to strike. She is, furthermore, more prone to despondency and less hopeful than the man, more void of shame or self-respect, more false of speech, more deceptive, and of more retentive memory. She is also more wakeful, more shrinking, more difficult to rouse to action, and requires a smaller quantity of nutriment.

14

As was previously stated, the male is more courageous than the female, and more sympathetic in the way of standing by to help. Even in the case of molluscs, when the cuttle-fish is struck with the trident the male stands by to help the female; but when the male is struck the female runs away.

[Aristotle's bookshelves may have been as bare as the cupboard of Mother Hubbard; he gives us no hint of the style of hair-do, sandals, or wrap-arounds worn by the 660th great-grandmother of today's American girl, but he delineates his ideas of the psychology of the female of the Grecian species in no uncertain terms. As to the authenticity of these characteristics, and the direct comparison of lady and cuttle-fish, I must remind the reader that of this anthology I am only the editor.]

Among insects and fishes, some cases are found wholly devoid of the duality of sex. For instance, the eel is neither male nor female, and can engender nothing. In fact, those who assert that eels are at times found with hair-like or worm-like progeny attached, make only random assertions from not having carefully noticed the locality of such attachments. For no eel nor animal of this kind is ever viviparous unless previously oviparous; and no eel was ever yet seen with an egg. . . . When people rest duality of sex in the eel on the assertion that the head of the male is bigger and longer, and the head of the female smaller and more snubbed, they are taking diversity of species for diversity of sex.

[Aristotle's comprehension of the life history of the eel was as uncertain as the physical grasp of his hand upon the living creature. But we must remember that it was less than four decades ago that we learned with certainty the truly amazing story of these creatures (see page 478) . We can feel only the sincerest respect for the doubts of this great Grecian naturalist as to the "worm-like progeny," especially as even today many an American farm boy believes implicitly that horsehairs will turn into snakes and eels.

Gilbert White spoke truth more than a century and a half ago, when he said: "It is the hardest thing in the world to shake off superstitious prejudices: they are sucked in as it were with our mother's milk; and growing up with us at a time when they take the fastest hold and make the most lasting impressions, become so interwoven into our very constitutions, that the strongest good sense is required to disengage ourselves from them. No wonder, therefore, that the lower people retain them their whole lives through, since their minds are not invigorated by a liberal education, and therefore not enabled to make any efforts adequate to the occasion."]

The cuckoo, as has been said elsewhere, makes no nest, but deposits its eggs in an alien nest, generally in the nest of the ring-dove, or on the ground in the nest of the hypolais or lark, or on a tree in the nest of the green linnet. It lays only one egg and does not hatch it itself, but the mother-bird in whose nest it has deposited it hatches and rears it; and, as they say, this mother bird, when the young cuckoo has grown big, thrusts her own brood out of the nest and lets them perish; others say that this mother-bird kills her own brood and gives them to the alien to devour, despising her own young owing to the beauty of the cuckoo. Personal observers agree in telling most of these stories, but are not in agreement as to the destruction of the young. Some say that the mother-cuckoo comes and devours the brood of the rearing mother; others say that the young cuckoo from its superior size snaps up the food brought before the smaller brood have a chance, and that in consequence the smaller brood die of hunger; others say that, by its superior strength, it actually kills the other ones whilst it is being reared up with them. The cuckoo shows great sagacity in the disposal of its progeny; the fact is, the mother-cuckoo is quite conscious of her own cowardice and of the fact that she could never help her young one in an emergency, and so, for the security of the young one, she makes of him a supposititious child in an alien nest. The truth is, this bird is pre-eminent among birds in the way of cowardice; it allows itself to be pecked at by little birds, and flies away from their attacks.

[Here again we have a striking example of the honesty of Aristotle. He is certain of the parasitic habit of the cuckoo, but as to details he offers only conjectures based on *it-is-saids* and *they-say*. The truth in this case is quite as remarkable as the imagination of the observers of old. The young cuckoo, when still a sprawling blind changeling, by means of a deep hollow in its back and unusual strength guided by a horribly certain instinct, pushes itself beneath the rightful eggs or nestlings of its host and heaves them up and over the rim of the nest.]

With regard to the sleeping and waking of animals, all creatures that are red-blooded and provided with legs give sensible proof that they go to sleep and that they waken up from sleep; for as a matter of fact, all animals that are furnished with eyelids shut them up when they go to sleep. Furthermore, it would appear that not only do men dream, but horses also, and dogs, and oxen; aye, and sheep, and goats, and all viviparous quadrupeds; and dogs show their dreaming by barking in their sleep. With regard to oviparous animals we cannot be sure that they dream, but most undoubtedly they sleep. And the same may be said of water animals, such as fishes, molluscs, crustaceans, to wit crawfish and the like. These animals sleep without doubt, although their sleep is of very short duration. The proof of their sleeping cannot be got from the condition of their eyes — for none of these creatures are furnished with eyelids — but can be obtained only from their motionless repose.

It is from the following facts that we may more reasonably infer that fishes sleep. Very often it is possible to take a fish off its guard so far as to catch hold of it or to give it a blow unawares; and all the while that you are preparing to catch or strike it, the fish is quite still but for a slight motion of the tail. And it is quite obvious that the animal is sleeping, from its movements if any disturbance be made during its repose; for it moves just as you would expect in a creature suddenly awakened. Further, owing to their being asleep, fish may be captured by torchlight. The watchmen in the tunny-fishery often take advantage of the fish being asleep to envelop them in a circle of nets; and it is quite obvious that they were thus sleeping by their lying still and allowing the glistening under-parts of their bodies to become visible, while the capture is taking place. They sleep in the night-time more than during the day; and so soundly at night that you may cast the net without making them stir. Fish, as a general rule, sleep close to the ground, or to the sand or to a stone at the bottom, or after concealing themselves under a rock or the ground. Flat fish go to sleep in the sand; and they can be distinguished by the outlines of their shapes in the sand, and are caught in this position by being speared with pronged instruments. The basse, the chrysophrys or gilthead, the mullet, and fish of the like sort are often caught in the daytime by the prong owing to their having been surprised when sleeping; for it is scarcely probable that such fish could be pronged while awake. Cartilaginous fish sleep at times so soundly that they may be caught by hand. The dolphin and the whale, and all such as are furnished with a blow-hole, sleep with the blow-hole over the surface of the water, and breathe through the blow-hole while they keep up a quiet flapping of their fins; indeed, some mariners assure us that they have actually heard the dolphin snoring.

Molluscs sleep like fishes, and crustaceans also. It is plain also that

17

insects sleep; for there can be no mistaking their condition of motionless repose. In the bee the fact of its being asleep is very obvious; for at night-time bees are at rest and cease to hum. But the fact that insects sleep may be very well seen in the case of common everyday creatures; for not only do they rest at nighttime from dimness of vision (and, by the way, all hard-eyed creatures see but indistinctly), but even if a lighted candle be presented they continue sleeping quite as soundly.

[Aristotle's knowledge of the sleep of animals is too astonishingly accurate for critical comment. Only this: as I sit by the open fire and watch the twitching limbs and quivering lips of the sleeping dog at my feet, as I listen to the low whimpering coming from some exciting chase in his dream world, Aristotle seems very close, the years between seem few. At midnight, when by the light of a flash lamp I reach into Pacific tide-pools, lift and replace fish immobile in wide-eyed slumber, I think again of the great Grecian naturalist.

Time, measured by human history, has little power over the love of a man for his dog, or the unceasing devotion of some of us to the study of the lives of the wild creatures that share our little planet.]

PLINY (A.D. 23–79)

Disagreements between Elephants and Dragons; The Birth of Pearls

From *Historia Naturalis*. A.D. 77

[PLINY THE ELDER, or, as his neighbors knew him, Caius Plinius Secundus, was a well-born Roman gentleman, pre-eminent as a compiler of ancient writings and famed as the author of the most "popular" natural history ever written. He was not a philosopher, and he lacked critical judgment and accuracy of observation, but his literary labors were indefatigable and he rescued a vast amount of material, both good and bad, from oblivion. As Aristotle's work stands for careful reporting of nature, any quotation from Pliny is notable for its curious mixture of fact and myth.

Three selections are chosen from the first English translation, the quaint but accurate rendition of Philomel Holland, published in London in 1601. The first is a brief eulogy of Aristotle and an account of his amazing zoological expedition, initiated and financed by Alexander the Great.]

ARISTOTLE [is] a man whom I cannot name, but with great honour and reverence, and whome in the historie and report of these matters I meane for the most part to follow. And in very truth King Alexander the great, of an ardent desire that he had to know the natures of all living creatures, gave this charge to Aristotle, a man singular and accomplished in all kind of science and learning, to search into this matter, and to set the same downe in writing: and to this effect commanded certaine thousands of men, one or other, throughout all the tract, as well of Asia as Greece, to give their Attendance, & obey him: to wit, all Hunters, Faulconers, Fowlers, and Fishers, that lived by those professions. Item, all Forresters, Park-keepers, and Wariners: all such as had the keeping of heards and flockes of cattell: of bee-hives, fish-pooles, stewes, and ponds: as also those that kept up foule, tame or wild, in mew, those that fed poultrie in barton or coupe: to the end that he should be ignorant of nothing in this behalfe, but be advertised by them, according to his commission, of all things in the world. I beseech the readers . . . to make a short start abroad with me, and in a breefe discourse by mine owne paines and diligence digested, to see all.

[Even in 1601 Englishmen had the gift of understatement, and so we find Philomel Holland writing of the *"Disagreement"* of the Dragons and the Elephants in introducing another chapter. By way of Holland's quill Pliny has this to say:]

Elephants breed in that part of Affricke which lyeth beyond the deserts and wildernesse of the Syrtes: also in Mauritania: they are found also among the Aethiopians and Troglodites, as hath been said: but India bringeth forth the biggest: as also the dragons, that are continually at variance with them, and evermore fighting, and those of such greatnesse, that they can easily claspe and wind around about the Elephants, and withal tye them fast with a knot. In this conflict they die, both the one and the other: the Elephant falls downe dead as conquered, and with his heavie weight crusheth and squeaseth the dragon that is wound and wreathed about him.

Wonderful is the wit and subtiltie that dumb creatures have & how they shift for themselves and annoy their enemies: which is the only difficultie that they have to arise and grow to so great an heigth and excessive bignesse. The dragon therefore espying the Elephant when he goeth to releese, assaileth him from an high tree and launceth himselfe upon him; but the Elephant knowing well enough he is not able to withstand his windings and knittings about him, seeketh to come close to some trees or hard rockes, and so for to crush and squise the dragon between him and them: the dragons ware hereof, entangle and snarle his feet and legges first with their taile: the Elephants on the other side, undoe those knots with their trunke as with a hand: but to prevent that againe, the dragons put in their heads into their snout, and so stop their breath, and withall, fret and gnaw the tenderest parts that they find there. Now in case these two mortall enemies chaunce to reencounter upon the way, they bristle and bridle one against another, and addresse themselves to fight; but the principall thing the dragons make at, is the eye: whereby it cometh to passe, that many times the Elephants are found blind, pined for hunger, and worne away, and after much languishing, for very anguish and sorrow die of their venime. What reason should a man alledge of this so mortall warre betweene them, if it be not a verie sport of Nature and pleasure that shee takes, in matching these two so great enemies togither, and so even and equall in every respect?

But some report this mutuall war between them after another sort: and that the occasion thereof ariseth from a naturall cause. For (say they) the Elephants bloud is exceeding cold, and therefore the dragons be wonderfull desirous thereof to refresh and coole themselves therewith, during the parching and hote season of the yeere. And to this purpose they lie under the water, waiting their time to take the Elephants at a vantage when they are drinking. Where they catch fast hold first of their trunke: and they have not so soone clasped and entangled it with their taile, but they set their venomous teeth in the Elephants eare, (the onely part of all their bodie, which they cannot reach unto with their trunke) and so bite it hard. Now these dragons are so big withall, that they be able to receive all the Elephants bloud. Thus are they sucked drie, untill they fall down dead: and the dragons again, drunken with their bloud, are squised under them, and die both together.

In Aethyopia there be as great dragons bred, as in India, namely, twentie cubites long. But I marvell much at this one thing, why king Juba should thinke that they were crested. They are bred most in a country of Aethyopia, where the Asachaei inhabite. It is reported, that upon their coasts they are enwrapped foure or five of them together, one within another, like to a hurdle or lattise worke, and thus passe the seas, for to find better pasturage in Arabia, cutting the waves, and bearing up their heads aloft, which serve them instead of sailes.

Megasthenes writeth, that there be serpents among the Indians grown to that bignesse, that they are able to swallow stags or buls all whole. Metrodorus saith, That about the river Rhyndacus in Pontus, there be Serpents that catch and devour the foules of the aire, bee they never so good and flight of wings, and sore they never so high. Well knowne it is, that Attilius Regulus, Generall under the Romanes, during the warres against the Carthaginians, assailed a Serpent neere the river Bagrada, which caried in length 1120 foot: and before he could conquer him, was driven to discharge upon him arrowes, quarrels, stones, bullets, and such like shot, out of brakes, slings, and other engines of artillerie, as if he had given the assault to some strong towne of warre. And the proofe of this was to be seene by the markes remaining in his skin and chaws, which, untill the warre of Numantia remained in a temple or conspicuous place of Rome. And that in the daies of the Emperour Claudius there was one of them killed in the Vaticane, within the bellie whereof there was found an infant all whole. This Serpent liveth at the first of kines milke, and whereupon taketh the name of Boae. As for other beasts, which ordinarily of late are brought from all parts into Italie, and oftentimes have there been scene, needlesse it is for me to describe their formes in particular curiously.

[Dr. Philomel Holland has no adverse comments to make on these tragic tales of the dragons and elephants, but more than two centuries and a half later, in a very stodgy translation of the same passage, John Bostock, M.D., and Fellow of the Royal Society, indignantly objects to two items in the account — the frigid character of the blood of the elephant, and the belief of crests in dragons.

As a pleasing antithesis to these sanguine encounters let us consider in a few lines Pliny's idea of the birth of a pearl.]

The richest merchandise of all, and the most soveraigne commoditie throughout the whole world, are Pearles. The Indian Ocean is chiefe for sending them: and yet to come by them, we must goe and search amongst those huge and terrible monsters of the sea, We must passe over so many seas, and saile into farre countries so remote and come into those parts where the heat of the sunne is so excessive and extreame: and when all is done, wee may perhaps misse of them.

This shell-fish which is the mother of Pearle, differeth not much in the manner of breeding and generation, from the oysters: for when the season of the yeere requireth that they should engender, they seeme to yawne and gape, and so doe open wide; and then (by report) they conceive a certaine moist dew as seed, wherewith they swell and grow bigge; and when the time commeth, the fruit of these shell-fishes are the pearles, better or worse, great or small, according to the qualitie and quantitie of the dew which they received. For if the dew were pure and cleare which went into them, then are the pearles white, faire and orient: if grosse and troubled, the pearles likewise are dimme, foule, and duskith; pale (I say) they are, if the weather were close, darke, and threatning raine, in the time of their conception. Whereby no doubt it is apparent and plaine, that they participate more of the aire and skie, than of the water and the sea. . . . The pearle is soft and tender so long as it is in the water, take it forth once and presently it hardeneth. As touching the shell that is the mother of Pearle, as soone as it perceiveth and feeleth a mans hand within it, by and by she shutteth, and by that meanes hideth and covereth her riches within: for well woteth she that therefore she is sought for. But let the fisher looke well to his fingers, for if she catch his hand betweene, off it goeth: so trenchant and sharpe an edge she carrieth, that is able to cut it quite a two. And verily this is a just punishment for the theefe, and none more: albeit shee be furnished and armed with other meanes of revenge. For they keepe for the most part about craggie rockes, and are there found: and if they be in the deepe, accompanied lightly they are with curst Sea-dogs. And yet all this will not serve to skare men away from fishing after them: for why? our dames and gentlewomen must have their eares behanged with them, there is no remedie.

ABBOT THEOBALDUS OF MONTE CASSINO
(ELEVENTH CENTURY)

The Whale

From *Physiologus: A Metrical Bestiary of Twelve Chapters.* CIRCA 1022–1035

[THROUGHOUT the first ten centuries after the birth of Christ, when natural history had reverted almost to the Stone Age, any writer who let himself go in the slightest degree covered his tracks with an appendix of morals. An example is the Abbot Theobaldus, who some time between 1022 and 1035 wrote a *Physiologus* or Bestiary. This is typical of the extensive drought in human mental history, so I present *The Whale* in full.]

GREATEST OF ALL is the Whale, of the beasts which live in the
  waters,
Monster indeed he appears, swimming on top of the waves,
Looking at him one thinks, that there in the sea is a mountain,
Or that an island had formed, here in the midst of the sea,
He also sometimes his hunger (which worries him often most greatly),
Wishes at once to relieve, warm is his wide open mouth,
Whence he then sends forth breaths of odours as sweet as the flowers,
By which to him he attracts fishes of sizes quite small,
Small ones indeed, we must say, because any fishes of great size,
Nor full grown can he eat, nor can eject from his mouth,
All little fishes he gladly retains to guard against hunger,
Not in the way that he did, swallowing Jonah of old,
On the approach of a storm, or fearing the heat of the summer,
All of the herd quick depart, troubling the depths of the sea,
Often again the Whale, rising up to the top of the waters,
Just like an arm of the land, seems he to those on the sea,
Hasten the seamen to this, and tie their ship, fearing a tempest,
They having made it secure, jump from the ship to the shore,
Kindle they then a hot fire, which by them is carried on shipboard,
That they themselves may be warmed, while cooking quickly some
  food,
The Whale now feeling the heat, at once plunges under the waters,

24

Thence to the place whence he came. Thus ship and all are destroyed.

[At the end of this sad tale, which in meter recalls Longfellow's *Hiawatha,* and in action the adventures of Sindbad, Theobaldus, not to be caught in the act of promulgating knowledge for knowledge' sake, disarms any religious critics by the following appendix:]

As is the Whale with his great bulk, so is the devil with all men,
Those, he has trained by his craft, men of great magic appear,
By him, through all the world the minds of all men are changed,
For them he hungers and thirsts, and when he can he destroys,
Those weak in faith he attracts, and with sweetness of words he
    entices,
Those who are strong in the faith, over these casts he no spell,
Whoso confides in the devil, to whom all his hopes are entrusted,
Quickly is dragged down to hell. So sorely is he deceived.

FREDERICK II (1194–1250)

The Migration of Birds

From *De Arte Venandi cum Avibus* or *The Art of Falconry.* 1245

[IN THE splendid isolation of that mental vacuum, the thirteenth century (and for centuries before and after), the position of the Emperor Frederick II was unique. As I said in the Introduction, he translated, quoted, and corrected Aristotle. In addition, inspired by a fanatical love of falconry, he wrote at length, originally and with considerable accuracy, of the habits and anatomy of birds in general. I regret only that space prevents the quoting of more than two of the eleven chapters on the migration of birds.]

WITH A PROPHETIC instinct for the proper time to migrate, birds as a rule anticipate the storms that usually prevail on their way to and from a warmer climate. They are conscious of the fact that autumn follows summer (when they are strongest and their plumage is at its best) and that after these seasons comes the winter — the time they dread most. They are instinctively aware of the proper date of departure for avoiding the winds to which they may be exposed in their wanderings and for eluding the local rains and hailstorms. They usually are able to choose a period of mild and favoring winds. North winds, either lateral or from the rear, are favorable, and they wait for them with the same sagacity that sailors exhibit when at sea. With such helpful breezes progress and steering in the air are made easy. With these to help them on their way, they reach, with comparative comfort, the distant lands of heart's desire. When they fly before the wind they can rest on an even keel, still maintaining progress, especially when propelled in a proper direction. When becalmed they do not fly so satisfactorily, for they must exert themselves all the more. With head winds there is a threefold difficulty in attempting to float, to fly forward, and to overcome direct aerial obstacles.

Among flight obstructions there are also to be considered not only contrary winds but local rains, hailstorms, and other forms of bad weather that may affect both air and sea, so that some birds fall into the ocean and others, when possible, fly on board a ship (where they are easily caught), preferring that fate to certain death or to continued exposure to the rigors and dangers of oceanic storms.

We notice also that when a favoring wind springs up, whether by day or night, migrating birds generally hasten to take advantage of it and even neglect food and sleep for this important purpose. We have observed that migrating birds of prey, that have begun to devour food we have thrown to them, will abandon it to fly off if a favorable wind begins to blow. They would rather endure and travel day and night than forgo such an advantageous opportunity.

The calls of migrating cranes, herons, geese, and ducks may be recognized flying overhead even during the night, and not, as Aristotle claims, as a part of their efforts in flight; they are the call notes of one or more birds talking to their fellows. For example, they understand wind and weather so thoroughly that they know when meteorologic conditions are favorable and are likely to remain so long enough to enable them to reach their intended haven. Weak fliers postpone their journey until they are sure of a prolonged period of good weather sufficient for their migrating venture, but hardy aviators take advantage of the first propitious period to begin their flight.

The slower migrants begin their departure early. For example, the smaller birds, as well as storks and herons, remain until the end of summer and leave the last of August so that they may not be embarrassed by changeable weather or early (autumn) storms. The more robust species and better fliers remain until the beginning of harvest (about the middle of September). Among the latter are the larger and smaller cranes. At that date strong fliers can readily defy the early winds and rains. There are, moreover, still better and swifter fliers who postpone their departure until the end of the autumnal season, say, until November. These include certain ducks and geese who do not fear high winds and heavy rains because of their skill in flight and because their plumage protection against cold is adequate. This rule also applies to the smaller geese who may remain behind in the sixth and seventh climatic zones the whole winter through, inasmuch as they can find there the herbage on which they feed. The larger geese also possess unusual meteorological instincts and avian alertness. In years when there are short summers, i.e., when the winter threatens to set in early, they migrate much sooner than usual.

Certain birds, cranes for example, who pass the summer in the far north (where winter comes on early) on account of the longer journey before them, migrate sooner than others of their species who, having nested farther south, prolong their northern visit, since their winter comes later and they have a shorter journey to make. When autumnal winds are favorable, these birds resume their southern flight and, traveling without intermission, quickly accomplish the voyage. Inclement weather, however, may delay the flight of species that have hatched their young in more southern localities until the storm has passed. Those nearest the equator begin their migration last.

The order of migration may be summed up as follows: not all shore birds depart pell-mell, like the disorderly land birds; the latter do not seem to care what birds lead the van or which form the rear guard of the migrating flocks. Water birds, on the contrary, preserve the following order: one forms the apex of advance, and all the others in the flock follow successively in a double row, one to the left and one to the right. Sometimes there are more in one series than in the other, but the two rows, meeting at an angle, form a pyramidal figure. Occasionally there is a single line.

This order they maintain not only when migrating to distant points and returning but, as has been explained, in going to and from their local feeding grounds.

One member of the flock usually acts as leader and, especially in the case of cranes, does this not because he alone knows the goal they seek but that he may be ever on the lookout for danger, of which he warns his companions; he also notifies them of any change to be made in the direction of flight. The whole flock is thus entirely under control of their leader or guide. When the latter becomes fatigued from the performance of this important work, his place in front is taken and his duties are assumed by another experienced commander, and the former leader retires to a rear rank. It is not true, as Aristotle asserts, that the same leader heads the migrant column during the whole of their journey.

[The best efforts of modern ornithologists have no better to offer in the way of presentation, investigation, and rejection of a fact and theory than was penned by the Emperor in the year 1245 in regard to the myth of the barnacle goose. Although at this early date he clearly explained and demolished it, this absurd tale continued to find credence until early in the nineteenth century.]

There is, also, a small species known as the barnacle goose, arrayed in motley plumage (it has in certain parts white and in others black, circular markings), of whose nesting haunts we have no certain knowledge. There is, however, a curious popular tradition that they spring from dead trees. It is said that in the far north old ships are to be found in whose rotting hulls a worm is born that develops into the barnacle goose. This goose hangs from the dead wood by its beak until it is old and strong enough to fly. We have made prolonged research into the origin and truth of this legend and even sent special envoys to the North with orders to bring back specimens of those mythical timbers for our inspection. When we examined them we did observe shell-like formations clinging to the rotten wood, but these bore no resemblance to any avian body. We therefore doubt the truth of this legend in the absence of corroborating evidence. In our opinion this superstition arose from the fact that barnacle geese breed in such remote latitudes that men, in ignorance of their real nesting places, invented this explanation.

CONRAD GESNER (1516–1565)

The Rhinoceros and the Mouse

From *Historiæ Animalium de Quadrupedibus viviparis*, 1551,
as translated and annotated by Edward Topsell
in *The Historie of Foure-footed Beastes*, 1607

[CONRAD GESNER was born in Switzerland in 1516, which, like most
sixteenth-century dates, means little until we tie it up with something more definite; think of it as being only twenty-four years after
the discovery of America. Gesner has been called the German Aristotle and the Swiss Pliny, and indeed his efforts lie somewhat between
the two. He traveled widely in Switzerland and Germany and through
extensive correspondence gathered a very great mass of facts on natural history, many of them correct and original. He also combed
ancient literature for the contents of his five folio volumes. He included such monstrosities as seven-headed dragons, but acknowledged
them as such.

To birds and plants Gesner gave the most reliable treatment. But
the illustrations are the outstanding feature of his work, mostly
drawn under his supervision by the best masters of woodcutting of his
age. Proof of their superiority to anything else extant is that they
were used over and over in subsequent publications by other authors.

Edward Topsell, whose translation is used here, gave to Gesner's
unadorned facts a warmth and personal touch which were quite
lacking in the original author's encyclopedic method of treatment.
Topsell's writing has something of the majestic swing of the style of
the King James Bible. At his dullest, Gesner's original Latin is considerably better reading than that of his contemporaries Aldrovandus,
Rondelet, and Salviani. He was most certainly the greatest naturalist
of the Renaissance.]

OF THE RHINOCEROS

WE ARE NOW to discourse of the second wonder in nature,
namely of a beast every way admirable, both for the outward
shape, quantity, and greatnesse, and also for the inward courage, disposition, and mildnes. For as the Elephant was the first wonder, of whom we have already discoursed, so this beast next unto the

Elephant filleth up the number, being every way as admirable as he, if he doe not excede him, except in quantity or height of stature; And being now come to the story of theis beast, I am hartily sorry, that so strange an outside, yealding no doubt through the omnipotent power of the creator, an answerable inside, and infinite testimonies of worthy and memorable vertues comprized in it, should through the ignorance of men, lye unfoulded and obscured before the Readers eyes: for he that shall but see our stories of the Apes, of the Dogs, of the Mice, & of other small beasts, and consider how larg a treatise we have collected together out of many writers, for the illustration of their natures and vulgar conditions, he cannot chuse but expect some rare and strange matters, as much unknowne to his minde about the storie of this Rhinoceros, as the outward shape and picture of him, appeareth rare and admirable to his eies: differing in every part from all other beasts, from the top of his nose to the tip of his taile, the eares and eies excepted, which are like Beares. But gentle Reader as thou art a man, so thou must consider since *Adam* went out of *Paradice,* ther was never any that was able perfectly to describe the universall conditions of all sorts of beasts, and it hath bin the counsell of the almighty himselfe, for the instruction of man, concerning his fall and naturall weakenesse, to keep him from the knowledge of many devine things, and also humane, which is of birds and beasts, Fishes and foule, that so he might learne, the difference betwixt his generations, & his degeneration, and consider how great a losse unto him was his fall in Paradice; who before that time knew both God himselfe and al creatures, but since that time neither knoweth God as he should know him, nor himselfe as he shall know it, nor the creatures as hee did know them.

But for my part which write the English story, I acknowledge that no man must looke for that at my hands, which I have not received from some other: for I would bee unwilling to write anything untrue, or uncertaine out of mine owne invention; and truth on every part is so deare unto mee, that I will not lie to bring any man in love and admiration with God and his works, for God needeth not the lies of men: To conclude therfore this Praeface, as the beast is strange and never seene in our countrey, so my eye-sight cannot adde any thing to the description: therefore harken unto that which I have observed out of other writers.

First of all that there is such a beast in the world, both *Pliny, Solinus, Diodorus, Aelianus, Lampridius,* and others, doe yeald ere-frigable testimony. *Heliogabalus* had one of them at Rome. *Pompey* the great, in his publicke spectacles did likewise produce a Rhinoc erot (As *Seneca* writeith). When *Augustus* rode triumphing for *Cleopatra,* he brought forth to the people a seahorse and a Rhinocerot which was the first time that ever a Rhinocerot was seene at Rome (as *Caelius writeth*). *Antoninus Pius* the Emperor, did give many gifts unto the people, amongst which were both Tigers and Rhinocerots, (saith *Iulius Capitalinus* in his life). *Martiall* also celebrateth an excellent epigram of a Rhinocerot, which in the presence of *Caesar Domitian* did cast up a Bull into the aire with his horne, as if he had bin a tenyce ball.

Lastly to put it out of all question that there is such a beast as this Rhinocerot, a picture was taken by *Gesner* from a beast alive at *Lysbon* in Portugale, before many witnesses, both Marchants and others; so that we have the Testimony both of antiquity and of the present age, for the Testimony of the forme and fashion of this beast, and that it is not the invention of man, but a worke of God in nature, first created into the beginning of the World, and ever since continued to this present day.

Concerning the name of this beast, the Graecians because of the horne in his Nose call him *Rhinoceros,* that is a Nose-horned-beast, and the Latins also have not altered that invention, for although there be many beastes that have but one horne, yet is there none that have that one horne growing out their Nose but this alone: All the residue have the horne growing out of their foreheads. There is some that have taken this *Rhinoceros,* for the *Monoceros* the Unicorne, because of this one horne, but they are deceived, taking the generall for the speciall which is a note of ignorance in them, and occasion of errour unto others. . . .

In quantity it is not much bigger than an *Orix: Pliny* maketh it equall in length to an Elephant, and some make it longer than an Elephant, but withall they say it is lower, and hath shorter Legges . . . (as *Artemidorus* said) he saw by one that was at *Alexandria,* the colour thereof was not like a Box-tree, but rather like an Elephantes, his quantity greater than a Buls, or as the greatest Bull, but his outward forme and proportion like a wilde Boares, especiallye in his mouth, except that out of his Nose groweth a horne, harder then any bones, which he useth in stead of armes, even as a Boare doth his teeth; hee hath also two girdles upon his body like the wings of a Dragon, comming from his backe downe to his belly, one toward his necke or mane, and the other toward his loines and hinder parts. . . .

When they are to fight they whet their horne upon a stone, and there is not only a discord betwixt these beasts and Elephants for their food, but a naturall description and enmity: for it is confidently affirmed, that when the *Rhinocerot* which was at *Lisborne,* was brought into the presence of an Elephant, the Elephant ran away from him. How and in what place he overcommeth the Elephant we have shewed already in his story, namely, how he fastneth his horne in the soft part of the *Elephantes* belly. Hee is taken by the same meanes that the *Unicorne* is taken, for it is said by *Albertus, Isidorus,* and *Alunnus,* that above all other creatures they love Virgins, and that unto them they will come be they never so wilde, and fall a sleepe before them, so being asleepe they are easily taken and carried away.

OF THE VULGAR LITTLE MOUSE

As we have handled the natures, and delivered the figures of the great Beasts, so also must we not disdaine in a perfect Hystory to touch the smallest: For Almighty God which hath made them al, hath disseminated in every kind both of great and smal beasts, seeds of his wisedome, maiesty, and glory. The little mouse therefore is iustly tearmed *Incola domus nostrae,* an inhabitant in our own houses, *Et rosor omnium rerum,* and a knawer of al things. . . . The Epithets of myce are these; short, small, fearful, peaceable, ridiculous, rustik, or country mouse, urbane, or citty mouse, greedy, wary, unhappy, harmefull, blacke, obscene, little, whiner, biter and earthly mouse. . . . Now to come to theyr generall nature and significations. First of all concerning their colour. It is divers, for although *Color murinus* be a common tearme for a mouse colour of Asses, yet notwithstanding Mice are sometimes blackish, sometimes white, sometims yellow, sometimes broune and sometimes ashe colour. There are White Mice amonge the people of *Savoy,* and Dolphin in France called *Alaubroges,* which the inhabitants of the country do beleev that they feede upon snow. . . .

There is no creature that heareth more perfectly then a Mouse, they dwell in houses of men, especially neare supping and dyning roomes, kitchins or larders, where any meat is stirring. And they make themselves places of aboade by gnawing with their teeth, if they finde not convenient lodginges prepared to their hand, and thy love the hollow places of wals, or the roofes of houses. . . . In the day time they lye still, so long as they either see or heare a man, or any other beast harmeful unto them, for they discerne their enemies, not fearing an Oxe, though they run away from a Cat.

They are very desirous of bread, and delight in all those meats which are made of fruit, for the nourishment of men. It is a creature very diligent & exquisite, both to compasse, seeke out and chuse the same, so that therefore it doth often endanger and loose his owne life: and finding any cubbards, wood, or such like hard matter, to withstand his purpose, and hinder his passage, it ceaseth not to weary it selfe with gnawing, untill it obtaine the purpose. All kinds of Mice love grain and corne, and prefer the hard before the soft, they love also cheese, and if they come to many cheeses together they tast all, but thy eate of the best. And therefore the Egyptians in the *Hyrogliphicks* do picture a mouse, to signifie sound judgement and good choice. . . .

And as their wisedome is admirable in this provision, so also is their love to be commended one to another, for falling into a vessell of Water or other deepe thing, out of which they cannot ascend againe of themselves, they help one another, by letting downe their tailes, and if their tailes be to short, then they lengthen them by this meanes, they take one anothers taile in their mouth, and so hang two or 3 in length untill the Mouse which was fallen downe take hold on the neathermost, which being performed, they al of them draw her out.

ANTONY VAN LEEUWENHOEK (1632–1723)

"Little Animals"

From *"Concerning Little Animals observed . . . in Rain- Well- Sea-
and Snow-Water."* In the *Philosophical Transactions of the
Royal Society of London.* 1677

[ANTONY VAN LEEUWENHOEK was born in Delft, Holland, about the
time when Galileo was being hailed before the Inquisition. He died
in the same city ninety years later. Holding some minor official post,
he spent all of his spare time throughout his life in making numerous
simple microscopic lenses and through them observing and describing everything small enough that came to hand. He never wrote a
scientific paper, but sent all his discoveries, written in old-fashioned
Dutch, in the form of letters to the Royal Society in London. There
they were translated and published in English, a language of which
Leeuwenhoek knew not a single word.

The charm of the diction of his letters and the delightful discursive
style in which his researches were couched are unique in the history
of science. A portion of his most famous letter will reveal the single-
mindedness of this man, devoting himself with no thought of personal fame or gain to the single-celled animals living in his rain
barrel.

His life was concentrated on everything about him too small to be
distinctly seen with the naked eye. He can stand for us as the most
human and one of the very ablest of pioneer microscopists.]

IN THE YEAR 1675, about half-way through September (being busy
with studying air, when I had much compressed it by means of
water), I discovered living creatures in rain, which had stood but
a few days in a new tub, that was painted blue within. This observation provoked me to investigate this water more narrowly; and
especially because these little animals were, to my eye, more than ten
thousand times smaller than the animalcule which Swammerdam
has portrayed, and called by the name of Water-flea, or Water-louse,
which you can see alive and moving in water with the bare eye.

Of the first sort that I discovered in the said water, I saw, after
divers observations, that the bodies consisted of 5, 6, 7, or 8 very clear

globules, but without being able to discern any membrane or skin that held these globules together, or in which they were inclosed. When these animalcules bestirred 'emselves, they sometimes stuck out two little horns, which were continually moved, after the fashion of a horse's ears. The part between these little horns was flat, their body else being roundish, save only that it ran somewhat to a point at the hind end; at which pointed end it had a tail, near four times as long as the whole body, and looking as thick, when viewed through my microscope, as a spider's web. At the end of this tail there was a pellet, of the bigness of one of the globules of the body; and this tail I could not perceive to be used by them for their movements in very clear water. These little animals were the most wretched creatures that I have ever seen; for when, with the pellet, they did but hit on any particles or little filaments (of which there are many in water, especially if it hath but stood some days), they stuck intangled in them; and then pulled their body out into an oval, and did struggle, by stretching themselves, to get their tail loose; whereby their whole body then sprang back towards the pellet of the tail, and their tails then coiled up serpent-wise, after the fashion of a copper or iron wire that, having been wound close about a round stick, and then taken off, kept all its windings. This motion, of stretching out and pulling together the tail, continued; and I have seen several hundred animalcules, caught fast by one another in a few filaments, lying within the compass of a coarse grain of sand.

I also discovered a second sort of animalcules, whose figure was an oval; and I imagined that their head was placed at the pointed end. These were a little bit bigger than the animalcules first mentioned. Their belly is flat, provided with divers incredibly thin little feet, or little legs, which were moved very nimbly, and which I was able to discover only after sundry great efforts, and wherewith they brought off incredibly quick motions. The upper part of their body was round, and furnished inside with 8, 10, or 12 globules: otherwise these animalcules were very clear. These little animals would change their body into a perfect round, but mostly when they came to lie high and dry. Their body was also very yielding: for if they so much as brushed against a tiny filament, their body bent in, which bend also presently sprang out again; just as if you stuck your finger into a bladder full of water, and then, on removing the finger, the inpitting went away. Yet the greatest marvel was when I brought any of the animalcules on a dry place, for I then saw them change themselves at last into a round, and then the upper part of the body rose up pyramid-like, with a point jutting out in the middle; and after having thus lain moving with their feet for a little while, they burst asunder, and the globules and a watery humour flowed away on all sides, without my being able to discern even the least sign of any skin wherein these globules and the liquid had, to all appearance, been inclosed; and at such times I could discern more globules than when they were alive. This bursting asunder I figure to myself to happen thus: imagine, for example, that you have a sheep's bladder filled with shot, peas, and water; then, if you were to dash it apieces on the ground, the shot, peas, and water would scatter themselves all over the place.

Furthermore, I discovered a third sort of little animals, that were about twice as long as broad, and to my eye quite eight times smaller than the animalcules first mentioned: and I imagined, although they were so small, that I could yet make out their little legs, or little fins. Their motion was very quick, both roundabout and in a straight line.

The fourth sort of animalcules, which I also saw a-moving, were so small, that for my part I can't assign any figure to 'em. These little animals were more than a thousand times less than the eye of a full-grown louse (for I judge the diameter of the louse's eye to be more than ten times as long as that of the said creature), and they surpassed in quickness the animalcules already spoken of. I have divers times seen them standing still, as 'twere, in one spot, and twirling themselves round with a swiftness such as you see in a whip-top a-spinning before your eyes; and then again they had a circular motion, the circumference whereof was no bigger than that of a small sand-grain; and anon they would go straight ahead, or their course would be crooked. . . .

On June 9th, collected rain-water betimes in a dish, and put it in a clean wine-glass, in my closet; and on examining it, I descried no animalcules. (*Note*. My closet standeth towards the north-east, and is partitioned off from my antechamber with pine-wood, very close joined, having no other opening than a slit an inch and a half high and 8 inches long, through which the wooden spring of my lathe passeth. 'Tis furnished towards the street with four windows, whereof the two lowermost can be opened from within, and which by night are closed outside with two wooden shutters; so that little or no air comes in from without, unless it chance that in making my observations I use a candle, when I draw up one casement a little, lest the candle inconvenience me; and I also then pull a curtain almost right across the panes.)

The 10th of June, observing this foresaid rain-water, which had

37

now stood about 24 hours in my closet, I perceived some few very little living creatures, to which, because of their littleness, no figure can be ascribed; and among others, I discovered a little animal that was a bit bigger, and that I could perceive to be oval. (*Note.* When I say that I have observed the water, I mean I have examined no more than 3, 4, or 5 drops thereof, which I also then throw away; and in narrowly scrutinizing 3 or 4 drops I may do such a deal of work, that I put myself into a sweat.)

RENÉ ANTOINE FERCHAULT DE RÉAUMUR
(1683–1757)

Why Ants Have Wings

From *Histoire des Fourmis*. Eighteenth century

[THE FRENCHMAN RÉAUMUR possessed great ability as a naturalist, but so did others who came before and after him and whose names I am forced to omit; it is perhaps partly sentiment that impels his inclusion. When fifty-four years old, he published the first of twelve volumes on the *Natural History of Insects,* a work that he was destined never to complete.

His life was passed amid the welter of eighteenth-century scientists, such as Buffon and Cuvier, together with other protagonists of the French scene, such as Louis XV and the Pompadour. Although he was rich and noble-born, and had a splendid mind, or perhaps because of this equipment, his disparagement by his contemporaries was due to the obnoxious sight of a gentleman devoting his time and interest to such ignoble creatures as ants and bees. In the face of all this, his unquenchable enthusiasm for scientific discovery and his surprisingly modern method of logical thinking are remarkable. His best epitaph is that his researches were appreciated by such men as Linnæus and Fabre.

At Réaumur's death in 1757 his remaining notes on insects were condemned by Cuvier as worthless. One hundred and sixty-eight years later, in 1926, William Morton Wheeler examined the contents of several dusty boxes in the archives of the Académie des Sciences, found them of pre-eminent value, and published a translation of Réaumuriana which places the author in the forefront of his eighteenth-century confrères.]

A T CERTAIN SEASONS in the majority of formicaries very large winged individuals, contrasting with the bulk of wingless individuals, may be observed. Moreover, at such times smaller individuals, which are also winged, may be detected. I know not by what chance it happened that these escaped the scrutiny of the attentive observer [Swammerdam], though he did not fail to observe certain wingless ants whose abdomen equalled or surpassed in volume that of the largest winged individuals. He thought that the large winged ants were the males, that they were to the formicaries what the drones are to the hives of bees, that the large wingless ants were the females and, finally, that all the other ants were sexless, that they were charged with all the work of the formicary as is the case with the majority of bees and that therefore they could be rightly designated by the name of "workers." Dissection confirmed him in the notion which analogy and the volume of the abdomen had suggested to him in regard to the large wingless ants. It showed him that their abdomens were full of eggs. . . .

But Swammerdam's inference in regard to the large winged ants, when he makes them out to be males, apparently because he is unable to make anything else of them, this conclusion, I say, is not so certain. If in order to ascertain their sex he had only resorted to dissection, or simply to strong pressure, he would have forced out of their bodies eggs, sometimes precisely similar to those which he extracted from the abdomens of the large wingless ants, but sometimes less distinct or smaller. In a word, he would have convinced himself that the large winged ants are themselves females.

Are there then in a formicary two kinds of females, winged and wingless? No, there is only one kind; both of them are the same individuals seen at different seasons. There constantly happens to the winged ants what happens to no other known animal of the class of those that bear wings: they lose theirs. A bird that should shed its wings would seem to us to be in a pitiable state; the ant which has four and at a certain season is glad to have them, at another season is better satisfied to possess them no longer. If we compare, therefore, a large wingless ant with a large winged ant, almost no differences, apart from the wings, can be detected between them, although Swammerdam believed that he could perceive other differences. The most real of those that he cites is a difference in colour, which reduces itself to a paler hue in the one than in the other. But it is usual to observe such variations in the same animal when examined at different ages.

The ancient naturalists, instead of telling us of the interesting fact in the history of ants that there comes a time when the winged individuals lose their wings, have unanimously maintained the very contrary. They have assured us that, having had no wings in their youth and at the time of their full vigour, they acquire them during their old age; that the wings sprout forth when the insects become decrepit and are about to die. One would have little ground to deny a statement so generally made if one could oppose to it only the apparent improbability that wings were given to ants merely in order that they might perish with greater dignity, or, as Cardano would have it, as a consolation in their old age. It does seem improbable that they would be bestowed on ants that are no longer in a condition to function for the good of their species. And in this case, what apparently should not be, is not. When we follow the ants through their various stages we see that those that are born without wings pass their lives without having them, whereas those that are to be winged have wings from the moment of their birth, like the other flies and the butterflies; that is, after the transformation which enables us to recognize them as ants.

After observing in a formicary nothing but wingless ants, and then after a certain time the occurrence of winged individuals, authors too hastily inferred that the ants that had wings were of the number that had formerly lacked them. There is no danger of drawing a similarly false conclusion from the observations that go to prove that the ants naturally lose their wings. But before describing the manner of making these observations, which is as easy as it is simple, it will be well to recount the circumstances that are most favourable to making them. And in order to understand these we must ascertain how the females are fecundated. . . .

Many reasons, which I shall not yet present in detail, conspire to prove that the small winged ants are the males. But to possess incontrovertible proof of this it is necessary to have surprised the female and male in the act of copulation. Hundreds and hundreds of times I vainly sought for them in this condition. The formicaries which I kept in the transparent jars showed me nothing more in this regard than the field nests, the interior of which I was unable to observe after they had been disturbed. Nevertheless, it is not only very certain that the small winged ants are the males, but it is equally certain that their mating, which has remained concealed during a long series of centuries, is more easily witnessed than that of any other insect. Not only does this occur in brightly illumined places, but often under one's very eyes, notwithstanding the fact that the ancients have eulogized these little insects on account of their modesty, because they believed that they exchanged the caresses that lead to the multiplication of their species only in darkness.

Perhaps I should have learned nothing from the first chance opportunity of seeing two ants mating, if I had been less familiar with the ways of these small insects. Being on the road to Poitou and finding myself on the levée of the Loire, very near Tours, on one of the first days of the month of September, 1731, I descended from my berlin,

enticed to stroll about by the beauty of the spot and the mild temperature of the air, which was the more agreeable because the earlier
hours of the day had been warm. The sun was within about an hour
of setting. During my stroll I noticed a lot of small mounds of sandy
and earthy particles rising above the openings that led the ants to
their subterranean abode. Many of them were at that time out of
doors; they were red, or rather reddish, of medium size. I stopped to
examine several of these earthen monticules and noticed on each
among the wingless ants a number of winged ones of two very different sizes. Some of them had abdomens no larger than those of the
wingless ants, and to judge from unaided vision one of the larger
winged individuals must have weighed more than two or three times
as much as one of the smaller. Over the beautiful levée, where I was
enjoying my walk, there appeared in the air in places not very far
apart small clouds of large flies which flew about in circling paths.
They might have been taken for gnats or craneflies or may-flies. Often
the small cloud hung in the air at a height within reach of the hand.
I used one of mine to capture some of these flies and succeeded repeatedly in doing so. All I secured were without difficulty recognized
for what they were, for they were winged ants like those I had found
at every step on the small mounds of earth. But I observed — and the
observation was as important as it was easy to make — that I almost
invariably captured them in pairs. Not only did I almost always find
in my hand one large and one small ant, but most frequently I took
them copulating and held them for some time before they separated. . . .

It is therefore in mid-air that the nuptials must be celebrated of
those ants that pass the greater portion of their lives underground
and the remainder of their lives crawling on its surface or at most
on walls, plants or trees. I have sometimes stood near a formicary,
part of whose inhabitants were winged, at about two or three o'clock
in the afternoon while it was still being warmed by the sun's rays.
Then the winged individuals of the two different sizes issued from
the earth, betook themselves, so so speak, to the roof of their abode
and there, after being thoroughly warmed, strolled about in various
directions, without, so far as I could see, any teasing of the large by
the small winged individuals; that is, without any tender preludes to
mating. Then one by one both the large and the small ones took
flight. Soon the surrounding air was seen to be filled with them, and
the large ones were seen each to have a small one dangling from its
posterior end. Not only, therefore, do they remain aloft while they

are copulating, but they actually begin the act in the air. Usually the female does not long remain alone. I have reason to believe this because along with the female and attached male which I expected to capture, and which I captured without their separating from each other, I have sometimes secured at the same time two or three additional males which, jealous of the good fortune of the first, apparently wished to supplant him, or were, perhaps, waiting till he left vacant the place that was the object of their desires.

Furthermore, I have always seen the ants return one by one to their formicary as they left it. Thus it is in the air that mating begins and continues. Then it is the task of the female to support the male, contrary to what is found in the flies called demoiselles, among which the male carries the female. Nature seems to have varied her combinations in all possible ways. . . . It has therefore been established that the wings are necessary to the ants, both male and female, in order that they may mate, and it would seem that these organs have been given them solely for this purpose. At least it is certain that the females do not long retain their wings after they have been fecundated. . . .

I will describe some of the observations, as easily made as they are decisive, that taught me that the wings, necessary throughout life to other animals provided with them, are necessary to the ants only during a portion of their existence, in order that the observations may be repeated by those who care to do so. On one of the last days of the month of August, towards the hour of sunset, having noticed in the air a lot of ants belonging to one of the largest species of this country, I captured three that passed within reach of my hand. Two were mating, but the third was a female that was at the time without a male. I confined all three in a small box. They were very vigorous, and their wings were in perfect condition. When at about nine or ten o'clock on the following morning I opened the box to examine my ants, I saw that one of the large ones — that is, one of the two females — no longer bore wings, but that she had lost none of her vigour and vivacity. She then appeared to me quite like those wingless females that one finds in small numbers in the formicary and are conspicuous by their size. I believed therefore, I had reason to suppose that the latter had all possessed wings. I found the four wings in the box. They were in such good condition and so complete that I had no reason to believe that they had been torn off by the other ants. In order to remain so perfect they must have fallen off naturally, as a leaf falls from a tree when the end of its petiole dries because the nutrient sap no longer enters it in sufficient quantity. Nothing was lacking to these four wings, which should have been torn in divers places had they been removed by repeated pulling. Finally, there remained not the minutest vestige attached to the ant's thorax; it was difficult to find even the cavities in which their basal ends had been articulated.

CARL LINNÆUS (1707–1778)

From Bats and Manatees to Cats and Camels

From *Systema Naturæ*. Tenth edition, 1758

[Linnæus, the famous Swedish naturalist, made several short expeditions, such as the one through Lapland, but the published results offer little that is quotable. The most striking fact about them is that he was able to travel forty-six hundred miles at a total cost of what today would be one hundred dollars.

By providing a two-word name in Latin for every known living animal and plant, Linnæus furnished the scientists of all nations the foundation of a basic, common nomenclature. The eyes of a Brazilian and an English scientist who know not a single word of each other's language will light up in mutual understanding at the mention of *Homo sapiens,* or *Canis familiaris.*

Linnæus's *Systema Naturæ,* which ensured his eternal fame among scientists, at sight promises as little readable material as would selections from a dictionary. But hidden at the bottom of the synopses of some of the animals is a concentrated résumé that, when done into English, satisfies requirements of interest, knowledge, humor, and myth.]

VAMPIRE BAT
Vespertilio Vampyrus
At night sucks in the blood of the sleeping, the combs of cocks and the juice of palm trees.

MANATEE
Trichechus Manatus
Lives in American seas. Eats vegetable matter; becomes tame; is delighted by music. Flesh edible.

SLOTH
Bradypus Tridactylus
Body very hairy, tenaceous of life. Climbs easily; walks with difficulty and exceedingly slowly, turns its head as though in astonishment; call, an exciting senarius; noise frightful, tears pitiful.

CAT

Felis Catus

Habits like allied forms, quiet, purrs (*ore molat*), erects tail; when roused is most agile, climbs, when angry emits an ambrosial odor; the lion of mice, she moves her tail when intent on prey; her eyes shine in the night, when she desires prey she devours it eagerly; makes love wretchedly with yowling and squabbling. Eats meat but disdains vegetables; in the next instant washes her face with her hand; when thrown from a high place, falls on her feet. Doesn't have fleas.

CAMEL

Camelus Dromedarius

A second, chambered stomach for pure water providing for a long time in the thirsty desert. Carries burdens, makes haste slowly, when weary lies down on its breast.

GILBERT WHITE (1720–1823)

Concerning Birds, Bats, and a Tortoise

From *The Natural History and Antiquities of Selborne.* 1789

[This gentle English curate was perhaps the first naturalist who clothed his observations in a real literary style. Although he corresponded with such scientific men as Thomas Pennant, yet he himself never went far from home, and only through others did he endeavor to make his little discoveries concrete parts of science.

Among such minor additions to knowledge is the fact that the real tail of the peacock is composed of short feathers which act as mechanical struts and supports to stiffen and hold erect the flimsy but enormously elongated and beautiful plumes of the train, which sprout from the lower back. And his keen eyes detected the little bats that fluttered around his home using their tail membranes as a creel to hold their prey while they secured a fresh grip.

The power of the written word has kept alive and vivid the recounting of these simple facts throughout a century and a half and in many editions.]

SELBORNE, *September* 9, 1767

I WAS much entertained last summer with a tame bat, which would take flies out of a person's hand. If you gave it anything to eat, it brought its wings round before the mouth, hovering and hiding its head in the manner of the birds of prey when they feed. The adroitness it showed in shearing off the wings of the flies, which were always rejected, was worthy of observation, and pleased me much. Insects seemed to be most acceptable, though it did not refuse raw flesh when offered: so that the notion, that bats go down chimneys and gnaw men's bacon, seems no improbable story. While I amused myself with this wonderful quadruped, I saw it several times confute the vulgar opinion, that bats when down upon a flat surface cannot get on the wing again, by rising with great ease from the floor. It ran, I observed, with more dispatch than I was aware of; but in a most ridiculous and grotesque manner.

SELBORNE, *April* 18, 1768

Dear Sir, — (As I had set my mind on the pleasure of y^r conversation, so I was in proportion disappointed when I found that you could not come. But as y^r business may be over now I shall still live in hopes of seeing you at this beautiful season, when every hedge and field abounds with matter of entertainment for the curious. If you could come down at the end of this week, or the beginning of next, I should be ready to partake with you in a post-chaise back to town on the second of May.)

The history of the stone-curlew, *charadrius aedicnemus,* is as follows. It lays its eggs, usually two, never more than three, on the bare ground, without any nest, in the field; so that the countryman, in stirring his fallows, often destroys them. The young run immediately from the egg, like partridges, etc., and are withdrawn to some flinty field by the dam, where they skulk among the stones, which are their best security; for their feathers are so exactly of the colour of our grey spotted flints, that the most exact observer, unless he catches the eye of the young bird, may be eluded. The eggs are short and round; of a dirty white, spotted with dark, bloody blotches. Though I might not be able, just when I pleased, to procure you a bird, yet I could show you them almost any day; and any evening you may hear them round the village, for they make a clamour which may be heard a mile. *Oedicnemus* is a most apt and expressive name for them, since their legs seem swoln like those of a gouty man. After harvest I have shot them before the pointers in turnip-fields.

The grasshopper-lark began his sibilous note in my fields last Saturday. Nothing can be more amusing than the whisper of this little bird, which seems to be close by though at an hundred yards distance; and, when close at your ear, is scarce any louder than when a great way off. Had I not been a little acquainted with insects, and known that the grasshopper kind is not yet hatched, I should have hardly believed but that it had been a *locusta* whispering in the bushes. The country people laugh when you tell them that it is the note of a bird. It is a most artful creature, skulking in the thickest part of a bush; and will sing at a yard distance, provided it be concealed. I was obliged to get a person to go on the other side of the hedge where it haunted, and then it would run, creeping like a mouse, before us for an hundred yards together, through the bottom of the thorns; yet it would not come into fair sight: but in a morning early, and when undisturbed, it sings on the top of a twig, gaping and shivering with its wings.

November 28, 1786

And here will be the properest place to mention, while I think of it, an anecdote which was told me by a gentleman, that, in a warren joining to his outlet, many daws (*corvi monedulae*) build every year in the rabbit-burrows under ground. The way he and his brothers used to take their nests, while they were boys, was by listening at the mouths of the holes; and if they heard the young ones cry, they twisted the nest out with a forked stick. Some water-fowls (*viz.* the puffins) breed, I know, in that manner; but I should never have suspected the daws of building in holes on the flat ground.

Another very unlikely spot is made use of by daws as a place to breed in, and that is *Stonehenge*. These birds deposit their nests in the interstices between the upright and the impost stones of that amazing work of antiquity: which circumstance alone speaks the prodigious height of the upright stones, that they should be tall enough to secure those nests from the annoyance of shepherd-boys, who are always idling round that place.

Happening to make a visit to my neighbour's peacocks, I could not help observing that the trains of those magnificent birds appear by no means to be their tails; those long feathers growing not from their *uropygium*, but all up their backs. A range of short brown stiff feathers, about six inches long, fixed in the *uropygium*, is the real tail, and serves as the fulcrum to prop the train, which is long and top-heavy, when set on end. When the train is up, nothing appears of the bird before but it's head and neck; but this would not be the case were those long feathers fixed only in the rump, as may be seen by the turkey-cock when in a strutting attitude. By a strong muscular vibration these birds can make the shafts of their long feathers clatter like the swords of a sword-dancer; they then trample very quick with their feet, and run backwards towards the females.

[A paragraph under date of March 19, 1772 makes very real one of the contemporaries of Gilbert White.]

When I came to London I found a long letter from Linnaeus to my Bro: John lying in Fleet-street, occasioned by an epistle & some phials of insects sent by the latter to the former. The old arch-naturalist writes with spirit still; & is very open and communicative, acknowledging that several of the Insects were new to him. He languishes to see a *pranticola,* being conscious that it belongs not to the genus of *hirundo.*

SELBORNE, *March* 15th, 1773

Dear Sir, — By my journal for last autumn it appears that the house-martins bred very late, and stayed very late in these parts; for, on the first of October, I saw young martins in their nest nearly fledged; and again on the twenty-first of October, we had at the next house a nest full of young martins just ready to fly; and the old ones were hawking for insects with great alertness. The next morning the brood forsook their nest, and were flying round the village. From this day I never saw one of the swallow kind till November the third; when twenty, or perhaps thirty, house-martins were playing all day long by the side of the hanging wood, and over my field. Did these small weak birds, some of which were nestlings twelve days ago, shift their quarters at this late season of the year to the other side of the northern tropic? Or rather, is it not more probable that the next church, ruin, chalk-cliff, steep covert, or perhaps sandbank, lake or pool (as a more northern naturalist would say), may become their hybernaculum, and afford them a ready and obvious retreat?

April 12, 1772

On the first of November I remarked that the old tortoise, formerly mentioned, began first to dig the ground in order to the forming its hybernaculum, which it had fixed on just beside a great tuft of hepaticas. It scrapes out the ground with its fore-feet, and throws it up over its back with its hind; but the motion of its legs is ridiculously slow, little exceeding the hour-hand of a clock; and suitable to the composure of an animal said to be a whole month in performing one feat of copulation. Nothing can be more assiduous than this creature night and day in scooping the earth, and forcing its great body into the cavity; but as the noons of that season proved unusually warm and sunny, it was continually interrupted, and called forth by the heat in the middle of the day; and though I continued there till the thirteenth of November, yet the work remained unfinished. Harsher weather, and frosty mornings, would have quickened its operations. No part of its behavior ever struck me more than the extreme timidity it always expresses with regard to rain; for though it has a shell that would secure it against the wheel of a loaded cart, yet does it discover as much solicitude about rain as a lady dressed in all her best attire, shuffling away on the first sprinklings, and running its head up in a corner. If attended to, it becomes an excellent weather-glass; for as sure as it walks elate, and as it were on tiptoe, feeding with great earnestness in a morning, so sure will it rain before night. It is totally a diurnal animal, and never pretends to stir after it becomes dark. The tortoise, like other reptiles, has an arbitrary stomach as well as lungs; and can refrain from eating as well as breathing for a great part of the year. When first awakened it eats nothing; nor again in the autumn before it retires: through the height of the summer it feeds voraciously, devouring all the food that comes in its way. I was much taken with its sagacity in discerning those that do it kind offices: for as soon as the good old lady comes in sight who has waited on it for more than thirty years, it hobbles toward its benefactress with awkward alacrity; but remains inattentive to strangers. Thus not only "the ox knoweth his owner, and the ass his master's crib," but the most abject reptile and torpid of beings distinguishes the hand that feeds it, and is touched with the feelings of gratitude.

[Under December 9 we find this note:]

The old tortoise that I have mentioned in a former letter, still continues in this garden; and retired underground about the twentieth of November, and came out again for one day on the thirtieth: it lies now buried in a wet swampy border under a wall facing to the south, and is enveloped at present in mud and mire!

[Eight years later, under date of April 21, 1780, we read as follows:]

The old Sussex tortoise, that I have mentioned to you so often, is become my property. I dug it out of it's winter dormitory in March last, when it was enough awakened to express it's resentments by hissing; and, packing it in a box with earth, carried it eighty miles in post-chaises. The rattle and hurry of the journey so perfectly roused it that, when I turned it out on a border, it walked twice down to the bottom of my garden; however, in the evening, the weather being cold, it buried itself in the loose mould, and continues still concealed.

As it will be under my eye, I shall now have an opportunity of enlarging my observations on it's mode of life, and propensities; and perceive already that, towards the time of coming forth, it opens a breathing place in the ground near it's head, requiring, I conclude, a freer respiration as it becomes more alive. The creature not only goes under the earth from the middle of November to the middle of April, but sleeps a great part of the summer: for it goes to bed in the longest days at four in the afternoon, and often does not stir in the morning till late. Besides, it retires to rest for every shower, and does not move at all in wet days.

When one reflects on the state of this strange being, it is a matter of wonder to find that Providence should bestow such a profusion of days, such a seeming waste of longevity, on a reptile that appears to relish it so little as to squander more than two-thirds of it's existence in a joyless stupor, and be lost to all sensation for months together in the profoundest of slumbers.

While I was writing this letter, a moist and warm afternoon, with the thermometer at 50, brought forth troops of shell-snails; and, at the same juncture, the tortoise heaved up the mould and put out it's head; and the next morning came forth, as it were raised from the dead; and walked about till four in the afternoon. This was a curious coincidence! a very amusing occurrence! to see such a similarity of feelings between both the shell-snail and the tortoise.

Because we call this creature an abject reptile, we are too apt to undervalue his abilities, and depreciate his powers of instinct. Yet he is, as Mr. Pope says of his lord,

" — Much too wise to walk into a well:"

and has so much discernment as not to fall down an haha, but to stop and withdraw from the brink with the readiest precaution.

Though he loves warm weather he avoids the hot sun; because his thick shell, when once heated, would, as the poet says of solid armour, "scald with safety." He therefore spends the more sultry hours under the umbrella of a large cabbage-leaf, or amidst the waving forests of an asparagus-bed.

But, as he avoids heat in the summer, so, in the decline of the year, he improves the faint autumnal beams, by getting within the reflection of a fruit-wall; and, though he never has read that planes inclining to the horizon receive a greater share of warmth, he inclines his shell, by tilting it against the wall, to collect and admit every feeble ray.

Pitiable seems the condition of this poor embarrassed reptile; to be cased in a suit of ponderous armour, which he cannot lay aside; to be imprisoned, as it were, within his own shell, must preclude, we should suppose, all activity and disposition for enterprise. Yet there is a season of the year (usually at the beginning of June) when his exertions are remarkable. He then walks on tiptoe, and is stirring by five in the morning; and, traversing the garden, examines every wicket and interstice in the fences, through which he will escape if possible; and often has eluded the care of the gardener, and wandered to some distant field. The motives that impell him to undertake these rambles seem to be of the amorous kind; his fancy then becomes intent on sexual attachments, which transports him beyond his usual gravity, and induces him to forget for a time his ordinary solemn deportment.

WILLIAM BARTRAM (1739–1823)

The Rattle Snake

From *Travels*. 1791

[WILLIAM was the son of John Bartram, that charming "natural" botanist, that wholly inarticulate lover and collector of rare plants. To an enthusiasm equal to his father's he added a pleasing gift of words. His description of scenes at the time of the early settlement of the states, leavened by accounts of birds and insects, fishes and alligators, is uninspiring but irreplaceable. Read in the blinding light of science, I fear that occasional chaff would be discerned among the kernels of his account of the habits of wild creatures. His "rattle snake," however, although somewhat tarnished with inaccuracies, is too good to omit.]

THE RATTLE SNAKE [is] a wonderful creature, when we consider his form, nature and disposition. It is certain that he is capable by a puncture or scratch of one of his fangs, not only to kill the largest animal in America, and that in a few minutes time, but to turn the whole body into corruption; but such is the nature of this dreadful reptile, that he cannot run or creep faster than a man or child can walk, and he is never known to strike until he is first assaulted or fears himself in danger, and even then always gives the earliest warning by the rattles at the extremity of the tail. I have in the course of my travels in the Southern States (where they are the largest, most numerous and supposed to be the most venomous and vindictive) stept unknowingly so close as almost to touch one of them with my feet, and when I perceived him he was already drawn up in circular coils ready for a blow. But however incredible it may appear, the generous, I may say magnanimous creature lay as still and motionless as if inanimate, his head crouched in, his eyes almost shut. I precipitately withdrew, unless when I have been so shocked with surprise and horror as to be in a manner rivetted to the spot, for a short time not having strength to go away; when he often slowly extends himself and quietly moves off in a direct line, unless pursued, when he erects his tail as far as the rattles extend, and gives the warning alarm by intervals. But if you pursue and overtake him with a show of enmity, he instantly throws himself into the spiral coil; his tail by the rapidity of its motion appears like a vapour, making a quick tremulous sound; his whole body swells through rage, continually rising and falling as a bellows; his beautiful parti-coloured skin becomes speckled and rough by dilatation; his head and neck are flattened, his cheeks swollen and his lips constricted, discovering his mortal fangs; his eyes red as burning coals, and his brandishing forked tongue of the colour of the hottest flame, continually menaces death and destruction, yet never strikes unless sure of his mark.

The rattle snake is the largest serpent yet known to exist in North America. I have heard of their having been found formerly, at the first settling of Georgia, seven, eight and even ten feet in length, and six or eight inches diameter; but there are none of that size now to be seen; yet I have seen them above six feet in length, and above six inches in thickness, or as large as a man's leg; but their general size is four, five, and six feet in length. They are supposed to have the power of fascination in an eminent degree, so as to inthral their prey. It is generally believed that they charm birds, rabbits, squirrels and other animals, and by steadfastly looking at them possess them with infatuation: be the cause what it may, the miserable creatures undoubtedly strive by every possible means to escape, but alas! their endeavours are in vain, they at last lose the power of resistance, and flutter or move slowly, but reluctantly, towards the yawning jaws of their devourers, and creep into their mouths, or lie down and suffer themselves to be taken and swallowed.

Since, within the circle of my acquaintance, I am known to be an advocate or vindicator of the benevolent and peaceable disposition of animal creation in general, not only towards mankind, whom they seem to venerate, but also towards one another, except where hunger or the rational and necessary provocations of the sensual appetite interfere, I shall mention a few instances, amongst many, which I have had an opportunity of remarking during my travels, particularly with regard to the animal I have been treating of. I shall strictly confine myself to facts.

When on the sea coast of Georgia, I consented, with a few friends, to make a party of amusement at fishing and fowling on Sapello, one of the sea coast islands. We accordingly descended the Alatamaha, crossed the sound and landed on the North end of the island, near the inlet, fixing our encampment at a pleasant situation, under the shade of a grove of Live Oaks and Laurels [1] on the high banks of a creek which we ascended, winding through a salt marsh, which had its source from a swamp and savanna in the island: our situation elevated and open, commanded a comprehensive landscape; the great ocean, the foaming surf breaking on the sandy beach, the snowy breakers on the bar, the endless chain of islands, checkered sound and high continent all appearing before us. The diverting toils of the day were not fruitless, affording us opportunities of furnishing ourselves plentifully with a variety of game, fish and oysters for our supper.

About two hundred yards from our camp was a cool spring, amidst a grove of the odoriferous Myrica: the winding path to this salubrious fountain led through a grassy savanna. I visited the spring several times in the night, but little did I know, or any of my careless drowsy companions, that every time we visited the fountain we were in imminent danger, as I am going to relate. Early in the morning, excited by unconquerable thirst, I arose and went to the spring; and having, thoughtless of harm or danger, nearly half past the dewy vale, along the serpentine foot path, my hasty steps were suddenly stopped by the sight of a hideous serpent, the formidable rattle snake, in a high spiral coil, forming a circular mound half the height of my knees, within six inches of the narrow path. As soon as I recovered my senses and strength from so sudden a surprise, I started back out of his reach, where I stood to view him: he lay quiet whilst I surveyed him, appearing no way surprised or disturbed, but kept his half-shut eyes fixed on me. My imagination and spirits were in a tumult, almost equally divided betwixt thanksgiving to the supreme Creator and preserver, and the dignified nature of the generous though terrible creature, who had suffered us all to pass many times by him during the night, without injuring us in the least, although we must have touched him, or our steps guided therefrom by a supreme guardian spirit. I hastened back to acquaint my associates, but with a determination to protect the life of the generous serpent. I presently brought my companions to the place, who were, beyond expression, surprised and terrified at the sight of the animal, and in a moment acknowledged their escape from destruction to be miraculous; and I am proud to assert, that all of us, except one person, agreed to let him lie undisturbed, and that that person at length was prevailed upon to suffer him to escape.

Again, when in my youth, attending my father on a journey to

1 Magnolia grandiflora, called by the inhabitants the Laurel.

the Catskill Mountains, in the government of New-york; having nearly ascended the peak of Giliad, being youthful and vigorous in the pursuit of botanical and novel objects, I had gained the summit of a steep rocky precipice, a-head of our guide; when just entering a shady vale, I saw at the root of a small shrub, a singular and beautiful appearance, which I remember to have instantly apprehended to be a large kind of Fungus which we call Jews ears, and was just drawing back my foot to kick it over; when at the instant, my father being near, cried out, a rattlesnake my son! and jerked me back, which probably saved my life. I had never before seen one. This was of the kind which our guide called a yellow one, it was very beautiful, speckled and clouded. My father pleaded for his life, but our guide was inexorable, saying he never spared the life of a rattle snake, and killed him; my father took his skin and fangs.

Some years after this, when again in company with my father on a journey into East Florida, on the banks of St. Juan, at Fort Picolata, attending the congress of a treaty between that government and the Creek Nation, for obtaining a territory from that people to annex to the new government; after the Indians and a detachment from the garrison of St. Augustine had arrived and encamped separately, near the fort, some days elapsed before the business of the treaty came on, waiting the arrival of a vessel from St. Augustine, on board of which were the presents for the Indians. My father employed this time of leisure in little excursions round about the fort; and one morning, being the day the treaty commenced, I attended him on a botanical excursion. Some time after we had been rambling in a swamp about a quarter of a mile from the camp, I being a-head a few paces, my father bid me observe the rattle snake before and just at my feet. I stopped and saw the monster formed in a high spiral coil, not half his length from my feet; another step forward would have put my life in his power, as I must have touched if not stumbled over him. The fright and perturbation of my spirits at once excited resentment; at that time I was entirely insensible to gratitude or mercy. I instantly cut off a little sapling and soon dispatched him: this serpent was about six feet in length, and as thick as an ordinary man's leg. The rencounter deterred us from proceeding on our researches for that day. So I cut off a long tough withe or vine, which fastening round the neck of the slain serpent, I dragged him after me, his scaly body sounding over the ground, and entering the camp with him in triumph, was soon surrounded by the amazed multitude, both Indians and my countrymen. The adventure soon reached the ears of the commander, who sent an officer to request that, if the snake had not bit himself, he might have him served up for his dinner. I readily delivered up the body of the snake to the cooks, and being that day invited to dine at the governor's table, saw the snake served up in several dishes; governor Grant being fond of the flesh of the rattle snake. I tasted of it but could not swallow it. I, however, was sorry after killing the serpent, when cooly recollecting every circumstance. He certainly had it in his power to kill me almost instantly, and I make no doubt but that he was conscious of it. I promised myself that I should never again be accessary to the death of a rattle snake, which promise I have invariably kept to. This dreaded animal is easily killed; a stick no thicker than a man's thumb is sufficient to kill the largest at one stroke, if well directed, either on the head or across the back; nor can they make their escape by running off, nor indeed do they attempt it when attacked.

ALEXANDER VON HUMBOLDT (1769–1859)

Jungle River

From *Views of Nature*. 1808

[BARON VON HUMBOLDT, born in Berlin, gained and maintained a pre-eminent place among the scientists and travelers of the world throughout a full life of ninety years. His first journey to South America is the one of most interest to naturalists, but even here geology and physical problems are usually given dominant treatment. Accounts of the zoological aspects of his expedition are somewhat diminished in value by his credulity in accepting the stories of Indians. But all this is compensated by his wonderful breadth of view, the connected sequence of his observations, and the beauty of his language. Charles Darwin, after himself traversing tropical jungles, wrote: "From what I have seen, Humboldt's glorious descriptions are and will forever be unparalleled."]

ON LEAVING the Island del Diamante, where the Zambos, who speak Spanish, cultivate the sugar-cane, we entered into a grand and wild domain of nature. The air was filled with countless flamingoes (*Phoenicopterus*) and other water-fowl, which seemed to stand forth from the blue sky like a dark cloud in ever-varying outlines. The bed of the river had here contracted to less than 1000 feet, and formed a perfectly straight canal, which was inclosed on both sides by thick woods. The margin of the forest presents a singular spectacle. In front of the almost impenetrable wall of colossal trunks of Caesalpinia, Cedrela, and Desmanthus, there rises with the greatest regularity on the sandy bank of the river, a low hedge of Sauso, only four feet high; it consists of a small shrub, *Hermesia castanifolia,* which forms a new genus of the family of Euphorbiaceae. A few slender, thorny palms, called by the Spaniards Piritu and Corozo (perhaps species of *Martinezia* or *Bactris*) stand close alongside; the whole resembling a trimmed garden hedge, with gate-like openings at considerable distances from each other, formed undoubtedly by the large four-footed animals of the forests, for convenient access to the river. At sunset, and more particularly at break of day, the American Tiger, the Tapir, and the Peccary (*Pecari, Dicotyles*) may be seen coming forth from these openings accompanied by their young, to give them drink. When they are disturbed by a passing Indian canoe, and are about to retreat into the forest, they do not attempt to rush violently through these hedges of Sauso, but proceed deliberately along the bank, between the hedge and river, affording the traveller the gratification of watching their motions for sometimes four or five hundred paces, until they disappear through the nearest opening. During a seventy-four days' almost uninterrupted river navigation of 1420 miles up the Orinoco, to the neighborhood of its sources, and along the Cassiquiare, and the Rio Negro — during the whole of which time we were confined to a narrow canoe — the same spectacle presented itself to our view at many different points, and, I may add, always with renewed excite* ment. There came to drink, bathe, or fish, groups of creatures belonging to the most opposite species of animals; the larger mammalia with many-coloured herons, palamedeas with the proudly-strutting curassow (*Crax Alector, C. Pauxi*). "It is here as in Paradise" (*es como en el Paradiso*), remarked with pious air our steersman, an old Indian, who had been brought up in the house of an ecclesiastic. But the gentle peace of the primitive golden age does not reign in the paradise of these American animals, they stand apart, watch, and avoid each other. The Capybara, a cavy (or river-hog) three or four feet long (a colossal repetition of the common Brazilian cavy, (*Cavia Aguti*), is devoured in the river by the crocodile, and on the shore by the tiger. They run so badly, that we were frequently able to overtake and capture several from among the numerous herds.

Below the mission of Santa Barbara de Arichuna we passed the night as usual in the open air, on a sandy flat, on the bank of the Apure, skirted by the impenetrable forest. We had some difficulty in finding dry wood to kindle the fires with which it is here customary to surround the bivouac, as a safeguard against the attacks of the Jaguar. The air was bland and soft, and the moon shone brightly. Several crocodiles approached the bank; and I have observed that fire attracts these creatures as it does our crabs and many other aquatic animals. The oars of our boats were fixed upright in the ground, to support our hammocks. Deep stillness prevailed, only broken at intervals by the blowing of the fresh-water dolphins, which are peculiar to the river net-work of the Orinoco (as, according to Colebrooke, they are also to the Ganges, as high up the river as Benares) ; they followed each other in long tracks.

After eleven o'clock, such a noise began in the contiguous forest, that for the remainder of the night all sleep was impossible. The wild cries of animals rung through the woods. Among the many voices which resounded together, the Indians could only recognize those which, after short pauses, were heard singly. There was the monotonous, plaintive, cry of the Aluates (howling monkeys), the whining, flute-like notes of the small sapajous, the grunting murmur of the striped nocturnal ape (*Nyctipithecus trivirgatus,* which I was the first to describe), the fitful roar of the great tiger, the Cuguar or maneless American lion, the peccary, the sloth, and a host of parrots, parraquas (*Ortalides*), and other pheasant-like birds. Whenever the tigers approached the edge of the forest, our dog, who before had barked incessantly, came howling to seek protection under the hammocks. Sometimes the cry of the tiger resounded from the branches of a tree, and was then always accompanied by the plaintive piping tones of the apes, who were endeavouring to escape from the unwonted pursuit.

If one asks the Indians why such a continuous noise is heard on certain nights, they answer, with a smile, that "the animals are rejoicing in the beautiful moonlight, and celebrating the return of the full moon." To me the scene appeared rather to be owing to an accidental, long-continued, and gradually increasing conflict among the animals. Thus, for instance, the jaguar will pursue the peccaries and the tapirs, which, densely crowded together, burst through the barrier of tree-like shrubs which opposes their flight. Terrified at the confusion, the monkeys on the tops of the trees join their cries with those of the larger animals. This arouses the tribes of birds who build their nests in communities, and suddenly the whole animal world is in a state of commotion. Further experience taught us that the voices were loudest during violent storms of rain, or when the thunder echoed and the lightning flashed through the depths of the woods. The good-natured Franciscan monk who (notwithstanding the fever from which he had been suffering for many months), accompanied us through the cataracts of Atures and Maypures to San Carlos, on the Rio Negro, and to the Brazilian coast, used to say, when apprehensive of a storm at night, "May Heaven grant a quiet night both to us and to the wild beasts of the forest!"

CHARLES WATERTON (1782–1865)

The Sloth

From *Wanderings in South America*. 1825

[WATERTON fills an appreciable gap in the writings of naturalists. His several visits to the Guianas in South America covered an extent of several years, and in his *Wanderings* and several volumes of essays he contributed much to our knowledge of the animal life of the tropical jungle. His explorations occurred at almost the same time as those of Humboldt, but the latter was concerned chiefly with physical geography, meteorology, and botany. So until the coming of Wallace and Bates into neighboring territory the writings of Waterton provide our only first-hand information.

In the appraisal of Waterton I can do no better than quote a paragraph from a review of the *Wanderings* by Sydney Smith, written in 1826: "Now, what shall we say, after all, of Mr. Waterton? That he has spent a great part of his life in wandering in the wild scenes he describes, and that he describes them with entertaining zeal and real feeling. His stories draw largely sometimes on our faith; but a man who lives in the woods of Cayenne must do many odd things, and see many odd things — things utterly unknown to the dwellers in Hackney and Highgate. We do not want to rein up Mr. Waterton too tightly — because we are convinced he goes best with his head free. But a little less of apostrophe, and some faint suspicion of his own powers of humour, would improve this gentleman's style."

His account of riding upon an enormous hooked crocodile and that of capturing a good-sized boa with his bare hands, while related with extravagant phrases, are very likely true, but they have been used to cast doubt on the authenticity of some of his other writing. His facts are, in the main, perfectly reliable, and his diction, while slightly stilted in the fashion of his time, is far more readable than that of some of his successors.

The selection on sloths herewith presented is a just refutation of the opinion of the great savant Buffon, who, in spite of his genius, fell into most grievous error in his estimation of a sloth. He says: "The inertia of this animal is not so much due to laziness as to wretchedness; it is the consequence of its faulty structure. Inactivity, stupidity, and even habitual suffering result from its strange and ill-constructed conformation. Having no weapons for attack or defense, no mode of refuge even by burrowing, its only safety is in flight. . . . Everything about it shows its wretchedness and proclaims it to be one of those defective monsters, those imperfect sketches, which Nature has sometimes formed, and which, having scarcely the faculty of existence, could only continue for a short time and have since been removed from the catalogue of living beings. They are the last possible term amongst creatures of flesh and blood, and any further defect would have made their existence impossible." If we imagine the dignified French savant himself, naked, and dangling from a lofty jungle branch in the full heat of the tropic sun, without water, and with the prospect of nothing but coarse leaves for breakfast, dinner, and all future meals, an impartial onlooker who was ignorant of man's normal haunts and life could very truthfully apply to the unhappy scientist Buffon's own comments. All of his terms of opprobrium would come home to roost with him. The sloth, as Waterton so well points out, is as perfectly adapted to its mode of life and, as far as we can tell, as unconsciously content with existence as any other tropical organism.]

L ET US NOW turn our attention to the Sloth, whose native haunts have hitherto been so little known, and probably little looked into. Those who have written on this singular animal, have remarked that he is in a perpetual state of pain, that he is proverbially slow in his movements, that he is a prisoner in space, and that as soon as he has consumed all the leaves of the tree upon which he had mounted, he rolls himself up in the form of a ball, and then falls to the ground. This is not the case.

If the naturalists who have written the history of the sloth had gone into the wilds, in order to examine his haunts and economy, they would not have drawn the foregoing conclusions; they would have learned, that though all other quadrupeds may be described while resting upon the ground, the sloth is an exception to this rule, and that his history must be written while he is in the tree.

This singular animal is destined by nature to be produced, to live and to die in the trees; and to do justice to him, naturalists must examine him in this his upper element. He is a scarce and solitary animal, and being good food, he is never allowed to escape. He inhabits remote and gloomy forests, where snakes take up their abode, and where cruelly stinging ants and scorpions, and swamps, and innumerable thorny shrubs and bushes, obstruct the steps of civilized man. Were you to draw your own conclusions from the descriptions which have been given of the sloth, you would probably suspect, that no naturalist has actually gone into the wilds with the fixed determination to find him out and examine his haunts and see whether nature has committed any blunder in the formation of this extraordinary creature, which appears to us so forlorn and miserable, so ill put together, and so totally unfit to enjoy the blessings which have been so bountifully given to the rest of animated nature; for, as it has formerly been remarked, he has no soles to his feet, and he is evidently ill at ease when he tries to move on the ground, and it is then that he looks up in your face with a countenance that says, "Have pity on me, for I am in pain and sorrow."

It mostly happens that Indians and Negroes are the people who catch the sloth, and bring it to the white man: hence it may be conjectured that the erroneous accounts we have hitherto had of the sloth, have not been penned down with the slightest intention to mislead the reader, or give him an exaggerated history, but that these errors have naturally arisen by examining the sloth in those places where nature never intended that he should be exhibited.

However, we are now in his own domain. Man but little frequents these thick and noble forests, which extend far and wide on every side of us. This, then, is the proper place to go in quest of the sloth. We will first take a near view of him. By obtaining a knowledge of his anatomy, we shall be enabled to account for his movements hereafter, when we see him in his proper haunts. His fore-legs, or, more correctly speaking, his arms, are apparently much too long, while his hind-legs are very short, and look as if they could be bent almost to the shape of a corkscrew. Both the fore and hind legs, by their form, and by the manner in which they are joined to the body, are quite incapacitated from acting in a perpendicular direction, or in supporting it on the earth, as the bodies of other quadrupeds are supported, by their legs. Hence, when you place him on the floor, his belly touches the ground. Now, granted that he supported himself on his legs like other animals, nevertheless he would be in pain, for he has no soles to his feet, and his claws are very sharp and long, and curved; so that, were his body supported by his feet it would be by their extremities, just as your body would be, were you to throw yourself on all fours, and try to support it on the ends of your toes and fingers — a trying position. Were the floor of glass, or of a polished surface, the sloth would actually be quite stationary; but as the ground is generally rough, with little protuberances upon it, such as stones, or roots of grass, &c., this just suits the sloth, and he moves his fore-legs in all directions, in order to find something to lay hold of; and when he has succeeded, he pulls himself forward, and is thus enabled to travel onwards, but at the same time in so tardy and awkward a manner, as to acquire him the name of Sloth.

Indeed his looks and his gestures evidently betray his uncomfortable situation; and as a sigh every now and then escapes him, we may be entitled to conclude that he is actually in pain.

Some years ago I kept a sloth in my room for several months. I often took him out of the house and placed him upon the ground, in order to have an opportunity of observing his motions. If the ground were rough, he would pull himself forwards, by means of his fore-legs, at a pretty good pace; and he invariably immediately shaped his course towards the nearest tree. But if I put him upon a smooth and well-trodden part of the road, he appeared to be in trouble and distress: his favourite abode was the back of a chair: and after getting all his legs in a line upon the topmost part of it, he would hang there for hours together, and often with a low and inward cry, would seem to invite me to take notice of him.

The sloth, in its wild state, spends its whole life in trees, and never leaves them but through force or by accident. An all-ruling Providence has ordered man to tread on the surface of the earth, the eagle to soar in the expanse of the skies, and the monkey and squirrel to inhabit the trees: still these may change their relative situations without feeling much inconvenience: but the sloth is doomed to spend his whole life in the trees; and, what is more extraordinary, not *upon* the branches, like the squirrel and the monkey, but *under* them. He moves suspended from the branch, he rests suspended from it, and he sleeps suspended from it. To enable him to do this, he must have a very different formation from that of any other known quadruped.

Hence his seemingly bungled conformation is at once accounted for; and in lieu of the sloth leading a painful life and entailing a melancholy and miserable existence on its progeny, it is but fair to surmise that it just enjoys life as much as any other animal, and that its extraordinary formation and singular habits are but further proofs to engage us to admire the wonderful works of Omnipotence.

It must be observed, that the sloth does not hang head-downwards like the vampire. When asleep, he supports himself from a branch parallel to the earth. He first seizes the branch with one arm, and then with the other; and after that, brings up both his legs, one by one, to the same branch; so that all four are in a line: he seems perfectly at rest in this position. Now, had he a tail, he would be at a loss to know what to do with it in this position: were he to draw it up within his legs, it would interfere with them; and were he to let it hang down, it would become the sport of the winds. Thus his deficiency of tail is a benefit to him; it is merely an apology for a tail, scarcely exceeding an inch and a half in length.

I observed, when he was climbing, he never used his arms both together, but first one and then the other, and so on alternately. There is a singularity in his hair, different from that of all other animals, and, I believe, hitherto unnoticed by naturalists; his hair is thick and coarse at the extremity, and gradually tapers to the root, where it becomes fine as a spider's web. His fur has so much the hue of the moss which grows on the branches of the trees, that it is very difficult to make him out when he is at rest.

The male of the three-toed sloth has a longitudinal bar of very fine black hair on his back, rather lower than the shoulder-blades; on each side of this black bar there is a space of yellow hair, equally fine; it has the appearance of being pressed into the body, and looks exactly as if it had been singed. If we examine the anatomy of his fore-legs, we shall immediately perceive by their firm and muscular texture, how very capable they are of supporting the pendent weight of his body, both in climbing and at rest; and, instead of pronouncing them a bungled composition, as a celebrated naturalist has done, we shall consider them as remarkably well calculated to perform their extraordinary functions.

As the sloth is an inhabitant of forests within the tropics, where the trees touch each other in the greatest profusion, there seems to be no reason why he should confine himself to one tree alone for food, and entirely strip it of its leaves. During the many years I have ranged the forests, I have never seen a tree in such a state of nudity; indeed, I would hazard a conjecture, that, by the time the animal had finished the last of the old leaves, there would be a new crop on the part of the tree he had stripped first, ready for him to begin again, so quick is the process of vegetation in these countries.

There is a saying amongst the Indians, that when the wind blows, the sloth begins to travel. In calm weather he remains tranquil, probably not liking to cling to the brittle extremity of the branches, lest they should break with him in passing from one tree to another; but as soon as the wind rises, the branches of the neighboring trees become interwoven, and then the sloth seizes hold of them, and pursues his journey in safety. There is seldom an entire day of calm in these forests. The trade-wind generally sets in about ten o'clock in the morning, and thus the sloth may set off after breakfast, and get a considerable way before dinner. He travels at a good round pace; and were you to see him pass from tree to tree, as I have done, you would never think of calling him a sloth.

Thus, it would appear that the different histories we have of this quadruped are erroneous on two accounts: first, that the writers of them, deterred by difficulties and local annoyances, have not paid sufficient attention to him in his native haunts; and secondly, they have described him in a situation in which he was never intended by nature to cut a figure; I mean on the ground. The sloth is as much at a loss to proceed on his journey upon a smooth and level floor, as a man would be who had to walk a mile in stilts upon a line of feather beds.

One day, as we were crossing the Essequibo, I saw a large two-toed sloth on the ground upon the bank; how he had got there nobody could tell: the Indian said he had never surprised a sloth in such a situation before: he would hardly have come there to drink, for both above and below the place, the branches of the trees touched the water, and afforded him an easy and safe access to it. Be this as it may, though the trees were not above twenty yards from him, he could not make his way through the sand in time enough to escape before we landed. As soon as we got up to him he threw himself upon his back, and defended himself in gallant style with his fore-legs. "Come, poor fellow," said I to him, "if thou hast got into a hobble today, thou shalt not suffer for it: I'll take no advantage of thee in misfortune; the forest is large enough both for thee and me to rove in: go thy ways up above, and enjoy thyself in these endless wilds; it is more than probable thou wilt never have another interview with man. So fare thee well." On saying this, I took a long stick which was lying there, held it for him to hook on, and then conveyed him to a high and stately mora. He ascended with wonderful rapidity, and in about a minute he was almost at the top of the tree. He now went off in a side direction, and caught hold of the branch of a neighbouring tree; he then proceeded towards the heart of the forest. I stood looking on, lost in amazement at his singular mode of progress. I followed him with my eye till the intervening branches closed in betwixt us; and then I lost sight for ever of the two-toed sloth. I was going to add, that I never saw a sloth take to his heels in such earnest; but the expression will not do, for the sloth has no heels.

JOHN JAMES AUDUBON (1785–1851)

The Wild Turkey

From *The Birds of America.* 1827–1838

[AUDUBON's early life, begun in Haiti, was that of a well-to-do sports-man and country gentleman, but after several unsuccessful business ventures he devoted himself to a great work on American birds, upon which rests his chief reason for fame. He was forty-two years of age when the first of the four elephant folios appeared, to be completed with 435 magnificent, hand-colored plates. This classic, in both pictures and text, suffers like some old masters in comparison with more recent work, but under the early, almost primitive conditions of its conception and completion it set a wholly new standard. For the first time· birds were shown in natural positions, as compared, for instance, with the woodeny profiles in Wilson's illustrations.

As the plate of the Wild Turkey is almost the best known of Audubon's illustrations, I have chosen a selection from the account of this bird. The writing is quite devoid of "beauty of form or emotional effect," but is valuable as one of the earliest, factual, first-hand accounts of the bird life of our country.]

THE TURKEY is irregularly migratory, as well as irregularly gre-garious. With reference to the first of these circumstances, I have to state, that whenever the *mast* of one portion of the country happens greatly to exceed that of another, the Turkeys are insensibly led toward that spot, by gradually meeting in their haunts with more fruit the nearer they advance towards the place where it is most plentiful. In this manner flock follows after flock, until one district is entirely deserted, while another is, as it were, overflowed by them. But as these migrations are irregular, and extend over a vast expanse of country, it is necessary that I should describe the manner in which they take place.

About the beginning of October, when scarcely any of the seeds and fruits have yet fallen from the trees, these birds assemble in flocks, and gradually move towards the rich bottom lands of the Ohio and Mississippi. The males, or, as they are more commonly called, the *gobblers,* associate in parties of from ten to a hundred, and search for food apart from the females; while the latter are seen either advancing singly, each with its brood of young, then about two-thirds grown, or in connexion with other families, forming parties often amounting to seventy or eighty individuals, all intent on shunning the old cocks, which, even when the young birds have attained this size, will fight with, and often destroy them by repeated blows on the head. Old and young, however, all move in the same course, and on foot, unless their progress be interrupted by a river, or the hunter's dog force them to take wing. When they come upon a river, they betake themselves to the highest eminences, and there often remain a whole day, or sometimes two, as if for the purpose of consultation. During this time the males are heard *gobbling,* calling, and making much ado, and are seen strutting about, as if to raise their courage to a pitch befitting the emergency. Even the females and young assume something of the same pompous demeanour, spread out their tails, and run around each other, *purring* loudly, and performing extravagant leaps. At length, when the weather appears settled, and all round is quiet, the whole party mount to the tops of the highest trees, whence, at a signal, consisting of a single *cluck,* given by a leader, the flock takes flight for the opposite shore. The old and fat birds easily get over, even should the river be a mile in breadth; but the younger and less robust frequently fall into the water, — not to be drowned, however, as might be imagined. They bring their wings close to their body, spread out their tail as a support, stretch forward their neck, and striking out their legs with great vigour, proceed rapidly toward the shore; on approaching which, should they find it too steep for landing, they cease their exertions for a few moments, float down the stream until they come to an accessible part, and by a violent effort generally extricate themselves from the water. It is remarkable that immediately after thus crossing a large stream, they ramble about for some time, as if bewildered. In this state, they fall an easy prey to the hunter.

When the Turkeys arrive in parts where the mast is abundant, they separate into smaller flocks, composed of birds of all ages and both sexes, promiscuously mingled, and devour all before them. This happens about the middle of November. So gentle do they sometimes become after these long journeys, that they have been seen to approach the farm-houses, associate with the domestic fowls, and enter the stables and corn-cribs in quest of food. In this way, roaming about the forests, and feeding chiefly on mast, they pass the autumn and part of the winter.

As early as the middle of February, they begin to experience the impulse of propagation. The females separate, and fly from the males. The latter strenuously pursue, and begin to gobble or to utter the notes of exultation. The sexes roost apart, but at no great distance from each other. When a female utters a call-note, all the gobblers within hearing return the sound, rolling note after note with as much rapidity as if they intended to emit the last and the first together, not with spread tail, as when fluttering round the females on the ground, or practising on the branches of the trees on which they have roosted for the night, but much in the manner of the domestic Turkey, when an unusual or unexpected noise elicits its singular hubbub. If the call of the female comes from the ground, all the males immediately fly towards the spot, and the moment they reach it, whether the hen be in sight or not, spread out and erect their tail, draw the head back on the shoulders, depress their wings with a quivering motion, and strut pompously about, emitting at the same time a succession of puffs from the lungs, and stopping now and then to listen and look. But whether they spy the female or not, they continue to puff and strut, moving with as much celerity as their ideas of ceremony seem to admit. While thus occupied, the males often encounter each other, in which case desperate battles take place, ending in bloodshed, and often in the loss of many lives, the weaker falling under the repeated blows inflicted upon their head by the stronger.

I have often been much diverted, while watching two males in fierce conflict, by seeing them move alternately backwards and forwards, as either had obtained a better hold, their wings drooping, their tails partly raised, their body-feathers ruffled, and their heads covered with blood. If, as they thus struggle, and gasp for breath, one of them should lose his hold, his chance is over, for the other, still holding fast, hits him violently with spurs and wings, and in a few minutes brings him to the ground. The moment he is dead, the conqueror treads him under foot, but, what is strange, not with hatred, but with all the motions which he employs in caressing the female. . . .

About the middle of April, when the season is dry, the hens begin to look out for a place in which to deposit their eggs. This place requires to be as much as possible concealed from the eye of the Crow, as that bird often watches the Turkey when going to her nest, and, waiting in the neighbourhood until she has left it, removes and eats the eggs. The nest, which consists of a few withered leaves, is placed on the ground, in a hollow scooped out, by the side of a log, or in the fallen top of a dry leafy tree, under a thicket of sumach or briars, or a few feet within the edge of a canebrake, but always in a dry place. The eggs, which are of a dull cream colour, sprinkled with red dots, sometimes amount to twenty, although the more usual number is from ten to fifteen. When depositing her eggs, the female always approaches the nest with extreme caution, scarcely ever taking the same course twice; and when about to leave them, covers them carefully with leaves, so that it is very difficult for a person who may have seen the bird to discover the nest. Indeed, few Turkeys' nests are found unless the female has been suddenly started from them, or a cunning Lynx, Fox, or Crow has sucked the eggs and left their shells scattered about. . . .

When an enemy passes within sight of a female, while laying or sitting, she never moves, unless she knows that she has been discovered, but crouches lower until he has passed. I have frequently approached within five or six paces of a nest, of which I was previously aware, on assuming an air of carelessness, and whistling or talking to myself, the female remaining undisturbed; whereas if I went cautiously towards it, she would never suffer me to approach within twenty paces, but would run off, with her tail spread on one side, to a distance of twenty or thirty yards, when assuming a stately gait she would walk about deliberately, uttering every now and then a cluck. They seldom abandon their nest, when it has been discovered by men; but, I believe, never go near it again when a snake or other animal has sucked any of the eggs. If the eggs have been destroyed or carried off, the female soon yelps again for a male; but, in general, she rears only a single brood each season. Several hens sometimes associate together, I believe for their mutual safety, deposit their eggs in the same nest, and rear their broods together. I once found three sitting on forty-two eggs. In such cases, the common nest is always watched by one of the females, so that no Crow, Raven, or perhaps even Pole-cat, dares approach it.

A mother will not leave her eggs, when near hatching, under any circumstances, while life remains. She will even allow an enclosure to be made around her, and thus suffer imprisonment rather than abandon them. I once witnessed the hatching of a brood of Turkeys, which I watched for the purpose of securing them together with the parent. I concealed myself on the ground within a very few feet, saw her raise herself half the length of her legs, look anxiously upon the eggs, cluck with a sound peculiar to the mother on such occasions, carefully remove each half-empty shell, and with her bill caress and dry the young birds, that already stood tottering and attempting to make their way out of the nest. Yes, I have seen this, and have left mother and young to better care than mine could have proved, — to the care of their Creator and mine. I have seen them all emerge from the shell, and, in a few moments after, tumble, roll, and push each other forward, with astonishing and inscrutable instinct.

HENRY DAVID THOREAU (1817–1862)

The Pond

From *Walden.* 1854

[THOREAU was born only forty-one years after the signing of the Declaration of Independence. Yet whenever I read him, the lead dust from his home-made pencils seems yet unbrushed from his writings. Everything in *Walden* could happen tomorrow. Volumes have been written about Thoreau's philosophy, his lack of science, his style, mysticism, and inner meanings, but the simplicity and directness of his natural-history observations elude all classification and diagraming. Either you agree with the pince-nez-on-a-black-ribbon estimate of an English poet and say "Thoreau isn't much," or you read, and at intervals reread, *Walden* and his story of the seasons, and become less and less articulate but more quietly, completely satisfied.]

SOMETIMES, having had a surfeit of human society and gossip, and worn out all my village friends, I rambled still farther westward than I habitually dwell, into yet more unfrequented parts of the town, "to fresh woods and pastures new," or, while the sun was setting, made my supper of huckleberries and blueberries on Fair Haven Hill, and laid up a store for several days. The fruits do not yield their true flavor to the purchaser of them, nor to him who raises them for the market. There is but one way to obtain it, yet few take that way. If you would know the flavor of huckleberries, ask the cow-boy or the partridge. It is a vulgar error to suppose that you have tasted huckleberries who never plucked them. A huckleberry never reaches Boston; they have not been known there since they grew on her three hills. The ambrosial and essential part of the fruit is lost with the bloom which is rubbed off in the market cart, and they become mere provender. As long as Eternal Justice reigns, not one innocent huckleberry can be transported thither from the country's hills.

Occasionally, after my hoeing was done for the day, I joined some impatient companion who had been fishing on the pond since morning, as silent and motionless as a duck or a floating leaf, and, after practising various kinds of philosophy, had concluded commonly,

by the time I arrived, that he belonged to the ancient sect of Coeno-
bites. There was one older man, an excellent fisher and skilled in all
kinds of woodcraft, who was pleased to look upon my house as a
building erected for the convenience of fishermen; and I was equally
pleased when he sat in my doorway to arrange his lines. Once in a
while we sat together on the pond, he at one end of the boat, and I
at the other; but not many words passed between us, for he had
grown deaf in his later years, but he occasionally hummed a psalm,
which harmonized well enough with my philosophy. Our intercourse
was thus altogether one of unbroken harmony far more pleasing to
remember than if it had been carried on by speech. When, as was
commonly the case, I had none to commune with, I used to raise
the echoes by striking with a paddle on the side of my boat, filling
the surrounding woods with circling and dilating sound, stirring
them up as the keeper of a menagerie his wild beasts, until I elicited
a growl from every wooded vale and hill-side.

In warm evenings I frequently sat in the boat playing the flute,
and saw the perch, which I seem to have charmed, hovering around
me, and the moon travelling over the ribbed bottom, which was
strewed with the wrecks of the forest. Formerly I had come to this
pond adventurously, from time to time, in dark summer nights, with
a companion, and making a fire close to the water's edge, which we
thought attracted the fishes, we caught pouts with a bunch of worms
strung on a thread, and when we had done, far in the night, threw
the burning brands high in the air like skyrockets, which, coming
down into the pond, were quenched with a loud hissing, and we were
suddenly groping in total darkness. Through this, whistling a tune,
we took our way to the haunts of men again. But now I had made
my home by the shore.

Sometimes, after staying in a village parlor till the family had all
retired, I have returned to the woods, and, partly with a view to the
next day's dinner, spent the hours of midnight fishing from a boat
by moonlight, serenaded by owls and foxes, and hearing, from time
to time, the creaking note of some unknown bird close at hand.
These experiences were very memorable and valuable to me, — anchored in forty feet of water, and twenty or thirty rods from the
shore, surrounded sometimes by thousands of small perch and shiners, dimpling the surface with their tails in the moonlight, and communicating by a long flaxen line with mysterious nocturnal fishes
which had their dwelling forty feet below, or sometimes dragging
sixty feet of line about the pond as I drifted in the gentle night
breeze, now and then feeling a slight vibration along it, indicative

of some life prowling about its extremity, of dull uncertain blundering purpose there, and slow to make up its mind. At length you slowly raise, pulling hand over hand, some horned pout squeaking and squirming to the upper air. It was very queer, especially in dark nights, when your thoughts had wandered to vast and cosmogonal themes in other spheres, to feel this faint jerk, which came to interrupt your dreams and link you to Nature again. It seemed as if I might next cast my line upward into the air, as well as downward into this element, which was scarcely more dense. Thus I caught two fishes as it were with one hook.

The scenery of Walden is on a humble scale, and, though very beautiful, does not approach to grandeur, nor can it much concern one who has not long frequented it or lived by its shore; yet this pond is so remarkable for its depth and purity as to merit a particular description. It is a clear and deep green well, half a mile long and a mile and three quarters in circumference, and contains about sixty-one and a half acres; a perennial spring in the midst of pine and oak woods, without any visible inlet or outlet except by the clouds and evaporation. The surrounding hills rise abruptly from the water to the height of forty to eighty feet, though on the south-east and east they attain to about one hundred and one hundred and fifty feet respectively, within a quarter and a third of a mile. They are exclusively woodland. All our Concord waters have two colors at least; one when viewed at a distance, and another, more proper, close at hand. The first depends more on the light, and follows the sky. In clear weather, in summer, they appear blue at a little distance, especially if agitated, and at a great distance all appear alike. In stormy weather they are sometimes of a dark slate color. The sea, however, is said to be blue one day and green another without any perceptible change in the atmosphere. I have seen our river, when, the landscape being covered with snow, both water and ice were almost as green as grass. Some consider blue "to be the color of pure water, whether liquid or solid." But, looking directly down into our waters from a boat, they are seen to be of very different colors. Walden is blue at one time and green at another, even from the same point of view. Lying between the earth and the heavens, it partakes of the color of both. Viewed from a hill-top it reflects the color of the sky; but near at hand it is of a yellowish tint next the shore where you can see the sand, then a light green, which gradually deepens to a uniform dark green in the body of the pond. In some lights, viewed even from a hill-top, it is of a vivid green next the shore. Some have referred this to the reflection of the verdure; but it is equally green there against the railroad sand-bank, and in the spring, before the leaves are expanded, and it may be simply the result of the prevailing blue mixed with the yellow of the sand. Such is the color of its iris. This is that portion also, where in the spring, the ice being warmed by the heat of the sun reflected from the bottom, and also transmitted through the earth, melts first and forms a narrow canal about the still frozen middle. Like the rest of our waters, when much agitated, in clear weather, so that the surface of the waves may reflect the sky at the right angle, or because there is more light mixed with it, it appears at a little distance of a darker blue than the sky itself; and at such a time, being on its surface, and looking with divided vision, so as to see the reflection, I have discerned a matchless and indescribable light blue, such as watered or changeable silks and sword blades suggest, more cerulean than the sky itself, alternating with the original dark green on the opposite sides of the waves, which last appeared but muddy in comparison. It is a vitreous greenish blue, as I remember it, like those patches of the winter sky seen through cloud vistas in the west before sundown. Yet a single glass of its water held up to the light is as colorless as an equal quantity of air. It is well known that a large plate of glass will have a green tint, owing, as the makers say, to its "body," but a small piece of the same will be colorless. How large a body of Walden water would be required to reflect a green tint I have never proved. The water of our river is black or a very dark brown to one looking directly down on it, and, like that of most ponds, imparts to the body of one bathing in it a yellowish tinge; but this water is of such crystalline purity that the body of the bather appears of an alabaster whiteness, still more unnatural, which, as the limbs are magnified and distorted withal, produces a monstrous effect, making fit studies for a Michael Angelo.

The water is so transparent that the bottom can easily be discerned at the depth of twenty-five or thirty feet. Paddling over it, you may see many feet beneath the surface the schools of perch and shiners, perhaps only an inch long, yet the former easily distinguished by their transverse bars, and you think that they must be ascetic fish that find a subsistence there. Once, in the winter, many years ago, when I had been cutting holes through the ice in order to catch pickerel, as I stepped ashore I tossed my axe back on to the ice, but, as if some evil genius had directed it, it slid four or five rods directly into one of the holes, where the water was twenty-five feet deep. Out of curiosity, I lay down on the ice and looked through the hole, until I saw the axe a little on one side, standing on its head, with its helve erect and gently swaying to and fro with the pulse of the pond; and there it might have stood erect and swaying till in the course of time the handle rotted off, if I had not disturbed it. Making another hole directly over it with an ice chisel which I had, and cutting down the longest birch which I could find in the neighborhood with my knife, I made a slip-noose, which I attached to its end, and, letting it down carefully, passed it over the knob of the handle, and drew it by a line along the birch, and so pulled the axe out again.

The shore is composed of a belt of smooth rounded white stones like paving-stones, excepting one or two short sand beaches, and is so steep that in many places a single leap will carry you into water over your head; and were it not for its remarkable transparency, that would be the last to be seen of its bottom till it rose on the opposite side. Some think it is bottomless. It is nowhere muddy, and a casual observer would say that there were no weeds at all in it; and of noticeable plants, except in the little meadows recently overflowed, which do not properly belong to it, a closer scrutiny does not detect a flag nor a bulrush, nor even a lily, yellow or white, but only a few small heart-leaves and potamogetons, and perhaps a water-target or two; all which however a bather might not perceive; and these plants are clean and bright like the element they grow in. The stones extend a rod or two into the water, and then the bottom is pure sand, except in the deepest parts, where there is usually a little sediment, probably from the decay of the leaves which have been wafted on to it so many successive falls, and a bright green weed is brought up on anchors even in midwinter. . . .

There have been caught in Walden pickerel, one weighing seven pounds, — to say nothing of another which carried off a reel with great velocity, which the fisherman safely set down at eight pounds because he did not see him, — perch and pouts, some of each weighing over two pounds, shiners, chivins or roach (*Leuciscus pulchellus*), a very few breams, and a couple of eels, one weighing four pounds, — I am thus particular because the weight of a fish is commonly its only title to fame, and these are the only eels I have heard of here; — also, I have a faint recollection of a little fish some five inches long, with silvery sides and a greenish back, somewhat dace-like in its character, which I mention here chiefly to link my facts to fable. Nevertheless, this pond is not very fertile in fish. Its pickerel, though not abundant, are its chief boast. I have seen at one time lying on the ice pickerel of at least three different kinds: a long and shallow one, steel-colored, most like those caught in the river; a bright golden kind, with greenish reflections and remarkably deep, which is the most common here; and another, golden-colored, and shaped like the last, but peppered on the sides with small dark brown or black spots, intermixed with a few faint blood-red ones, very much like a trout. The specific name *reticulatus* would not apply to this; it should be *guttatus* rather. These are all very firm fish, and weigh more than their size promises. The shiners, pouts, and perch also, and indeed all the fishes which inhabit this pond, are much cleaner, handsomer, and firmer fleshed than those in the river and most other ponds, as the water is purer, and they can easily be distinguished from them. Probably many ichthyologists would make new varieties of some of them. There are also a clean race of frogs and tortoises, and a few muscles in it; musk-rats and minks leave their traces about it, and occasionally a travel-ling mud-turtle visits it. Sometimes, when I pushed off my boat in the morning, I disturbed a great mud-turtle which had secreted himself under the boat in the night. Ducks and geese frequent it in the spring and fall, the white-bellied swallows (*Hirundo bicolor*) skim over it, and the peetweets (*Totanus macularius*) "teter" along its stony shores all summer. I have sometimes disturbed a fish-hawk sit-ting on a white-pine over the water; but I doubt if it is ever profaned by the wing of a gull, like Fair Haven. At most, it tolerates one an-nual loon. These are all the animals of consequence which frequent it now.

You may see from a boat, in calm weather, near the sandy eastern shore, where the water is eight or ten feet deep, and also in some other parts of the pond, some circular heaps half a dozen feet in diameter by a foot in height, consisting of small stones less than a hen's egg in size, where all around is bare sand. At first you wonder if the Indians could have formed them on the ice for any purpose, and so, when the ice melted, they sank to the bottom; but they are too regular and some of them plainly too fresh for that. They are similar to those found in rivers; but as there are no suckers nor lam-preys here, I know not by what fish they could be made. Perhaps they are the nests of the chivin. These lend a pleasing mystery to the bottom.

The shore is irregular enough not to be monotonous. I have in my mind's eye the western, indented with deep bays, the bolder northern, and the beautifully scolloped southern shore, where successive capes overlap each other and suggest unexplored coves between. The forest has never so good a setting, nor is so distinctly beautiful, as when seen from the middle of a small lake amid hills which rise from the water's edge; for the water in which it is reflected not only makes the best foreground in such a case, but, with its winding shore, the most natural and agreeable boundary to it. There is no rawness nor imperfection in its edge there, as where the axe has cleared a part, or a cultivated field abuts on it. The trees have ample room to expand on the water side, and each sends forth its most vigorous branch in that direction. There Nature has woven a natural selvage, and the eye rises by just gradations from the low shrubs of the shore to the highest trees. There are few traces of man's hand to be seen. The water laves the shore as it did a thousand years ago.

A lake is the landscape's most beautiful and expressive feature. It is earth's eye; looking into which the beholder measures the depths of his own nature. The fluviatile trees next the shore are the slender eyelashes which fringe it, and the wooded hills and cliffs around are its overhanging brows.

Standing on the smooth sandy beach at the east end of the pond, in a calm September afternoon, when a slight haze makes the opposite shore line indistinct, I have seen whence came the expression, "the glassy surface of a lake." When you invert your head, it looks like a thread of finest gossamer stretched across the valley, and gleaming against the distant pine woods, separating one stratum of the atmosphere from another. You would think that you could walk dry under it to the opposite hills, and that the swallows which skim over might perch on it. Indeed, they sometimes dive below the line, as it were by mistake, and are undeceived. As you look over the pond westward you are obliged to employ both your hands to defend your eyes against the reflected as well as the true sun, for they are equally bright; and if, between the two, you survey its surface critically, it is literally as smooth as glass, except where the skater insects, at equal intervals scattered over its whole extent, by their motions in the sun produce the finest imaginable sparkle on it, or, perchance, a duck plumes itself, or, as I have said, a swallow skims so low as to touch it. It may be that in the distance a fish describes an arc of three or four feet in the air, and there is one bright flash where it emerges, and another where it strikes the water; sometimes the whole silvery arc is revealed; or here and there, perhaps, is a thistle-down floating on its surface, which the fishes dart at and so dimple it again. It is like molten glass cooled but not congealed, and the few motes in it are pure and beautiful like the imperfections in glass. You may often detect a yet smoother and darker water, separated from the rest as if by an invisible cobweb, boom of the water nymphs, resting on it. From a hill-top you can see a fish leap in almost any part; for not a pickerel or shiner picks an insect from this smooth surface but it manifestly disturbs the equilibrium of the whole lake. It is wonderful with what elaborateness this simple fact is advertised, — this piscine murder will out, — and from my distant perch I distinguish the circling undulations when they are half a dozen rods in diameter. You can even detect a water-bug (*Gyrinus*) ceaselessly progressing over the smooth surface a quarter of a mile off; for they furrow the water slightly, making a conspicuous ripple bounded by two diverging lines, but the skaters glide over it without rippling it perceptibly. When the surface is considerably agitated there are no skaters nor water-bugs on it, but apparently, in calm days, they leave their havens and adventurously glide forth from the shore by short impulses till they completely cover it. It is a soothing employment, on one of these fine days in the fall when all the warmth of the sun is fully appreciated, to sit on a stump on such a height as this, over-looking the pond, and study the dimpling circles which are incessantly inscribed on its otherwise invisible surface amid the reflected skies and trees. Over this great expanse there is no disturbance but it is thus at once gently smoothed away and assuaged, as, when a vase of water is jarred, the trembling circles seek the shore and all is smooth again. Not a fish can leap or an insect fall on the pond but it is thus reported in circling dimples, in lines of beauty, as it were the constant welling up of its fountain, the gentle pulsing of its life, the heaving of its breast. The thrills of joy and thrills of pain are undistinguishable. How peaceful the phenomena of the lake! Again the works of man shine as in the spring. Ay, every leaf and twig and stone and cobweb sparkles now at mid-afternoon as when covered with dew in a spring morning. Every motion of an oar or an insect produces a flash of light; and if an oar falls, how sweet the echo!

In such a day, in September or October, Walden is a perfect forest mirror, set round with stones as precious to my eye as if fewer or rarer. Nothing so fair, so pure, and at the same time so large, as a lake, perchance, lies on the surface of the earth. Sky water. It needs no fence. Nations come and go without defiling it. It is a mirror which no stone can crack, whose quicksilver will never wear off, whose gilding Nature continually repairs; no storms, no dust, can dim its surface ever fresh; — a mirror in which all impurity presented to it sinks, swept and dusted by the sun's hazy brush, — this the light dustcloth, — which retains no breath that is breathed on it, but sends its own to float as clouds high above its surface, and be reflected in its bosom still.

A field of water betrays the spirit that is in the air. It is continually receiving new life and motion from above. It is intermediate in its nature between land and sky. On land only the grass and trees wave, but the water itself is rippled by the wind. I see where the breeze dashes across it by the streaks or flakes of light. It is remarkable that we can look down on its surface. We shall, perhaps, look down thus on the surface of air at length, and mark where a still subtler spirit sweeps over it.

The skaters and water-bugs finally disappear in the latter part of October, when the severe frosts have come; and then and in November, usually, in a calm day, there is absolutely nothing to ripple the surface. One November afternoon, in the calm at the end of a rain storm of several days' duration, when the sky was still completely overcast and the air was full of mist, I observed that the pond was remarkably smooth, so that it was difficult to distinguish its surface; though it no longer reflected the bright tints of October, but the sombre November colors of the surrounding hills. Though I passed over it as gently as possible, the slight undulations produced by my boat extended almost as far as I could see, and gave a ribbed appearance to the reflections. But, as I was looking over the surface, I saw here and there at a distance a faint glimmer, as if some skater insects which had escaped the frosts might be collected there, or, perchance, the surface, being so smooth, betrayed where a spring welled up from the bottom. Paddling gently to one of these places, I was surprised to find myself surrounded by myriads of small perch, about five inches long, of a rich bronze color in the green water, sporting there, and constantly rising to the surface and dimpling it, sometimes leaving bubbles on it. In such transparent and seemingly bottomless water, reflecting the clouds, I seemed to be floating through the air as in a balloon and their swimming impressed me as a kind of flight or hovering, as if they were a compact flock of birds passing just beneath my level on the right or left, their fins, like sails, set all around them. There were many such schools in the pond, apparently improving the short season before winter would draw an icy shutter over their broad skylight, sometimes giving to the surface an appearance as if a slight breeze struck it, or a few rain-drops fell there. When I approached carelessly and alarmed them, they made a sudden plash and rippling with their tails, as if one had struck the water with a brushy bough, and instantly took refuge in the depths. At length the wind rose, the mist increased, and the waves began to run, and the perch leaped much higher than before, half out of the water, a hundred black points, three inches long, at once above the surface. Even as late as the fifth of December, one year, I saw some dimples on the surface, and thinking it was going to rain hard immediately, the air being full of mist, I made haste to take my place at the oars and row homeward; already the rain seemed rapidly increasing, though I felt none on my cheek, and I anticipated a thorough soaking. But suddenly the dimples ceased, for they were produced by the perch, which the noise of my oars had scared into the depths, and I saw their schools dimly disappearing; so I spent a dry afternoon after all.

An old man who used to frequent this pond nearly sixty years ago, when it was dark with surrounding forests, tells me that in those days he sometimes saw it all alive with ducks and other water fowl, and that there were many eagles about it. He came here a-fishing, and used an old log canoe which he found on the shore. It was made of two white-pine logs dug out and pinned together, and was cut off square at the ends. It was very clumsy, but lasted a great many years before it became water-logged and perhaps sank to the bottom. He did not know whose it was; it belonged to the pond. He used to make a cable for his anchor of strips of hickory bark tied together. An old man, a potter, who lived by the pond before the Revolution, told him once that there was an iron chest at the bottom, and that he had seen it. Sometimes it would come floating up to the shore; but when you went toward it, it would go back into deep water and disappear. I was pleased to hear of the old log canoe, which took the place of an Indian one of the same material but more graceful construction, which perchance had first been a tree on the bank, and then, as it were, fell into the water, to float there for a generation, the most proper vessel for the lake. I remember that when I first looked into these depths there were many large trunks to be seen indistinctly lying on the bottom, which had either been blown over formerly, or left on the ice at the last cutting, when wood was cheaper; but now they have mostly disappeared.

When I first paddled a boat on Walden, it was completely surrounded by thick and lofty pine and oak woods, and in some of its coves grapevines had run over the trees next the water and formed bowers under which a boat could pass. The hills which form its shores are so steep, and the woods on them were then so high, that, as you looked down from the west end, it had the appearance of an amphitheatre for some kind of sylvan spectacle. I have spent many an hour, when I was younger, floating over its surface as the zephyr willed, having paddled my boat to the middle, and lying on my back across the seats, in a summer forenoon, dreaming awake, until I was aroused by the boat touching the sand, and I arose to see what shore my fates had impelled me to; days when idleness was the most attractive and productive industry. Many a forenoon have I stolen away, preferring to spend thus the most valued part of the day; for I was rich, if not in money, in sunny hours and summer days, and spent them lavishly; nor do I regret that I did not waste more of them in the workshop or the teacher's desk. But since I left those shores the woodchoppers have still further laid them waste, and now for many a year there will be no more rambling through the aisles of the wood, with occasional vistas through which you see the water. My Muse may be excused if she is silent henceforth. How can you expect the birds to sing when their groves are cut down?

Now the trunks of trees on the bottom, and the old log canoe, and the dark surrounding woods, are gone, and the villagers, who scarcely know where it lies, instead of going to the pond to bathe or drink, are thinking to bring its water, which should be as sacred as the Ganges at least, to the village in a pipe, to wash their dishes with! — to earn their Walden by the turning of a cock or drawing of a plug! That devilish Iron Horse, whose ear-rending neigh is heard throughout the town, has muddied the Boiling Spring with his foot, and he it is that has browsed off all the woods on Walden shore, that Trojan horse, with a thousand men in his belly, introduced by mercenary Greeks! Where is the country's champion, the Moore of Moore Hall, to meet him at the Deep Cut and thrust an avenging lance between the ribs of the bloated pest?

Nevertheless, of all the characters I have known, perhaps Walden wears best, and best preserves its purity. Many men have been likened to it, but few deserve that honor. Though the woodchoppers have laid bare first this shore and then that, and the Irish have built their sties by it, and the railroad has infringed on its border, and the ice-men have skimmed it once, it is itself unchanged, the same water which my youthful eyes fell on; all the change is in me. It has not acquired one permanent wrinkle after all its ripples. It is perennially young, and I may stand and see a swallow dip apparently to pick an insect from its surface as of yore. It struck me again to-night, as if I had not seen it almost daily for more than twenty years, — Why, here is Walden, the same woodland lake that I discovered so many years ago; where a forest was cut down last winter another is springing up by its shore as lustily as ever; the same thought is welling up to its surface that was then; it is the same liquid joy and happiness to itself and its Maker, ay, and it *may* be to me. It is the work of a brave man surely, in whom there was no guile! He rounded this water with his hand, deepened and clarified it in his thought, and in his will bequeathed it to Concord. I see by its face that it is visited by the same reflection; and I can almost say, Walden, is it you?

It is no dream of mine,
To ornament a line;
I cannot come nearer to God and Heaven
Than I live to Walden even.
I am its stony shore,
And the breeze that passes o'er;
In the hollow of my hand
Are its water and its sand
And its deepest resort
Lies high in my thought.

The cars never pause to look at it; yet I fancy that the engineers and firemen and brakemen, and those passengers who have a season ticket and see it often, are better men for the sight. The engineer does not forget at night, or his nature does not, that he has beheld this vision of serenity and purity once at least during the day. Though seen but once, it helps to wash out State-street and the engine's soot. One proposes that it be called "God's Drop."

THE BOOK OF
Naturalists

PART II

PART II

INTRODUCTION

FOR YOU to understand the reason for the majority of the selections in this second part of the Anthology I must explain what is meant by *naturalist* and *literary natural history*. This is difficult because of the dearth of exact English words that will express accurately what I have in mind. The words *science* and *scientist, nature* and *naturalist, evolution* and *evolutionist, field naturalist, laboratory* and *museum technical scientist* — none of these can be defined exactly; yet this is not entirely a drawback, for if they are worthy, all should in more ways than one grade and intergrade with one another.

My particular definition of natural history for the purposes of Part Two is that which is concerned with the observations of living animals and plants in their natural wild state, and the results (to quote part of the *Oxford English Dictionary* definition of literature) put into "writing which has claim to consideration on the ground of beauty of form or emotional effect," without the slightest deviation from the fundamental concomitant of absolute truth.

There has been almost as much disputatious writing concerning the so-called closet scientist on the one hand and the nature writer on the other, as on the also so-called conflict of science and religion. Yet each owes much to his fellow, and each should strive to use eclectically the methods of the other. The common bond is that any overstepping of the sheerest boundaries of truth, any dalliance, however ephemeral, with any doubtful theory, however attractive, any poetic lure of exaggeration, automatically damns forever *all* the work of any naturalist.

To sum up, I present an *ideal* equipment for a naturalist writer of literary natural history: Supreme enthusiasm, tempered with infinite patience and a complete devotion to truth; the broadest possible education; keen eyes, ears, and nose; the finest instruments; opportunity for observation; thorough training in laboratory technique; comprehension of known facts and theories, and the habit of giving full credit for these in the proper place; awareness of what is not known; ability to put oneself in the subject's place; interpretation and integration of observations; a sense of humor; facility in writing; an eternal sense of humbleness and wonder.

My friend William Morton Wheeler was an outstanding example

of one who fulfilled every requirement of scientist, investigator, naturalist, and literary master. How well he appreciated the various methods of presentation of zoological knowledge is shown by a sentence or two from the prefatory discussion of his own style in the writing of *Demons of the Dust:*

". . . With advancing years and a growing conviction of the value of all science as a *social* undertaking, I felt that a detailed account of the insects which had so long interested me might interest the general reader. . . . Some of my colleagues maintain that I have attempted the impossible, that there are only two ways of writing on such a subject as the one I have chosen. Either I should have written as an undefiled specialist in the refined jargon and with the ponderous documentation demanded by the ritual of my caste, or I should have turned to literature and presented the matter with the well-known devices and embellishments of rhetoric, which may perhaps delight but are sure to mislead the uncritical reader. Any intermediate course, they claimed, can only lead to a compromise distasteful alike to the high-brow scientist and the thrill-seeking, movie-fed public." Needless to say, Dr. Wheeler's diction embodies the best in all three styles.

In my ideal mental equipment for a literary naturalist I have intentionally made him responsible for many of the necessities of a technical scientist. This should be carried a step farther; by this I mean documentation in print. Lack of this is the cause of much of the criticism heaped by specialists upon nature writers. Wheeler, referring to Fabre and his success as a popular writer, says: "He almost never referred to contemporary entomologists, even to those who anticipated him in the publication of similar or identical observations, and he rarely cited even such early French authorities as Réaumur, Latreille and Dufour. Now the scholar and the scientist are extremely sensitive in regard to documentation, not only because it is so thrilling to find their own works cited by their friends and enemies, but also because it enables them to judge whether an author has a sufficient knowledge of previous achievements in a particular field, and has, or thinks he has, contributed any new facts, inferences or reflections."

This is absolutely reasonable and obligatory for any reliable work, but the exact facies of documentation must, in the case of the naturalist, be somewhat different from that of the specialist. One basic need of the former is to make his writing pleasing in form, and this would be seriously impaired were his pages filled with a spate of references and parenthetical allusions to sources and the work of others.

Both documentation and the scientific names of creatures written about on the literary page can always satisfactorily be included in an appendix, with exact reference to page and line of text, as Wheeler himself has done to perfection in his *Social Life among the Insects.*

Another quotation from the same discerning author defines natural history, recognizing its basic nature and essential qualities: "History shows that throughout the centuries, from Aristotle and Pliny to the present day, natural history constitutes the perennial root-stock or stolon of biological science and that it retains this character because it satisfies some of our most fundamental and vital interests in organisms as living individuals more or less like ourselves. From time to time the stolon had produced special disciplines which have grown into great, flourishing complexes. . . . Taxonomy, morphology, paleontology, physiology began to shoot up, branch and differentiate, becoming independent specialties, developing their own methods, fictions and hypotheses. . . . It has itself changed its name from time to time as the investigators of different periods have been impressed by different aspects of its fundamental tendencies. Aristotle wrote of the 'histories' of animals, the naturalists of more recent centuries spoke of their 'habits'; we have become more articulate and speak of their 'behavior.' Last and most recent, and in danger of being lost in its all-inclusiveness is the commonly used and convenient word substitute for natural history, 'Ecology.' "

Wheeler dealt in the main with ants, little creatures which are in themselves fascinating and dramatic, as well as being available for observation to the most sluggish of us. But even more of a miracle is worked by a master in a field of science alien to our volume, the astronomer Harlow Shapley, who can hold an audience spellbound by materializing pin-points of distant stars and heavenly smudges of galaxies into creative, understandable words that stir our deepest emotions.

With men such as these two embodying all we should like to have in each author, I have chosen a few of the best-known names in the realm of recent natural history, giving the preference, when possible, to the more articulately literate.

Concerning the thirty-four authors in Part Two I will make only the following distinctions: scientists of the first rank, Darwin, Wallace, Agassiz, T. H. Huxley, Osborn, and Wheeler; zoologists of note, Thomson, Fabre, Levick, Roule, Chapman, Haskins, and J. Huxley.

The remainder hold high place as articulate naturalists, excellent interpreters of science, rather than original investigators on their own, but all "afflicted with the necessary word." Men such as these had small part in the early development of natural history. In the first place there were too few facts available, and secondly there were almost no audiences to listen or to read. So while the ideal is to discover for oneself as well as to compile, yet in these days interpretation is of great merit — in fact, is absolutely essential. It too has its own ideals, and to excel it must be outstanding in distinction, in felicity of phrase and clarity of presentation.

A bare fact describing some character which goes back two hundred million years through evolutionary development needs no fine literary dressing for a zoologist or a naturalist of imagination. To the intelligent but unscientific reader, however, it must be put into language that is comprehensible, clear, and perhaps pointed by some apt simile out of his own experience.

Before we pass on to the actual words of the selected authors, here are two extreme examples of presentation.

In 1819, when our grandfathers were children, a delightfully naïve little book was published in London under the name of *Dialogues on Entomology*. The author will ever remain a mystery, but the dedication is to Miss Maria Edgeworth, "from whose writings the Youth of the Present Age have derived equal Entertainment and Advantage." I quote a few pages with no further comment:

LUCY. My dear mamma, are you very busy? If you can spare time, I wish you would come to the garden and explain something which appears very odd. Oh dear, I have run so fast!

MOTHER. I am rather busy; but, as I have great pleasure in explaining what you wish to know, I will go with you.

LUCY. Thank you, mamma: I am sure you will not call this one of my foolish interruptions.

MOTHER. I approve so much of your disposition to inquire into every thing of which you are ignorant, that I never think such interruptions troublesome. Well, now, here we are in the garden.

LUCY. See here, mamma, at the ivy wall, a strange thing like part of a moth, struggling among bits of leaves and sticks, look, look! it moves, it is growing larger — there are its legs. I really believe it is a moth: now it is entirely come out from among those scraps of dirt. Is it a moth, mamma?

MOTHER. Yes, it is a moth; but softly: watch how its wings unfold. You see they gradually spread open, with a little quivering motion. Could you have thought so large an animal had been concealed in that small case, which you see in the midst of those broken leaves?

90

It is now a perfect moth, and this is the first time it has appeared in that form.

LUCY. How much brighter the colours are than when I first saw its wings! They now look bright and clean, at first they had a wet crumpled look.

MOTHER. Yes, they were damp; an exceedingly small quantity of moisture surrounds the fly in its case, which preserves the joints in a state of flexibility, that is suppleness, and enables the wings to unfold to their full size; when they have done so, the moisture is no longer useful, and the air quickly dries it up.

LUCY. But, mamma, if this is the first time of its being a moth, how did it get into that case? how was it put there? My sister told me once, that caterpillars turned into moths and butterflies; do they, mamma?

MOTHER. Yes; every kind of caterpillar changes into some species of winged insect.

LUCY. Then, I suppose, all kinds of flies, gnats, and bees, have been caterpillars. What tiny things the caterpillars of flies must be.

MOTHER. They are very small indeed; so small that many of them are called maggots.

LUCY. Maggots! I always thought maggots were nasty and dirty.

MOTHER. As several kinds of caterpillars or maggots live in rotten wood, mud, carrion, dunghills, and other substances of which we have the most disagreeable notions, we are too apt to extend the disgusting or nasty idea of the place in which they live to the creature itself.

Leaving Lucy palpitating with her new discoveries among the lepidoptera let us quote a few lines by Julian Huxley from a chapter in his book *Evolution: The Modern Synthesis.*

"A quite different development, however, may occur if recessive lethal mutations occur in both interchange chromosome-groups. In that case, the homozygotes will be inviable and only the hybrid will survive. This is the condition of *balanced lethals.* Since lethal mutations are common, and since the heterozygote will enjoy increased advantage in various ways as soon as one homozygote has become inviable, we may expect this condition to develop out of segmental interchange at least as readily as that of differentiated prime types.

"The classical case is that of the evening primroses (*Oenothera*). Here abundant genetic and cytological evidence converge to show that almost all the species are balanced-lethal heterozygotes, the original pure types having disappeared. Elaborate subsidiary mechanisms ensure the production of the heterozygotes with the minimum of wastage."

This presentation is intended primarily for professional biologists, and is an excellent example of the necessary use of technical language for the clear elucidation of intricate genetic divergences. If the reader has an acquaintance with the basic principles of Mendelian heredity the passage is clear and logical. If not, then

> "A primrose by a river's brim
> A yellow primrose was to him,
> And it was nothing more."

My attempt in this Anthology has been to steer a middle course between Lucy and Julian. The possibility of this may be judged by the masterly essay of Dr. Huxley's (p. 395), written with admirable popular balance, which I have chosen for inclusion in the main body of Part Two.

We cannot begrudge a paragraph of appreciation of the literary joy to be found in the penumbra of nature, perhaps "unnatural history" is a good term, such as the delightful actors in *The Wind in the Willows,* and *The House at Pooh Corner.* I hope there are some "earnest seekers after truth" besides myself who like occasionally to turn from Whitehead and Einstein, Eddington and D'Arcy Thompson to Kaa, Bagheera, Rikki-tikki, the Walrus and the Gryphon, and back again without feeling they have degraded or scrambled their mental processes.

One other diverting aspect of our theme might be called Famous Unnaturalists. From a long and amusing list I select a few quotations.

Master François Villon: "Not all the birds that singen all the way from here to Babylon could induce me to spend one day amid the hard, sober fare of a country life."

Pepys: "And, among other rarities, a hive of bees, so as being hived in glass, you may see the bees making their honey and combs mighty pleasantly. . . ."

"I to Mrs. Turner's, in Salisbury Court, and carried her our Eagle, which she desired, and we glad to be rid of her, she fouling our office mightily. . . ."

"Dr. Williams did carry me into his garden, where he hath abundance of grapes; and did show me how a dog which he hath to kill all the cats that come thither to kill his pigeons, and do afterwards bury them; and do it with so much care that they shall be quite covered, that but if the tip of a tail hangs out he will take up the cat again and dig the hole deeper. . . .

"At table I had very good discourse with Mr. Ashmole, wherein he did assure me that frogs and many insects do often fall from the sky, ready formed."

Samuel Johnson: In 1786 it was written of the late Samuel Johnson: "He had indeed that strong aversion felt by all the lower ranks of people towards four-footed companions very completely, notwithstanding he had for many years a cat which he called Hodge, that kept always in his room at Fleet Street; but so exact was he not to offend the human species by superfluous attention to brutes, that when the creature was grown sick and old, and could eat nothing but oysters, Mr. Johnson always went out himself to buy Hodge's dinner that Francis the black's delicacy might not be hurt, at seeing himself employed for the convenience of a quadruped."

CHARLES DARWIN (1809–1882)

From Bahia and the Galápagos to the Origin of Species

From *Diary of the Voyage of H.M.S. Beagle,* 1831–1836, and *The Origin
of Species,* 1859

[To WRITE a volume about Darwin would be comparatively a simple
matter; to sum up anything about him in a few sentences is impu-
dent, futile, and hopeless. His simplicity eludes all attempt at con-
structive appraisal. He seems somehow the most natural of naturalists, the most completely amateur, the greatest scientist who derived
all his magnificent results from living animals, from observations
and meditations on their lives, homes, habits — their natural history.

At the age of twenty-two he went as naturalist on the *Beagle* from
England, around South America, and on to the Galápagos, Tahiti,
and Australia. The effect of this expedition on his life work is best
told by three sentences from his diary for 1837, not long after his
return; "In July opened first note-book on Transmutation of Species.
Had been greatly struck from about the month of previous March
on character of South American fossils, and species of Galapagos Archipelago. These facts (especially latter), origin of all my
views."

I once discovered a great expanse of water in the heart of Tower
Island, Galápagos, named it Darwin Bay, and longed with all my
heart to have the man himself there with me. What would not any
of us give to have walked with him eighty-eight years before, when
first he observed with growing wonder the strange animal life of this
archipelago! What a marvelous thing if we could listen to his comments *after* he had brought his theories to fruition!

What Linnæus did for the nomenclature of living beings, Darwin
achieved for their whys and hows of existence, the magnificent demonstration of the oneness and continuity of life on the planet, from
the lowliest of one-celled plants and animals which came into existence in the beginning, to living sequoias and man himself.

The following pages give examples of two extreme forms of Darwin's writings. The first is a selection from his recently published
original *Diary of the Voyage of the Beagle,* including his youthful
and very human reactions on his first sight of the tropics, and a part

94

of the account of his days in the Galápagos, which were to be so important both to himself and to the world. The second selection is from the concluding section of *The Origin of Species*.]

BAHIA

[1832.]

February 23rd, 24th & 25th. Since leaving Teneriffe the sea has been so calm, that it is hard to believe it the same element which tossed us about in the Bay of Biscay. This stillness is of great moment to the quantity of comfort which is attainable on ship-board. Hitherto I have been surprised how enjoyable life is in this floating prison. But the greatest & most constant drawback to this is the very long period which separates us from our return. Excepting when in the midst of tropical scenery, my greatest share of pleasure is in anticipating a future time when I shall be able to look back on past events; & the consciousness that this prospect is so distant never fails to be painful. To enjoy the soft & delicious evenings of the Tropic; to gaze at the bright band of stars, which stretches from Orion to the Southern Cross, & to enjoy such pleasures in quiet solitude, leaves an impression which a few years will not destroy.

26th. For the first time in my life I saw the sun at noon to the North: yesterday it was very near over our heads & therefore of course we are a little to the South of it. I am constantly surprised at not finding the heat more intense than it is; when at sea & with a gentle breeze blowing one does not even wish for colder weather. I am sure I have frequently been more oppressed by a hot summer's day in England.

27th. Quietly sailing. Tomorrow we shall reach Bahia.

February 28th. About 9 o'clock we were near to the coast of Brazil; we saw a considerable extent of it, the whole line is rather low & irregular, & from the profusion of wood & verdure of a bright green colour. About 11 o'clock we entered the bay of All Saints, on the Northern side of which is situated the town of Bahia or San Salvador. It would be difficult [to] imagine, before seeing the view, anything so magnificent. It requires, however, the reality of nature to make it so. If faithfully represented in a picture, a feeling of distrust would be raised in the mind, as I think is the case in some of Martins' views. The town is fairly embosomed in a luxuriant wood & situated on a steep bank overlooks the calm waters of the great bay of All Saints. The houses are white & lofty & from the windows being narrow & long have a very light & elegant appearance. Convent, Porticos & public buildings vary the uniformity of the houses: the bay is scattered over with large ships; in short the view is one of the finest in the Brazils. But these beauties are as nothing compared to the Vegetation; I believe from what I have seen Humboldt's glorious descriptions are & will for ever be unparalleled: but even he with his dark blue skies & the rare union of poetry with science which he so strongly displays when writing on tropical scenery, with all this falls far short of the truth. The delight one experiences in such times bewilders the mind; if the eye attempts to follow the flight of a gaudy butterfly, it is arrested by some strange tree or fruit; if watching an insect one forgets it in the stranger flower it is crawling over; if turning to admire the splendour of the scenery, the individual character of the foreground fixes the attention. The mind is a chaos of delight, out of which a world of future & more quiet pleasure will arise. I am at present fit only to read Humboldt; he like another sun illumines everything I behold.

29th. The day has passed delightfully: delight is however a weak term for such transports of pleasure: I have been wandering by myself in a Brazilian forest: amongst the multitude it is hard to say what set of objects is most striking; the general luxuriance of the vegetation bears the victory, the elegance of the grasses, the novelty of the parasitical plants, the beauty of the flowers, the glossy green of the foliage, all tend to this end. A most paradoxical mixture of sound & silence pervades the shady parts of the wood: the noise from the insects is so loud that in the evening it can be heard even in a vessel anchored several hundred yards from the shore: yet within the recesses of the forest a universal stillness appears to reign. To a person fond of Natural history such a day as this brings with it pleasure more acute than he ever may again experience. After wandering about for some hours, I returned to the landing place. Before reaching it I was overtaken by a Tropical storm. I tried to find shelter under a tree so thick that it would never have been penetrated by common English rain, yet here in a couple of minutes, a little torrent flowed down the trunk. It is to this violence we must attribute the verdure in the bottom of the wood: if the showers were like those of a colder clime, the moisture would be absorbed or evaporated before reaching the ground.

March 1st. I can only add raptures to the former raptures. I walked with the two Mids a few miles into the interior. The country is composed of small hills & each new valley is more beautiful than the last. I collected a great number of brilliantly coloured flowers, enough to make a florist go wild. Brazilian scenery is nothing more nor less than a view in the Arabian Nights, with the advantage of reality. The air is deliciously cool & soft; full of enjoyment one fervently desires to live in retirement in this new & grander world.

2nd & 3rd. I am quite ashamed at the very little I have done during these two days; a few insects & plants make up the sum total. My only excuse is the torrents of rain, but I am afraid idleness is the true reason. Yesterday Cap. Paget dined with us & made himself very amusing by detailing some of the absurdities of naval etiquette. To day Rowlett & myself went to the city & he performed the part of cicerone to me: in the lower part near to the wharfs, the streets are very narrow & the houses even more lofty than in the old town of Edinburgh. The smell is very strong & disagreeable, which is not to be wondered at, since I observe they have the same need of crying "gardez l'eau" as in Auld Reekie. All the labor is done by the black men, who stand collected in great numbers round the merchants' warehouses. The discussions which arise about the amount of hire are very animated; the negroes at all times use much gesticulation & clamor & when staggering under their heavy burthens, beat time & cheer themselves by a rude song. I only saw one wheel carriage; but the horses are by no means scarce; they are generally small & well shaped & are chiefly used for the merchants to ride. We paid a visit to one of the principal churches; we here found for a guide, a little Irish boy about 13 years old. His father was buried there two months ago, & was one of the unfortunate people whom Don Pedro enticed into the country under the pretence of settling [him] there. This little fellow contrives to support his mother & sister by the few Vintems which in the course of the day he earns by messages. Mr Gond, one of the principal merchants in the place, offered to lend us horses, if we would walk to his country house. We gladly accepted his offer & enjoyed a most delightful ride; one beautiful view after another opening upon us in endless succession.

March 4th. This day is the first of the Carnival; but Wickham, Sullivan & myself nothing (un)daunted were determined to face its dangers. These dangers consist in being unmercifully pelted by wax balls full of water & being wet through by large tin squirts. We found it very difficult to maintain our dignity whilst walking through the streets. Charles the V has said that he was a brave man who could snuff a candle with his fingers without flinching; I say it is he who can walk at a steady pace, when buckets of water on each side are ready to be dashed over him. After an hour's walking the gauntlet, we at length reached the country & there we were well determined to remain till it was dark. We did so, & had some difficulty in finding the road back again, as we took care to coast along the outside of the town. To complete our ludicrous miseries a heavy shower wet us to the skins; & at last gladly we reached the Beagle. It was the first time Wickham had been on shore, & he vowed if he was here for six months it should be [the] only one.

5th. King & myself started at 9 o'clock for a long naturalizing· walk. Some of the valleys were even more beautiful than any I have yet seen. There is a wild luxuriance in these spots that is quite enchanting. One of the great superiorities that Tropical scenery has over European is the wildness even of the cultivated ground. Cocoa Nuts, Bananas, Plantain, Oranges, Papaws are mingled as if by Nature, & between them are patches of the herbacious plants such as Indian corn, Yams & Cassava: & in this class of views, the knowledge that all conduces to the subsistence of mankind, adds much to the pleasure of beholding them. We returned to the ship about $\frac{1}{2}$ after 5 o'clock & during these eight hours we scarcely rested one. The sky was cloudless & the day very hot, yet we did not suffer much. It appears to me that the heat merely brings on indolence, & if there is any motive sufficient to overcome this it is very easy to undergo a good deal of fatigue. During the walk I was chiefly employed in collecting numberless small beetles & in geologising. King shot some pretty birds & I a most beautiful large lizard. It is a new & pleasant thing for me to be conscious that naturalizing is doing my duty, & that if I neglected that duty I should at same time neglect what has for some years given me so much pleasure.

THE GALAPAGOS

1835.

September 7th. The Beagle sailed for the Galapagos: (*15th*) on the 15th she was employed in surveying the outer coast of Chatham Island, the S. Eastern one of the Archipelago.

16th. The next day we ran near Hood's Isd & there left the Whale boat. In the evening the Yawl was also sent away on a surveying cruize of some length. The weather now & during the passage has continued as on the coast of Peru, a steady, gentle breeze of wind & gloomy sky. We landed for an hour on the N. W. end of Chatham Isd. These islands at a distance have a sloping uniform outline, excepting where broken by sundry paps & hillocks; the whole black Lava, *completely* covered by small leafless brushwood & low trees. The fragments of Lava where most porous, are reddish like cinders; the stunted trees show little signs of life. The black rocks heated by the rays of the Vertical sun, like a stove, give to the air a close & sultry feeling. The plants also smell unpleasantly. The country was compared to what we might imagine the cultivated parts of the Infernal regions to be.

This day, we now being only 40 miles from the Equator, has been the first warm one; up to this time all on board have worn cloth clothese, & although no one would complain of cold, still less would they of too much warmth. The case would be very different if we were cruizing on the Atlantic side of the continent.

September 17th. The Beagle was moved into St. Stephen's harbor. We found there an American Whaler & we previously had seen two at Hoods Island. The Bay swarmed with animals; Fish, Shark & Turtles were popping their heads up in all parts. Fishing lines were soon put overboard & great numbers of fine fish 2 & even 3 feet long were caught. This sport makes all hands very merry; loud laughter & the heavy flapping of the fish are heard on every side. After dinner a party went on shore to try to catch Tortoises, but were unsuccessful. These islands appear paradises for the whole family of Reptiles. Besides three kinds of Turtles, the Tortoise is so abundant that [a] single Ship's company here caught 500–800 in a short time. The black Lava rocks on the beach are frequented by large (2–3 ft.) most disgusting, clumsy Lizards. They are as black as the porous rocks over which they crawl & seek their prey from the Sea. Somebody çalls them "imps of darkness." They assuredly well become the land they inhabit. When on shore I proceeded to botanize & obtained 10 different flowers; but such insignificant, ugly little flowers, as would better become an Arctic than a Tropical country. The birds are Strangers to Man & think him as innocent as their countrymen the huge Tortoises. Little birds, within 3 or four feet, quietly hopped about the Bushes & were not frightened by stones being thrown at them. Mr King killed one with his hat & I pushed off a branch with the end of my gun a large Hawk. . . .

[Charles Island.]

September 26th & 27th. I industriously collected all the animals, plants, insects & reptiles from this Island. It will be very interesting to find from future comparison to what district or "centre of creation" the organized beings of this archipelago must be attached. . . .

THE ORIGIN OF SPECIES

[Note: Taken from the Sixth Edition, which contains some additions and corrections.]

I have now recapitulated the facts and considerations which have thoroughly convinced me that species have been modified, during a long course of descent. This has been effected chiefly through the natural selection of numerous successive, slight, favourable variations; aided in an important manner by the inherited effects of the use and disuse of parts; and in an unimportant manner, that is in relation to adaptive structures, whether past or present, by the direct action of external conditions, and by variations which seem to us in our ignorance to arise spontaneously. It appears that I formerly underrated the frequency and value of these latter forms of variation, as leading to permanent modifications of structure independently of natural selection. But as my conclusions have lately been much misrepresented, and it has been stated that I attribute the modification of species exclusively to natural selection, I may be permitted to remark that in the first edition of this work, and subsequently, I placed in a most conspicuous position — namely, at the close of the Introduction — the following words: "I am convinced that natural selection has been the main but not the exclusive means of modification." This has been of no avail. Great is the power of steady misrepresentation; but the history of science shows that fortunately this power does not long endure.

It can hardly be supposed that a false theory would explain, in so satisfactory a manner as does the theory of natural selection, the several large classes of facts above specified. It has recently been objected that this is an unsafe method of arguing; but it is a method used in judging of the common events of life, and has often been used by the greatest natural philosophers. The undulatory theory of light has thus been arrived at; and the belief in the revolution of the earth on its own axis was until lately supported by hardly any direct evidence. It is no valid objection that science as yet throws no light on the far higher problem of the essence or origin of life. Who can explain what is the essence of the attraction of gravity? No one now objects to following out the results consequent on this unknown element of attraction; notwithstanding that Leibnitz formerly accused Newton of introducing "occult qualities and miracles into philosophy."

I see no good reason why the views given in this volume should shock the religious feelings of any one. It is satisfactory, as showing how transient such impressions are, to remember that the greatest discovery ever made by man, namely, the law of the attraction of gravity, was also attacked by Leibnitz, "as subversive of natural, and inferentially of revealed, religion." A celebrated author and divine has written to me that "he has gradually learnt to see that it is just as noble a conception of the Deity to believe that He created a few original forms capable of self-development into other and needful forms, as to believe that He required a fresh act of creation to supply the voids caused by the action of His laws."

Why, it may be asked, until recently did nearly all the most eminent living naturalists and geologists disbelieve in the mutability of species. It cannot be asserted that organic beings in a state of nature are subject to no variation; it cannot be proved that the amount of variation in the course of long ages is a limited quantity; no clear distinction has been, or can be, drawn between species and well-marked varieties. It cannot be maintained that species when intercrossed are invariably sterile, and varieties invariably fertile; or that sterility is a special endowment and sign of creation. The belief that species were immutable productions was almost unavoidable as long as the history of the world was thought to be of short duration; and now that we have acquired some idea of the lapse of time, we are too apt to assume, without proof, that the geological record is so perfect that it would have afforded us plain evidence of the mutation of species, if they had undergone mutation.

But the chief cause of our natural unwillingness to admit that one species has given birth to clear and distinct species, is that we are always slow in admitting great changes of which we do not see the steps. The difficulty is the same as that felt by so many geologists, when Lyell first insisted that long lines of inland cliffs had been formed, and great valleys excavated, by the agencies which we see still at work. The mind cannot possibly grasp the full meaning of the term of even a million years; it cannot add up and perceive the full effects of many slight variations, accumulated during an almost infinite number of generations.

Although I am fully convinced of the truth of the views given in this volume under the form of an abstract, I by no means expect to convince experienced naturalists whose minds are stocked with a multitude of facts all viewed, during a long course of years, from a point of view directly opposite to mine. It is so easy to hide our ignorance under such expressions as the "plan of creation," "unity of design," &c., and to think that we give an explanation when we only re-state a fact. Any one whose disposition leads him to attach more weight to unexplained difficulties than to the explanation of a certain number of facts will certainly reject the theory. A few naturalists, endowed with much flexibility of mind, and who have already begun to doubt the immutability of species, may be influenced by this volume; but I look with confidence to the future, — to young and rising naturalists, who will be able to view both sides of the question with impartiality. Whoever is led to believe that species are mutable will do good service by conscientiously expressing his conviction; for thus only can the load of prejudice by which this subject is overwhelmed be removed. . . .

When the views advanced by me in this volume, and by Mr. Wallace, or when analogous views on the origin of species are generally admitted, we can dimly foresee that there will be a considerable revolution in natural history. Systematists will be able to pursue their labours as at present; but they will not be incessantly haunted by the shadowy doubt whether this or that form be a true species. This, I feel sure and I speak after experience, will be no slight relief. The endless disputes whether or not some fifty species of British brambles are good species will cease. Systematists will have only to decide (not that this will be easy) whether any form be sufficiently constant and distinct from other forms, to be capable of definition; and if definable, whether the differences be sufficiently important to deserve a specific name. This latter point will become a far more essential consideration than it is at present; for differences, however slight, between any two forms, if not blended by intermediate gradations, are looked at by most naturalists as sufficient to raise both forms to the rank of species.

Hereafter we shall be compelled to acknowledge that the only distinction between species and well-marked varieties is, that the latter are known, or believed, to be connected at the present day by intermediate gradations, whereas species were formerly thus connected. Hence, without rejecting the consideration of the present existence of intermediate gradations between any two forms, we shall be led to weigh more carefully and to value higher the actual amount of difference between them. It is quite possible that forms now generally acknowledged to be merely varieties may hereafter be thought worthy of specific names; and in this case scientific and common language will come into accordance. In short, we shall have to treat species in the same manner as those naturalists treat genera, who admit that genera are merely artificial combinations made for convenience. This may not be a cheering prospect; but we shall at least be free from the vain search for the undiscovered and undiscoverable essence of the term species.

The other and more general departments of natural history will rise greatly in interest. The terms used by naturalists, of affinity, relationship, community of type, paternity, morphology, adaptive characters, rudimentary and aborted organs, &c., will cease to be metaphorical, and will have a plain signification. When we no longer look at an organic being as a savage looks at a ship, as something wholly beyond his comprehension; when we regard every production of nature as one which has had a long history; when we contemplate every complex structure and instinct as the summing up of many contrivances, each useful to the possessor, in the same way as any great mechanical invention is the summing up of the labour, the experience, the reason, and even the blunders of numerous workmen; when we thus view each organic being, how far more interesting — I speak from experience — does the study of natural history become!

A grand and almost untrodden field of inquiry will be opened, on the causes and laws of variation, on correlation, on the effects of use and disuse, on the direct action of external conditions, and so forth. The study of domestic productions will rise immensely in value. A new variety raised by man will be a more important and interesting subject for study than one more species added to the infinitude of already recorded species. Our classifications will come to be, as far as they can be so made, genealogies; and will then truly give what may be called the plan of creation. The rules for classifying will no doubt become simpler when we have a definite object in view. We possess no pedigrees or armorial bearings; and we have to discover and trace the many diverging lines of descent in our natural genealogies, by characters of any kind which have long been inherited. Rudimentary organs will speak infallibly with respect to the nature of long-lost structures. Species and groups of species which are called aberrant, and which may fancifully be called living fossils, will aid us in forming a picture of the ancient forms of life. Embryology will often reveal to us the structure, in some degree obscured, of the prototypes of each great class.

When we can feel assured that all the individuals of the same species, and all the closely allied species of most genera, have within a not very remote period descended from one parent, and have migrated from some one birth-place; and when we better know the many means of migration, then, by the light which geology now throws, and will continue to throw, on former changes of climate and of the level of the land, we shall surely be enabled to trace in an admirable manner the former migrations of the inhabitants of the whole world. Even at present, by comparing the differences between the inhabitants of the sea on the opposite sides of a continent, and the nature of the various inhabitants on that continent in relation to their apparent means of immigration, some light can be thrown on ancient geography.

The noble science of Geology loses glory from the extreme imperfection of the record. The crust of the earth with its imbedded remains must not be looked at as a well-filled museum, but as a poor collection made at hazard and at rare intervals. The accumulation of each great fossiliferous formation will be recognized as having depended on an unusual concurrence of favourable circumstances, and the blank intervals between the successive stages as having been of vast duration. But we shall be able to gauge with some security the duration of these intervals by a comparison of the preceding and succeeding organic forms. We must be cautious in attempting to correlate as strictly contemporaneous two formations, which do not include many identical species, by the general succession of the forms of life. As species are produced and exterminated by slowly acting and still existing causes, and not by miraculous acts of creation; and as the most important of all causes of organic change is one which is almost independent of altered and perhaps suddenly altered physical conditions, namely, the mutual relation of organism to organism — the improvement of one organism entailing the improvement or the extermination of others; it follows, that the amount of organic change in the fossils of consecutive formations probably serves as a fair measure of the relative though not actual lapse of time. A number of species, however, keeping in a body might remain for a long period unchanged, whilst within the same period several of these species by migrating into new countries and coming into competition with foreign associates, might become modified; so that we must not overrate the accuracy of organic change as a measure of time.

In the future I see open fields for far more important researches. Psychology will be securely based on the foundation already well laid by Mr. Herbert Spencer, that of the necessary acquirement of each mental power and capacity by gradation. Much light will be thrown on the origin of man and his history.

Authors of the highest eminence seem to be fully satisfied with the view that each species has been independently created. To my mind it accords better with what we know of the laws impressed on matter by the Creator, that the production and extinction of the past and present inhabitants of the world should have been due to secondary causes, like those determining the birth and death of the individual. When I view all beings not as special creations, but as the lineal descendants of some few beings which lived long before the first bed of the Cambrian system was deposited, they seem to me to become ennobled. Judging from the past, we may safely infer that not one living species will transmit its unaltered likeness to a distant futurity. And of the species now living very few will transmit progeny of any kind to a far distant futurity; for the manner in which all organic beings are grouped, shows that the greater number of species in each genus, and all the species in many genera, have left no descendants, but have become utterly extinct. We can so far take a prophetic glance into futurity as to foretell that it will be the common and widely-spread species, belonging to the larger and dominant groups within each class, which will ultimately prevail and procreate new and dominant species. As all the living forms of life are the lineal descendants of those which lived long before the Cambrian epoch, we may feel certain that the ordinary succession by generation ·has never once been broken, and that no cataclysm has desolated the whole world. Hence we may look with some confidence to a secure future of great length. And as natural selection works solely by and for the good of each being, all corporeal and mental endowments will tend to progress towards perfection.

It is interesting to contemplate a tangled bank, clothed with many plants of many kinds, with birds singing on the bushes, with various insects flitting about, and with worms crawling through the damp earth, and to reflect that these elaborately constructed forms, so different from each other, and dependent upon each other in so complex a manner, have all been produced by laws acting around us. These laws, taken in the largest sense, being Growth with Reproduction; Inheritance which is almost implied by reproduction; Variability from the indirect and direct action of the conditions of life, and from use and disuse: a Ratio of Increase so high as to lead to a Struggle for Life, and as a consequence to Natural Selection, entailing Divergence of Character and the Extinction of less-improved forms. Thus, from the war of nature, from famine and death, the most exalted object which we are capable of conceiving, namely, the production of the higher animals, directly follows. There is grandeur in this view of life, with its several powers, having been originally breathed by the Creator into a few forms or into one; and that, whilst this planet has gone cycling on according to the fixed law of gravity, from so simple a beginning endless forms most beautiful and most wonderful have been, and are being evolved.

ALFRED RUSSEL WALLACE (1823–1913)

Mimicry, and Other Protective Resemblances among Animals

First published in the *Westminster Review,* July 1867; reprinted with
additions and corrections in *Natural Selection and Tropical
Nature,* 1895

[WALLACE lived a full life of four score years and ten, a life of travel,
thinking, and writing. His long collecting expeditions to the Amazon,
the Malay States, and the East Indies were productive of scientific
results of the first order. In Borneo, in 1858, he wrote an essay on
his theory of the origin of new species which struck him as he lay
in bed with a bout of fever. He sent it to Darwin, and it proved to be
almost the counterpart of the foundations of the evolutionary theory, which Darwin had arrived at twenty years before, but never
published.

Wallace showed a very clear realization of the distinction between
Lamarck's idea of conscious effort preceding structural change, and
the Darwinian belief of variation followed by adaptive habits. Wallace's name is closely associated with original concepts and proof,
taken from actual living specimens in the field, of geographical distribution, the effects of insular life, natural selection, and mimicry.
His books, while of great importance, are straightforward narratives
or lucid biological arguments, embellished with little or no attempt
at vividness, and giving few hints of his unending delight in tropical
wild life. Only in some of his letters, such as the following passage,
are there glimpses of the personality of the man himself:

"LOBO ROMAN, SUMATRA, *December* 22, 1861

"My Dear George:

"Between eight and nine years ago, when we were concocting that
absurd book 'Travels on the Amazon and Rio Negro,' you gave me
this identical piece of waste paper with sundry others, and now having scribbled away my last sheet of 'hot-pressed writing,' and being
just sixty miles from another, I send you back your gift, with interest; so you see that a good action, sooner or later, finds its sure
reward.

"I now write you a *letter,* I hope for the last time, for I trust our
future letters may be *viva voce,* as an Irishman would say, while our

epistolary correspondence will be confined to *notes*. I really do now think and believe that I am coming home, and as I am quite uncertain when I may be able to send you this letter, I may possibly arrive not very long after it. Some fine morning I expect to walk into 79, Pall Mall, and shall, I suppose, find things just the same as if I had walked out yesterday and come in to-morrow! There will you be seated on the same chair, at the same table, surrounded by the same account books, and writing upon paper of the same size and colour as when I last beheld you. I shall find your inkstand, pens, and pencils in the same places, and in the same beautiful order, which my idiosyncrasy compels me to admire, but forbids me to imitate. (Could you see the table at which I am now writing, your hair would stand on end at the reckless confusion it exhibits!) . . .

"I am here in one of the places unknown to the Royal Geographical Society, situated in the very centre of East Sumatra, about one hundred miles from the sea in three directions. It is the height of the wet season, and the rain pours down strong and steady, generally all night and half the day. Bad times for me, but I walk out regularly three or four hours every day, picking up what I can, and generally getting some little new or rare or beautiful thing to reward me. This is the land of the two-horned rhinoceros, the elephant, the tiger, and the tapir; but they all make themselves very scarce, and beyond their tracks and their dung, and once hearing a rhinoceros *bark* not far off, I am not aware of their existence. This, too, is the very land of monkeys; they swarm about the villages and plantations, long-tailed and short-tailed, and with no tail at all, white, black, and grey; they are eternally racing about the tree-tops, and gambolling in the most amusing manner. The way they jump is amazing. They throw themselves recklessly through the air, apparently sure, with one or other of their four hands, to catch hold of something. I estimated one jump by a long-tailed white monkey, at thirty feet horizontal, and sixty feet vertical, from a high tree to a lower one; he fell through, however, so great was his impetus, on to a lower branch, and then, without a moment's stop, scampered away from tree to tree, evidently quite pleased with his own pluck. When I startle a band, and one leader takes a leap like this, it is amusing to watch the others — some afraid and hesitating on the brink till at last they pluck up courage, take a run at it, and often roll over in the air with their desperate efforts. Then there are the long-armed apes, who never walk or run upon the trees, but travel altogether by their long arms, swinging themselves from bough to bough in the easiest and most graceful manner possible."]

THERE is no more convincing proof of the truth of a comprehensive theory than its power of absorbing and finding a place for new facts, and its capability of interpreting phenomena which had been previously looked upon as unaccountable anomalies. It is thus that the law of universal gravitation and the undulatory theory of light have become established and universally accepted by men of science. Fact after fact has been brought forward as being apparently inconsistent with them, and one after another these very facts have been shown to be the consequences of the laws they were at first supposed to disprove. A false theory will never stand this test. Advancing knowledge brings to light whole groups of facts which it cannot deal with, and its advocates steadily decrease in numbers, notwithstanding the ability and scientific skill with which it may have been supported. . . .

The course of a true [theory] is very different, as may be well seen by the progress of opinion on the subject of Natural Selection. In less than eight years *The Origin of Species* has produced conviction in the minds of a majority of the most eminent living men of science. New facts, new problems, new difficulties as they arise are accepted, solved, or removed by this theory; and its principles are illustrated by the progress and conclusions of every well established branch of human knowledge. It is the object of the present chapter to show how it has recently been applied to connect together and explain a variety of curious facts which had long been considered as inexplicable anomalies. . . .

The adaptation of the external colouring of animals to their conditions of life has long been recognised, and has been imputed either to an originally created specific peculiarity, or to the direct action of climate, soil, or food. Where the former explanation has been accepted it has completely checked inquiry, since we could never get any further than the fact of the adaptation. There was nothing more to be known about the matter. The second explanation was soon found to be quite inadequate to deal with all the varied phases of the phenomena, and to be contradicted by many well known facts. For example, wild rabbits are always of gray or brown tints well suited for concealment among grass and fern. But when these rabbits are domesticated, without any change of climate or food, they vary into white or black, and these varieties may be multiplied to any extent, forming white or black races. Exactly the same thing has occurred with pigeons; and in the case of rats and mice, the white variety has not been shown to be at all dependent on alteration of climate, food, or other external conditions. In many cases the wings of an insect not only assume the exact tint of the bark or leaf it is accustomed to rest on, but the form and veining of the leaf or the exact rugosity of the bark is imitated; and these detailed modifications cannot be reasonably imputed to climate or to food, since in many cases the species does not feed on the substance it resembles, and when it does, no reasonable connection can be shown to exist between the supposed cause and the effect produced. It was reserved for the theory of Natural Selection to solve all these problems, and many others which were not at first supposed to be directly connected with them. To make these latter intelligible, it will be necessary to give a sketch of the whole series of phenomena which may be classed under the head of useful or protective resemblances.

Concealment, more or less complete, is useful to many animals, and absolutely essential to some. Those which have numerous enemies from which they cannot escape by rapidity of motion find safety in concealment. Those which prey upon others must also be so constituted as not to alarm them by their presence or their approach, or they would soon die of hunger. Now it is remarkable in how many cases nature gives this boon to the animal, by colouring it with such tints as may best serve to enable it to escape from its enemies or to entrap its prey. Desert animals as a rule are desert-coloured. The lion is a typical example of this, and must be almost invisible when crouched upon the sand or among desert rocks and stones. Antelopes are all more or less sandy-coloured. The camel is pre-eminently so. The Egyptian cat and the Pampas cat are sandy or earth-coloured. The Australian kangaroos are of the same tints, and the original colour of the wild horse is supposed to have been a sandy or clay-colour.

The desert birds are still more remarkably protected by their assimilative hues. The stonechats, the larks, the quails, the goatsuckers and the grouse, which abound in the North African and Asiatic deserts, are all tinted and mottled so as to resemble with wonderful accuracy the average colour and aspect of the soil in the district they inhabit. The Rev. H. Tristam, in his account of the ornithology of North Africa in the first volume of the *Ibis* says: "In the desert, where neither trees, brushwood, nor even undulation of the surface afford the slightest protection to its foes, a modification of colour which shall be assimilated to that of the surrounding country is absolutely necessary. Hence *without exception* the upper plumage of *every bird,* whether lark, chat, sylvain, or sand-grouse, and also the fur of *all the smaller mammals,* and the skin of *all the snakes and lizards,* is of one uniform isabelline sand colour." After the testimony of so able an observer it is unnecessary to adduce further examples of the protective colours of desert animals.

Almost equally striking are the cases of arctic animals possessing the white colour that best conceals them upon snowfields and icebergs. The polar bear is the only bear that is white, and it lives constantly among snow and ice. The arctic fox, the ermine, and the alpine hare change to white in winter only, because in summer white would be more conspicuous than any other colour, and therefore a danger rather than a protection; but the American polar hare, inhabiting regions of almost perpetual snow, is white all the year round. Other animals inhabiting the same Northern regions do not, however, change colour. The sable is a good example, for throughout the severity of a Siberian winter it retains its rich brown fur. But its habits are such that it does not need the protection of colour, for it is said to be able to subsist on fruits and berries in winter, and to be so active upon the trees as to catch small birds among the branches. So also the woodchuck [otter] of Canada has a dark-brown fur; but then it lives in burrows and frequents river banks, catching fish and small animals that live in or near the water.

Among birds, the ptarmigan is a fine example of protective colouring. Its summer plumage so exactly harmonises with the lichen-coloured stones among which it delights to sit, that a person may walk through a flock of them without seeing a single bird; while in winter its white plumage is an almost equal protection. The snow-bunting, the jer-falcon, and the snowy owl are also white-coloured birds inhabiting the arctic regions, and there can be little doubt but that their colouring is to some extent protective.

Nocturnal animals supply us with equally good illustrations. Mice, rats, bats and moles possess the least conspicuous of hues, and must be quite invisible at times when any light colour would be instantly seen. Owls and goatsuckers are of those dark mottled tints that will assimilate with bark and lichen, and thus protect them during the day, and at the same time be inconspicuous in the dusk.

It is only in the tropics, among forests which never lose their foliage, that we find whole groups of birds whose chief colour is green. The parrots are the most striking example, but we have also a group of green pigeons in the East; and the barbets, leaf-thrushes, bee-eaters, white-eyes, turacos, and several smaller groups, have so much green in their plumage as to tend greatly to conceal them among the foliage. . . .

It is, however, in the insect world that this principle of the adaptation of animals to their environment is most fully and strikingly developed. In order to understand how general this is, it is necessary to enter somewhat into details, as we shall thereby be better able to appreciate the significance of the still more remarkable phenomena we shall presently have to discuss. It seems to be in proportion to their sluggish motions or the absence of other means of defence, that insects possess the protective colouring. In the tropics there are thousands of species of insects which rest during the day clinging to the bark of dead or fallen trees; and the greater portion of these are delicately mottled with gray and brown tints, which, though symmetrically disposed and infinitely varied, yet blend so completely with the usual colours of the bark, that at two or three feet distance they are quite indistinguishable. In some cases a species is known to frequent only one species of tree. This is the case with the common South American long-horned beetle (Onychocerus scorpio), which, Mr. Bates informed me, is found only on a rough-barked tree, called Tapiribá, on the Amazon. It is very abundant, but so exactly does it resemble the bark in colour and rugosity, and so closely does it cling to the branches, that until it moves it is absolutely invisible! An allied species (O. concentricus) is found only at Pará, on a distinct species of tree, the bark of which it resembles with equal accuracy. Both these insects are abundant, and we may fairly conclude that the protection they derive from this strange concealment is at least one of the causes that enable the race to flourish. . . .

The distribution of colour in butterflies and moths respectively is very instructive from this point of view. The former have all their brilliant colouring on the upper surface of all four wings, while the under surface is almost always soberly coloured, and often very dark and obscure. The moths on the contrary have generally their chief colour on the hind wings only, the upper wings being of dull, sombre, and often imitative tints, and these generally conceal the hind wings when the insects are in repose. This arrangement of the colours is therefore eminently protective, because the butterfly always rests with his wings raised so as to conceal the dangerous brilliancy of his upper surface. It is probable that if we watched their habits sufficiently we should find the under surface of the wings of butterflies very frequently imitative and protective. Mr. T. W. Wood has pointed out that the little orange-tip butterfly often rests in the evening on the green and white flower heads of an umbelliferous plant, the wild chervil, and that when observed in this position the beautiful green and white mottling of the under surface completely assimilates with the flower heads and renders the creature very difficult to be seen. It

111

is probable that the rich dark colouring of the under side of our pea-cock, tortoiseshell, and red-admiral butterflies answers a similar purpose. . . .

But the most wonderful and undoubted case of protective resemblance in a butterfly which I have ever seen, is that of the common Indian Kallima inachis, and its Malayan ally, Kallima paralekta. The upper surface of these insects is very striking and showy, as they are of a large size, and are adorned with a broad band of rich orange on a deep-bluish ground. The underside is very variable in colour, so that out of fifty specimens no two can be found exactly alike, but every one of them will be of some shade of ash or brown or ochre, such as are found among dead, dry, or decaying leaves. The apex of the upper wings is produced into an acute point, a very common form in the leaves of tropical shrubs and trees, and the lower wings are also produced into a short narrow tail. Between these two points runs a dark curved line exactly representing the midrib of a leaf, and from this radiate on each side a few oblique lines, which serve to indicate the lateral veins of a leaf. These marks are more clearly seen on the outer portion of the base of the wings, and on the inner side towards the middle and apex, and it is very curious to observe how the usual marginal and transverse striae of the group are here modified and strengthened so as to become adapted for an imitation of the venation of a leaf. We come now to a still more extraordinary part of the imitation, for we find representations of leaves in every stage of decay, variously blotched and mildewed and pierced with holes, and in many cases irregularly covered with powdery black dots gathered into patches and spots, so closely resembling the various kinds of minute fungi that grow on dead leaves that it is impossible to avoid thinking at first sight that the butterflies themselves have been attacked by real fungi.

But this resemblance, close as it is, would be of little use if the habits of the insect did not accord with it. If the butterfly sat upon leaves or upon flowers, or opened its wings so as to expose the upper surface, or exposed and moved its head and antennae as many other butterflies do, its disguise would be of little avail. We might be sure, however, from the analogy of many other cases, that the habits of the insect are such as still further to aid its deceptive garb; but we are not obliged to make any such supposition, since I myself had the good fortune to observe scores of Kallima paralekta, in Sumatra, and to capture many of them, and can vouch for the accuracy of the following details. These butterflies frequent dry forests and fly very swiftly. They were never seen to settle on a flower or a green leaf, but were many times lost sight of in a bush or tree of dead leaves. On such occasions they were generally searched for in vain, for while gazing intently at the very spot where one had disappeared, it would often suddenly dart out, and again vanish twenty or fifty yards farther on. On one or two occasions the insect was detected reposing, and it could then be seen how completely it assimilates itself to the surrounding leaves. It sits on a nearly upright twig, the wings fitting closely back to back, concealing the antennae and head, which are drawn up between their bases. The little tails of the hind wing touch the branch, and form a perfect stalk to the leaf, which is supported in its place by the claws of the middle pair of feet, which are slender and inconspicuous. The irregular outline of the wings gives exactly the perspective effect of a shrivelled leaf. We thus have size, colour, form, markings, and habits all combining together to produce a disguise which may be said to be absolutely perfect; and the protection which it affords is sufficiently indicated by the abundance of the individuals that possess it. . . .

The whole order of Orthoptera, grasshoppers, locusts, crickets, etc., are protected by their colours harmonising with that of the vegetation or the soil on which they live, and in no other group have we such striking examples of special resemblance. Most of the tropical Mantidae and Locustidae are of the exact tint of the leaves on which they habitually repose, and many of them in addition have the veinings of their wings modified so as exactly to imitate that of a leaf. This is carried to the furthest possible extent in the wonderful genus, Phyllium, the "walking leaf," in which not only are the wings perfect imitations of leaves in every detail, but the thorax and legs are flat, dilated, and leaf-like; so that when the living insect is resting among the foliage on which it feeds, the closest observation is often unable to distinguish between the animal and the vegetable.

The whole family of the Phasmidae, or spectres, to which this insect belongs, is more or less imitative, and a great number of the species are called "walking-stick insects," from their singular resemblance to twigs and branches. Some of these are a foot long and as thick as one's finger, and their whole colouring, form, rugosity, and the arrangement of the head, legs, and antennae are such as to render them absolutely identical in appearance with dead sticks. They hang loosely about shrubs in the forest, and have the extraordinary habit of stretching out their legs unsymmetrically, so as to render the deception more complete. One of these creatures obtained by myself in Borneo (Ceroxylus laceratus) was covered over with foliaceous excrescences of a clear olive green colour, so as exactly to re-

113

semble a stick grown over by a creeping moss or jungermannia. The Dyak who brought it me assured me it was grown over with moss although alive, and it was only after a most minute examination that I could convince myself it was not so.

We need not adduce any more examples to show how important are the details of form and of colouring in animals, and that their very existence may often depend upon their being by these means concealed from their enemies. This kind of protection is found apparently in every class and order, for it has been noticed wherever we can obtain sufficient knowledge of the details of an animal's life-history. It varies in degree, from the mere absence of conspicuous colour or a general harmony with the prevailing tints of nature, up to such a minute and detailed resemblance to inorganic or vegetable structures as to realise the talisman of the fairy tale, and to give its possessor the power of rendering itself invisible.

We will now endeavour to show how these wonderful resemblances have most probably been brought about. Returning to the higher animals, let us consider the remarkable fact of the rarity of white colouring in the mammalia or birds of the temperate or tropical zones in a state of nature. There is not a single white land-bird or quadruped in Europe, except the few arctic or alpine species, to which white is a protective colour. Yet in many of these creatures there seems to be no inherent tendency to avoid white, for directly they are domesticated white varieties arise, and appear to thrive as well as others. We have white mice and rats, white cats, horses, dogs and cattle, white poultry, pigeons, turkeys and ducks, and white rabbits. Some of these animals have been domesticated for a long period, others only for a few centuries; but in almost every case in which an animal has been thoroughly domesticated, parti-coloured and white varieties are produced and become permanent.

It is also well known that animals in a state of nature produce white varieties occasionally. Blackbirds, starlings, and crows are occasionally seen white, as well as elephants, deer, tigers, hares, moles, and many other animals; but in no case is a permanent white race produced. Now there are no statistics to show that the normal-coloured parents produce white offspring oftener under domestication than in a state of nature, and we have no right to make such an assumption if the facts can be accounted for without it. But if the colours of animals do really, in the various instances already adduced, serve for their concealment and preservation, then white or any other conspicuous colour must be hurtful, and must in most cases shorten an animal's life. A white rabbit would be more surely the prey of hawk or buzzard, and the white mole, or field mouse, could not long escape from the vigilant owl. So, also, any deviation from those tints best adapted to conceal a carnivorous animal would render the pursuit of its prey much more difficult, would place it at a disadvantage among its fellows, and in a time of scarcity would probably cause it to starve to death. On the other hand, if an animal spreads from a temperate to an arctic district, the conditions are changed. During a large portion of the year, and just when the struggle for existence is more severe, white is the prevailing tint of nature, and dark colours will be the most conspicuous. The white varieties will now have an advantage; they will escape from their enemies or will secure food, while their brown companions will be devoured or will starve; and as "like produces like" is the established rule in nature, the white race will become permanently established, and dark varieties, when they occasionally appear, will soon die out from their want of adaptation to their environment. In each case the fittest will survive, and a race will be eventually produced adapted to the conditions in which it lives.

We have here an illustration of the simple and effectual means by which animals are brought into harmony with the rest of nature. That slight amount of variability in every species, which we often look upon as something accidental or abnormal, or so insignificant as to be hardly worthy of notice, is yet the foundation of all those wonderful and harmonious resemblances which play such an important part in the economy of nature. Variation is generally very small in amount,[1] but it is all that is required, because the change in the external conditions to which an animal is subject is generally very slow and intermittent. When these changes have taken place too rapidly, the result has often been the extinction of species; but the general rule is, that climatal and geological changes go on slowly, and the slight but continual variations in the colour, form, and structure of all animals have furnished individuals adapted to these changes, and who have become the progenitors of modified races. Rapid multiplication, incessant slight variation, and survival of the fittest — these are the laws which ever keep the organic world in harmony with the inorganic, and with itself. These are the laws which we believe have produced all the cases of protective resemblance already adduced, as well as those still more curious examples we have yet to bring before our readers. . . .

It has been long known to entomologists that certain insects bear

[1] Later research has shown that variation is more frequent and of greater amount than at first supposed.

a strange external resemblance to others belonging to distinct genera, families, or even orders, and with which they have no real affinity whatever. The fact, however, appears to have been generally considered as dependent upon some unknown law of "analogy" — some "system of nature," or "general plan," which had guided the Creator in designing the myriads of insect forms, and which we could never hope to understand. In only one case does it appear that the resemblance was thought to be useful, and to have been designed as a means to a definite and intelligible purpose. The flies of the genus Volucella enter the nests of bees to deposit their eggs, so that their larvae may feed upon the larvae of the bees, and these flies are each wonderfully like the bee on which it is parasitic. Kirby and Spence believed that this resemblance or "mimicry" was for the express purpose of protecting the flies from the attacks of the bees, and the connection is so evident that it was hardly possible to avoid this conclusion. The resemblance, however, of moths to butterflies or to bees, of beetles to wasps, and of locusts to beetles, has been many times noticed by eminent writers; but scarcely ever till within the last few years does it appear to have been considered that these resemblances had any special purpose, or were of any direct benefit to the insects themselves. In this respect they were looked upon as accidental, as instances of the "curious analogies" in nature which must be wondered at but which could not be explained. Recently, however, these instances have been greatly multiplied; the nature of the resemblances has been more carefully studied, and it has been found that they are often carried out into such details as almost to imply a purpose of deceiving the observer. The phenomena, moreover, have been shown to follow certain definite laws, which again all indicate their dependence on the more general law of the "survival of the fittest," or, "the preservation of favoured races in the struggle for life." It will, perhaps, be as well here to state what these laws or general conclusions are, and then to give some account of the facts which support them.

The first law is, that in an overwhelming majority of cases of mimicry, the animals (or the groups) which resemble each other inhabit the same country, the same district, and in most cases are to be found together on the very same spot.

The second law is, that these resemblances are not indiscriminate, but are limited to certain groups, which in every case are abundant in species and individuals, and can often be ascertained to have some special protection.

The third law is, that the species which resemble or "mimic" these dominant groups are comparatively less abundant in individuals, and are often very rare.

These laws will be found to hold good in all the cases of true mimicry among various classes of animals to which we have now to call the attention of our readers.

As it is among butterflies that instances of mimicry are most numerous and most striking, an account of some of the more prominent examples in this group will first be given. There is in South America an extensive family of these insects, the Heliconidae, which are in many respects very remarkable. They are so abundant and characteristic in all the woody portions of the American tropics, that in almost every locality they will be seen more frequently than any other butterflies. They are distinguished by very elongate wings, body, and antennae, and are exceedingly beautiful and varied in their colours; spots and patches of yellow, red or pure white upon a black, blue or brown ground being most general. They frequent the forests chiefly, and all fly slowly and weakly; yet although they are so conspicuous, and could certainly be caught by insectivorous birds more easily than almost any other insects, their great abundance all over the wide region they inhabit shows that they are not so persecuted. It is to be especially remarked also, that they possess no adaptive colouring to protect them during repose, for the under side of their wings presents the same, or at least an equally conspicuous colouring as the upper side; and they may be observed after sunset suspended at the end of twigs and leaves, where they have taken up their station for the night, fully exposed to the attacks of enemies if they have any.

These beautiful insects possess, however, a strong pungent semi-aromatic or medicinal odour, which seems to pervade all the juices of their system. When the entomologist squeezes the breast of one of them between his fingers to kill it, a yellow liquid exudes which stains the skin, and the smell of which can only be got rid of by time and repeated washings. Here we have probably the cause of their immunity from attack, since there is a great deal of evidence to show that certain insects are so disgusting to birds that they will under no circumstances touch them. Mr. Stainton has observed that a brood of young turkeys greedily devoured all the worthless moths he had amassed in a night's "sugaring," yet one after another seized and rejected a single white moth which happened to be among them. Young pheasants and partridges which eat many kinds of caterpillars seem to have an absolute dread of that of the common currant moth, which they will never touch, and tomtits as well as other small birds appear never to eat the same species. In the case of the Heliconidae, however, we have some direct evidence to the same effect. In the Brazilian forests there are great numbers of insectivorous birds — as jacamars, trogons, and puffbirds — which catch insects on the wing, and that they destroy many butterflies is indicated by the fact that the wings of these insects are often found on the ground where their bodies have been devoured. But among these there are no wings of Heliconidae, while those of the large showy Nymphalidae, which have a much swifter flight, are often met with. Again, a gentleman who had recently returned from Brazil stated at a meeting of the Entomological Society that he once observed a pair of puffbirds catching butterflies, which they brought to their nest to feed their young; yet during half an hour they never brought one of the Heliconidae, which were flying lazily about in great numbers, and which they could have captured more easily than any others. It was this circumstance that led Mr. Belt to observe them so long, as he could not understand why the most common insects should be altogether passed by. Mr. Bates also tells us that he never saw them molested by lizards or predacious flies, which often pounce on butterflies.

If therefore, we accept it as highly probable (if not proved) that the Heliconidae are very greatly protected from attack by their peculiar odour and taste, we find it much more easy to understand their chief characteristics — their great abundance, their slow flight, their gaudy colours, and the entire absence of protective tints on their under surfaces. This property places them somewhat in the position of those curious wingless birds of oceanic islands, the dodo, the apteryx, and the moas, which are with great reason supposed to have lost the power of flight on account of the absence of carnivorous quadrupeds. Our butterflies have been protected in a different way, but quite as effectually; and the result has been that as there has been nothing to escape from, there has been no weeding out of slow flyers, and as there has been nothing to hide from, there has been no extermination of the bright-coloured varieties, and no preservation of such as tended to assimilate with surrounding objects.

Now let us consider how this kind of protection must act. Tropical insectivorous birds very frequently sit on dead branches of a lofty tree, or on those which overhang forest paths, gazing intently around, and darting off at intervals to seize an insect at a considerable distance, which they generally return to their station to devour. If a bird began by capturing the slow-flying conspicuous Heliconidae, and found them always so disagreeable that it could not eat them, it would after a very few trials leave off catching them at all; and their whole appearance, form, colouring and mode of flight are so peculiar that there can be little doubt birds would soon learn to distinguish them at a long distance, and never waste any time in pursuit of them. Under these circumstances, it is evident that any other butterfly of a group which birds were accustomed to devour would be almost equally well protected by closely resembling a Heliconian externally, as if it acquired also the disagreeable odour; always supposing that there were only a few of them among a great number of the Heliconias. If the birds could not distinguish the two kinds externally, and there were on the average only one eatable among fifty uneatable, they would soon give up seeking for the eatable ones, even if they knew them to exist. If, on the other hand, any particular butterfly of an eatable group acquired the disagreeable taste of the Heliconias while it retained the characteristic form and colouring of its own group, this would be really of no use to it whatever; for the birds would go on catching it among its eatable allies (compared with which it would rarely occur), it would be wounded and disabled, even if rejected, and its increase would thus be as effectually checked as if it were devoured. It is important, therefore, to understand that if any one genus of an extensive family of eatable butterflies were in danger of extermination from insect-eating birds, and if two kinds of variation were going on among them, some individuals possessing a slightly disagreeable taste, others a slight resemblance to the Heliconidae, this latter quality would be much more valuable than the former. The change in flavour would not at all prevent the variety from being captured as before, and it would almost certainly be thoroughly disabled before being rejected. The approach in colour and form to the Heliconidae, however, would be at the very first a positive, though perhaps a slight advantage; for although at short distances this variety would be easily distinguished and devoured, yet at a longer distance it might be mistaken for one of the uneatable group, and so be passed by and gain another day's life, which might in many cases be sufficient for it to lay a quantity of eggs and leave a numerous progeny, many of which would inherit the peculiarity which had been the safeguard of their parent.

Now this hypothetical case is exactly realised in South America. Among the white butterflies forming the family Pieridae (many of which do not greatly differ in appearance from our own cabbage butterflies) is a genus of rather small size (Leptalis), some species of which are white like their allies, while the larger number exactly resemble the Heliconidae in the form and colouring of the wings. It must always be remembered that these two families are as absolutely distinguished from each other by structural characters as are the carnivora and the ruminants among quadrupeds, and that an entomologist can always distinguish the one from the other by the structure of the feet, just as certainly as a zoologist can tell a bear from a buffalo by the skull or by a tooth. Yet the resemblance of a species of the one family to another species in the other family was often so great, that both Mr. Bates and myself were many times deceived at the time of capture, and did not discover the distinctness of the two insects till a closer examination detected their essential differences. During his residence of eleven years in the Amazon valley, Mr. Bates found a number of species or varieties of Leptalis, each of which was a more or less exact copy of one of the Heliconidae of the district it inhabited; and the results of his observations are embodied in a paper published in the *Linnaean Transactions,* in which he first explained the phenomena of "mimicry" as the result of natural selection, and showed its identity in cause and purpose with protective resemblance to vegetable or inorganic forms.

The imitation of the Heliconidae by the Leptalides is carried out to a wonderful degree in form as well as in colouring. The wings have become elongated to the same extent, and the antennae and abdomen have both become lengthened, to correspond with the unusual condition in which they exist in the former family. . . . As if to derive all the benefit possible from this protective mimicry, the habits have become so modified that the Leptalides generally frequent the very same spots as their models, and have the same mode of flight; and as they are always very scarce (Mr. Bates estimating their numbers at about one to a thousand of the group they resemble), there is hardly a possibility of their being found out by their enemies. It is also very remarkable that in almost every case the particular Ithomias and other species of Heliconidae which they resemble are noted as being very common species, swarming in individuals, and found over a wide range of country. This indicates antiquity and permanence in the species, and is exactly the condition most essential both to aid in the development of the resemblance and to increase its utility. . . .

In the preceding cases we have found Lepidoptera imitating other species of the same order, and such species only as we have good reason to believe were free from the attacks of many insectivorous creatures; but there are other instances in which they altogether lose the external appearance of the order to which they belong, and take on the dress of bees or wasps — insects which have an undeniable protection in their stings. The Sesiidae and Aegeriidae, two families of day-flying moths, are particularly remarkable in this respect, and a mere inspection of the names given to the various species shows how the resemblance has struck every one. We have apiformis, vespiforme, ichneumoniforme, scoliaeforme, sphegiforme (bee-like, wasplike, ichneumon-like, etc.), and many others, all indicating a resemblance to stinging Hymenoptera. . . .

Charis melipona, a South American Longicorn of the family Necydalidae, has been so named from its resemblance to a small bee of the genus Melipona. It is one of the most remarkable cases of mimicry, since the beetle has the thorax and body densely hairy like the bee, and the legs are tufted in a manner most unusual in the order Coleopters. Another Longicorn, Odontocera odyneroides, has the abdomen banded with yellow, and constricted at the base, and is altogether so exactly like a small common wasp of the genus Odynerus, that Mr. Bates informs us he was afraid to take it out of his net with his fingers for fear of being stung. Had Mr. Bates' taste for insects been less omnivorous than it was, the beetle's disguise might have saved it from his pin, as it had no doubt often done from the beak of hungry birds.

Perhaps the most wonderful case of all is the large caterpillar mentioned by Mr. Bates, which startled him by its close resemblance to a small snake. The first three segments behind the head were dilatable at the will of the insect, and had on each side a large black pupillated spot, which resembled the eye of the reptile. Moreover, it resembled a poisonous viper, not a harmless species of snake, as was proved by the imitation of keeled scales on the crown produced by the recumbent feet, as the caterpillar threw itself backward! . . .

Although such a store of interesting facts has been already accumulated, the subject we have been discussing is one of which comparatively little is really known. The natural history of the tropics has never yet been studied on the spot with a full appreciation of "what to observe" in this matter. The varied ways in which the colouring and form of animals serve for their protection, their strange disguises as vegetable or mineral substances, their wonderful mimicry of other beings, offer an almost unworked and inexhaustible field of discovery for the zoologist, and will assuredly throw much light on the laws and conditions which have resulted in the wonderful variety of colour, shade, and marking which constitutes one of the most pleasing characteristics of the animal world, but the immediate causes of which it has hitherto been most difficult to explain.

JEAN LOUIS RODOLPHE AGASSIZ (1807–1873)

The Aims of an Expedition

From *A Journey in Brazil*. 1867

[AGASSIZ is one of the best-known names in science. In the order mentioned, he was a great teacher, geologist, ichthyologist, embryologist, and comparative anatomist. At twenty-two he produced a monograph on Brazilian fishes, and then studied fossil fishes, invertebrates, and glaciers. He was born in Switzerland, but spent the second half of his life in the United States, where as a lecturer and teacher he became supreme. This was inevitable with a charming personality added to his illustrative ability. Before he had mastered the English language, and when at a loss for a word or phrase, he would go to the blackboard and begin to draw. In this way "he would lead his listeners along the successive phases of insect development, talking as he drew and drawing as he talked, till suddenly the winged creature stood declared upon the blackboard, almost as if it had burst then and there from the chrysalis, and the growing interest of his hearers culminated in a burst of delighted applause."

The following selection is from two lectures given on board ship en route to Brazil on his first expedition in 1865. Agassiz and his wife wrote the account together, she furnishing the narrative and he the various dissertations to his assembled staff. The instructions are of great interest as showing that even three-quarters of a century ago the same general problems held as today; even the warning about over-collecting is as true today as then. His account of the recapitulation in the embryo of ancestry and of development in lower groups of animals is suggestive of later discoveries.

As did Huxley in his early convictions, Agassiz believed in the original creation of a number of isolated forms of life or archetypes, each giving rise to diverging animal species. But, in spite of his clarity of reasoning, Agassiz could never accept the Darwinian idea of continuous evolution, his intensely religious mind still holding the belief that every organic creature was "a thought of God." Nevertheless his investigations added greatly to that same tree of life.

Any piecemeal quotations from Agassiz's writings, such as are

122

necessitated by the limitations of this volume, can but inadequately represent the genius of this illustrious scientist.]

April 6th. — . . . We are now fairly in the tropics. "The trades" blow heavily, and yesterday was a dreary day for those unused to the ocean; the beautiful blue water, of a peculiar metallic tint, as remarkable in color, it seemed to me, as the water of the Lake of Geneva, did not console us for the heavy moral and physical depression of sea-sick mortals. To-day the world looks brighter; there is a good deal of motion, but we are more accustomed to it. This morning the lecture had, for the first time, a direct bearing upon the work of the expedition. The subject was, "How to observe, and what are the objects of scientific explorations in modern times."

"My companions and myself have come together so suddenly and so unexpectedly on our present errand, that we have had little time to organize our work. The laying out of a general scheme of operations is, therefore, the first and one of the most important points to be discussed between us. The time for great discoveries is passed. No student of nature goes out now expecting to find a new world, or looks in the heavens for any new theory of the solar system. The work of the naturalist, in our day, is to explore worlds the existence of which is already known; to investigate, not to discover. The first explorers, in this modern sense, were Humboldt in the physical world, Cuvier in natural history, Lavoisier in chemistry, La Place in astronomy. They have been the pioneers in the kind of scientific work characteristic of our century. We who have chosen Brazil as our field must seek to make ourselves familiar with its physical features, its mountains and its rivers, its animals and plants. There is a change, however, to be introduced in our mode of work, as compared with that of former investigators. When less was known of animals and plants the discovery of new species was the great object. This has been carried too far, and is now almost the lowest kind of scientific work. The discovery of a new species as such does not change a feature in the science of natural history, any more than the discovery of a new asteroid changes the character of the problems to be investigated by astronomers. It is mere adding to the enumeration of objects. We should look rather for the fundamental relations among animals; the number of species we may find is of importance only so far as they explain the distribution and limitation of different genera and families, their relations to each other and to the physical conditions under which they live. Out of such investigations there looms up a deeper question for scientific men, the solution of which is to be the most important result of their work in coming generations. The origin of life is the great question of the day. How did the organic world come to be as it is? It must be our aim to throw some light on this subject by our present journey. How did Brazil come to be inhabited by the animals and plants now living there? Who were its inhabitants in past times? What reason is there to believe that the present condition of things in this country is in any sense derived from the past? The first step in this investigation must be to ascertain the geographical distribution of the present animals and plants. Suppose we first examine the Rio San Francisco. The basin of this river is entirely isolated. Are its inhabitants, like its waters, completely distinct from those of other basins? Are its species peculiar to itself, and not repeated in any other river of the continent? Extraordinary as this result would seem, I nevertheless expect to find it so. The next water-basin we shall have to examine will be that of the Amazons, which connects through the Rio Negro with the Orinoco. It has been frequently repeated that the same species of fish exist in the waters of the San Francisco and in those of Guiana and of the Amazons. At all events, our works on fishes constantly indicate Brazil and Guiana as the common home of many species; but this observation has never been made with sufficient accuracy to merit confidence. Fifty years ago the exact locality from which any animal came seemed an unimportant fact in its scientific history, for the bearing of this question on that of origin was not then perceived. To say that any specimen came from South America was quite enough; to specify that it came from Brazil, from the Amazons, the San Francisco, or the La Plata, seemed a marvellous accuracy in the observers. In the museum at Paris, for instance, there are many specimens entered as coming from New York or from Pará; but all that is absolutely known about them is that they were shipped from those sea-ports. Nobody knows exactly where they were collected. So there are specimens entered as coming from the Rio San Francisco, but it is by no means sure that they came exclusively from that water-basin. All this kind of investigation is far too loose for our present object. Our work must be done with much more precision; it must tell something positive of the geographical distribution of animals in Brazil. Therefore, my young friends who come with me on this expedition, let us be careful that every specimen has a label, recording locality and date, so secured that it shall reach Cambridge safely. It would be still better to attach two labels to each specimen, so that, if any mischance happens to one, our record may not be lost. We must try not to mix the fishes of different rivers, even though they flow into each other, but to keep our collections perfectly distinct. You will easily see the vast importance of thus ascertaining the limitation of species, and the bearing of the result on the great question of origin. . . .

"Our next aim, and with the same object, namely, its bearing upon the question of origin, will be the study of the young, the collecting of eggs and embryos. This is the more important, since museums generally know only adult specimens. As far as I know, the Zoological Museum at Cambridge is the only one containing large collections of embryological specimens from all the classes of the animal kingdom. One significant fact, however, is already known. In their earliest stages of growth all animals of the same class are much more alike than in their adult condition, and sometimes so nearly alike as hardly to be distinguished. Indeed, there is an early period when the resemblances greatly outweigh the differences. How far the representatives of different classes resemble one another remains to be ascertained with precision. There are two possible interpretations of these facts. One is that animals so nearly identical in the beginning must have been originally derived from one germ, and are but modifications or transmutations, under various physical conditions, of this primitive unit. The other interpretation, founded on the same facts, is, that since, notwithstanding this material identity in the beginning, no germ ever grows to be different from its parent, or diverges from the pattern imposed upon it at its birth, therefore some other cause besides a material one must control its development; and if this be so, we have to seek an explanation of the differences between animals outside of physical influences. Thus far both these views rest chiefly upon personal convictions and opinions. The true solution of the problem must be sought in the study of the development of the animals themselves, and embryology is still in its infancy; for, though a very complete study of the embryology of a few animals has been made, yet these investigations include so small a number of representatives from the different classes of the animal kingdom that they do not yet give a basis for broad generalizations. Very little is known of the earlier stages in the formation of hosts of insects whose later metamorphoses, including the change of the already advanced larva, first to the condition of a chrysalis and then to that of a perfect insect, have been carefully traced. It remains to be ascertained to what extent the caterpillars of different kinds of butterflies, for instance, resemble one another during the time of their formation in the egg. An immense field of observation is open in this order alone.

"I have, myself, examined over one hundred species of bird embryos, now put up in the museum of Cambridge, and found that, at a certain age, they all have bills, wings, legs, feet, &c., &c., exactly alike. The young robin and the young crow are web-footed, as well as the duck. It is only later that the fingers of the foot become distinct. How very interesting it will be to continue this investigation among the tropical birds! — to see whether, for instance, the toucan, with its gigantic bill, has, at a certain age, a bill like that of all other birds; whether the spoonbill ibis has, at the same age, nothing characteristic in the shape of its bill. No living naturalist could now tell you one word about all this; neither could he give you any information about corresponding facts in the growth of the fishes, reptiles, or quadrupeds of Brazil, not one of the young of these animals having ever been compared with the adult. In these lectures I only aim at showing you what an extensive and interesting field of investigation opens before us; if we succeed in cultivating even a few corners of it we shall be fortunate."

April 10th. — A rough sea to-day, notwithstanding which we had our lecture as usual, though I must say, that, owing to the lurching of the ship, the lecturer pitched about more than was consistent with the dignity of science. Mr. Agassiz returned to the subject of embryology, urging upon his assistants the importance of collecting materials for this object as a means of obtaining an insight into the deeper relations between animals.

"Heretofore classification has been arbitrary, inasmuch as it has rested mainly upon the interpretation given to structural differences by various observers, who did not measure the character and value of these differences by any natural standard. I believe that we have a more certain guide in these matters than opinion or the individual estimate of any observer, however keen his insight into structural differences. The true principle of classification exists in Nature herself, and we have only to decipher it. If this conviction be correct, the next question is, How can we make this principle a practical one in our laboratories, an active stimulus in our investigations? Is it susceptible of positive demonstration in material facts? Is there any method to be adopted as a correct guide, if we set aside the idea of originating systems of classification of our own, and seek only to read that already written in nature? I answer, Yes. The standard is to be found in the changes animals undergo from their first forma-tion in the egg to their adult condition.

"It would be impossible for me here and now to give you the de-tails of this method of investigation, but I can tell you enough to illustrate my statement. Take a homely and very familiar example, that of the branch of Articulates. Naturalists divide this branch into three classes, — Insects, Crustacea, and Worms; and most of them tell you that Worms are lowest, Crustacea next in rank, and that Insects stand highest, while others have placed the Crustacea at the head of the group. We may well ask why. Why does an insect stand above a crustacean, or, *vice versa*, why is a grasshopper or a butterfly struc-turally superior to a lobster or a shrimp? And indeed there must be a difference in opinion as to the respective standing of these groups so long as their classification is allowed to remain a purely arbitrary one, based only upon interpretation of anatomical details. One man thinks the structural features of Insects superior, and places them highest; another thinks the structural features of the Crustacea highest, and places them at the head. In either case it is only a ques-tion of individual appreciation of the facts. But when we study the gradual development of the insect, and find that in its earliest stages it is worm-like, in its second, or chrysalis stage, it is crustacean-like, and only in its final completion it assumes the character of a perfect insect, we have a simple natural scale by which to estimate the com-parative rank of these animals. Since we cannot suppose that there is a retrograde movement in the development of any animal, we must believe that the insect stands highest, and our classification in this instance is dictated by Nature herself. This is one of the most striking examples, but there are others quite as much so, though not as familiar. The frog, for instance, in its successive stages of develop-ment, illustrates the comparative standing of the orders composing the class to which it belongs. These orders are differently classified by various naturalists, according to their individual estimate of their structural features. But the growth of the frog, like that of the insects, gives us the true grade of the type. There are not many groups in which this comparison has been carried out so fully as in the insects and frogs; but wherever it has been tried it is found to be a perfectly sure test. Occasional glimpses of these facts, seen disconnectedly, have done much to confirm the development theory, so greatly in vogue at present, though under a somewhat new form. Those who sustain these views have seen that there was a gradation between animals, and have inferred that it was a material connection. But when we follow it in the growth of animals themselves, and find that, close as it is, no animal ever misses its true development, or grows to be anything but what it was meant to be, we are forced to admit that the gradation which unquestionably unites all animals is an intellectual, not a material one. It exists in the Mind which made them. As the works of a human intellect are bound together by mental kinship, so are the thoughts of the Creator spiritually united. I think that considerations like these should be an inducement for us all to collect the young of as many animals as possible on this journey. In so doing we may change the fundamental principles of classification, and confer a lasting benefit on science.

"It is very important to select the right animals for such investigations. I can conceive that a lifetime should be passed in embryological studies, and yet little be learned of the principles of classification. The embryology of the worm, for instance, would not give us the natural classification of the Articulates, because we should see only the first step of the series; we should not reach the sequence of the development. It would be like reading over and over again the first chapter of a story. The embryology of the Insects, on the contrary, would give us the whole succession of a scale on the lowest level of which the Worms remain forever. So the embryology of the frog will give us the classification of the group to which it belongs, but the embryology of the Cecilia, the lowest order in the group, will give us only the initiatory steps. In the same way the naturalist who, in studying the embryology of the reptiles, should begin with their lowest representatives, the serpents, would make a great mistake. But take the alligator, so abundant in the regions to which we are going. An alligator's egg in the earliest condition of growth has never been opened by a naturalist. The young have been occasionally taken from the egg just before hatching, but absolutely nothing is known of their first phases of development. A complete embryology of the alligator would give us not only the natural classification of reptiles as they exist now, but might teach us something of their history from the time of their introduction upon earth to the present day. For embryology shows us not only the relations of existing animals to each other, but their relations to extinct types also. One prominent result of embryological studies has been to show that animals in the earlier stages of their growth resemble ancient representatives of the same type belonging to past geological ages. The first reptiles were introduced in the carboniferous epoch, and they were very different from those now existing. They were not numerous at that period; but later in the world's history there was a time, justly called the 'age of reptiles,' when the gigantic Saurians, Plesiosaurians, and Ichthyosaurians abounded. I believe, and my conviction is drawn from my previous embryological studies, that the changes of the alligator in the egg will give us the clew to the structural relations of the Reptiles from their first creation to the present day, — will give us, in other words, their sequence in time as well as their sequence in growth. In the class of Reptiles, then, the most instructive group we can select with reference to the structural relations of the type as it now exists, and their history in past times, will be the alligator. We must therefore neglect no opportunity of collecting their eggs in as large numbers as possible.

"There are other animals in Brazil, low in their class to be sure, but yet very important to study embryologically, on account of their relation to extinct types. These are the sloths and armadillos, — animals of insignificant size in our days, but anciently represented in gigantic proportions. The Megatherium, the Mylodon, the Megalonyx, were some of these immense Mammalia. I believe that the embryonic changes of the sloths and armadillos will explain the structural relations of those huge Edentata and their connection with the present ones. South America teems with the fossil bones of these animals, which indeed penetrated into the northern half of the hemisphere as high up as Georgia and Kentucky, where their remains have been found. The living representatives of the family are also numerous in South America, and we should make it one of our chief objects to get specimens of all ages and examine them from their earliest phases upward. We must, above all, try not to be led away from the more important aims of our study by the diversity of objects. I have known many young naturalists to miss the highest success by trying to cover too much ground, — by becoming collectors rather than investigators. Bitten by the mania for amassing a great number and variety of species, such a man never returns to the general consideration of more comprehensive features. We must try to set before ourselves certain important questions, and give ourselves resolutely to the investigation of these points, even though we should sacrifice less important things more readily reached.

"Another type full of interest, from an embryological point of view, will be the monkeys. Since some of our scientific colleagues look upon them as our ancestors, it is important that we should collect as many facts as possible concerning their growth. Of course it would be better if we could make the investigation in the land of the Orangs, Gorillas, and Chimpanzees, — the highest monkeys and the nearest to man in their development. Still even the process of growth in the South American monkey will be very instructive. Give a mathematician the initial elements of a series, and he will work out the whole; and so I believe when the laws of embryological development are better understood, naturalists will have a key to the limits of these cycles of growth, and be able to appoint them their natural boundaries even from partial data."

THOMAS HENRY HUXLEY (1825–1895)

On a Piece of Chalk

A lecture delivered to the working men of Norwich, 1868

[BY A CURIOUS COINCIDENCE, Huxley's life work, like that of Darwin, was more or less crystallized by a long ocean voyage, in this instance on H.M.S. *Rattlesnake* when he was twenty-one years old. The trip, which extended to Torres Strait, between New Guinea and Australia, gave the young naturalist an opportunity to study the life of the ocean, and later to write important papers concerning it.

He was profoundly affected by Darwin's *Origin of Species,* accepted it completely, and defended it most successfully. Huxley was a master of English, but he kept it too much under control and after a burst of brilliance we sometimes wish that the succeeding pages were not so perfect, so concise.

If I had the opportunity of dining and enjoying a long evening of conversation with anyone from past years, Huxley would be almost first choice. What a treat it would be to sit and listen to Huxley and Wheeler for hours on end!

A paragraph from an essay on the natural history of *Yeast* might have been by either man, although Wheeler was only six years of age when it was written.

"It is highly creditable to the ingenuity of our ancestors that the peculiar property of fermented liquids, in virtue of which they 'make glad the heart of man,' seems to have been known in the remotest periods of which we have any record. All savages take to alcoholic fluids as if they were to the manner born. Our Vedic forefathers intoxicated themselves with the juice of the 'soma'; Noah, by a not unnatural reaction against a superfluity of water, appears to have taken the earliest practicable opportunity of qualifying that which he was obliged to drink; and the ghosts of the ancient Egyptians were solaced by pictures of banquets in which the wine-cup passes round, graven on the walls of their tombs. A knowledge of the process of fermentation, therefore, was in all probability possessed by the prehistoric populations of the globe; and it must have become a matter of great interest even to primeval wine-bibbers to study the methods by which fermented liquids could be surely manufactured. No doubt

it was soon discovered that the most certain, as well as the most ex-
peditious, way of making a sweet juice ferment was to add to it a
little of the scum, or lees, of another fermenting juice. And it can
hardly be questioned that this singular excitation of fermentation
in one fluid, by a sort of infection, or inoculation, of a little ferment
taken from some other fluid, together with the strange swelling,
foaming and hissing of the fermented substance, must have always
attracted attention from the more thoughtful. Nevertheless, the commencement of the scientific analysis of the phenomenon dates from
a period not earlier than the first half of the seventeenth century."

From the world of yeast is but a step to the tiny animals which
have built the white cliffs of Albion, and Huxley's classic *On a Piece
of Chalk* takes rank with the best of the writing of this grand scientist.]

I F A WELL were sunk at our feet in the midst of the city of Norwich,
the diggers would very soon find themselves at work in that white
substance almost too soft to be called rock, with which we are all
familiar as "chalk."

Not only here, but over the whole county of Norfolk, the well-
sinker might carry his shaft down many hundred feet without coming to the end of the chalk; and, on the sea-coast, where the waves
have pared away the face of the land which breasts them, the scarped
faces of the high cliffs are often wholly formed of the same material.
Northward, the chalk may be followed as far as Yorkshire; on the
south coast it appears abruptly in the picturesque western bays of
Dorset, and breaks into the Needles of the Isle of Wight; while on
the shores of Kent it supplies that long line of white cliffs to which
England owes her name of Albion.

Were the thin soil which covers it all washed away, a curved band
of white chalk, here broader, and there narrower, might be followed
diagonally across England from Lulworth in Dorset, to Flamborough
Head in Yorkshire — distance of over 280 miles as the crow flies. From
this band to the North Sea, on the east, and the Channel, on the
south, the chalk is largely hidden by other deposits; but, except in
the Weald of Kent and Sussex, it enters into the very foundation of
all the south-eastern counties.

Attaining, as it does in some places, a thickness of more than a
thousand feet, the English chalk must be admitted to be a mass of
considerable magnitude. Nevertheless, it covers but an insignificant
portion of the whole area occupied by the chalk formation of the
globe, much of which has the same general characters as ours, and

is found in detached patches, some less, and others more extensive, than the English. Chalk occurs in north-west Ireland; it stretches over a large part of France, — the chalk which underlies Paris being, in fact, a continuation of that London basin; it runs through Denmark and Central Europe, and extends southward to North Africa; while eastward, it appears in the Crimea and in Syria, and may be traced as far as the shores of the Sea of Aral, in Central Asia. If all the points at which true chalk occurs were circumscribed, they would lie within an irregular oval about 3,000 miles in long diameter — the area of which would be as great as that of Europe, and would many times exceed that of the largest existing inland sea — the Mediterranean.

Thus the chalk is no unimportant element in the masonry of the earth's crust, and it impresses a peculiar stamp, varying with the conditions to which it is exposed, on the scenery of the districts in which it occurs. The undulating downs and rounded coobs, covered with sweet-grassed turf, of our inland chalk country, have a peacefully domestic and mutton-suggesting prettiness, but can hardly be called either grand or beautiful. But on our southern coasts, the wall-sided cliffs, many hundred feet high, with vast needles and pinnacles standing out in the sea, sharp and solitary enough to serve as perches for the wary cormorant, confer a wonderful beauty and grandeur upon the chalk headlands. And, in the East, chalk has its share in the formation of some of the most venerable of mountain ranges, such as the Lebanon.

What is this wide-spread component of the surface of the earth? and whence did it come?

You may think this no very hopeful inquiry. You may not unnaturally suppose that the attempt to solve such problems as these can lead to no result, save that of entangling the inquirer in vague speculations, incapable of refutation and of verification. If such were really the case, I should have selected some other subject than a "piece of chalk" for my discourse. But, in truth, after much deliberation, I have been unable to think of any topic which would so well enable me to lead you to see how solid is the foundation upon which some of the most startling conclusions of physical science rest.

A great chapter of the history of the world is written in the chalk. Few passages in the history of man can be supported by such an overwhelming mass of direct and indirect evidence as that which testifies to the truth of the fragment of the history of the globe, which I hope to enable you to read, with your own eyes, to-night. Let me add, that few chapters of human history have a more profound significance for ourselves. I weigh my words well when I assert, that the man who should know the true history of the bit of chalk which every carpenter carries about in his breeches-pocket, though ignorant of all other history, is likely, if he will think his knowledge out to its ultimate results, to have a truer, and therefore a better, conception of this wonderful universe, and of man's relation to it, than the most learned student who is deep-read in the records of humanity and ignorant of those of Nature.

The language of the chalk is not hard to learn, not nearly so hard as Latin, if you only want to get at the broad features of the story it has to tell; and I propose that we now set to work to spell that story out together.

We all know that if we "burn" chalk the result is quick-lime. Chalk, in fact, is a compound of carbonic acid gas, and lime, and when you make it very hot the carbonic acid flies away and the lime is left. By this method of procedure we see the lime, but we do not see the carbonic acid. If, on the other hand, you were to powder a little chalk and drop it into a good deal of strong vinegar, there would be a great bubbling and fizzing, and, finally, a clear liquid, in which no sign of chalk would appear. Here you see the carbonic acid in the bubbles; the lime, dissolved in the vinegar, vanishes from sight. There are a great many other ways of showing that chalk is essentially nothing but carbonic acid and quick-lime. Chemists enunciate the result of all the experiments which prove this, by stating that chalk is almost wholly composed of "carbonate of lime."

It is desirable for us to start from the knowledge of this fact, though it may not seem to help us very far towards what we seek. For carbonate of lime is a widely-spread substance, and is met with under very various conditions. All sorts of limestones are composed of more or less pure carbonate of lime. The crust which is often deposited by waters which have drained through limestone rocks, in the form of what are called stalagmites and stalactites, is carbonate of lime. Or, to take a more familiar example, the fur on the inside of a tea-kettle is carbonate of lime; and, for anything chemistry tells us to the contrary, the chalk might be a kind of gigantic fur upon the bottom of the earth-kettle, which is kept pretty hot below.

Let us try another method of making the chalk tell us its own history. To the unassisted eye chalk looks simply like a very loose and open kind of stone, but it is possible to grind a slice of chalk down so thin that you can see through it — until it is thin enough, in fact, to be examined with any magnifying power that may be thought desirable. A thin slice of the fur of a kettle might be made in the same

134

way. If it were examined microscopically, it would show itself to be a more or less distinctly laminated mineral substance, and nothing more.

But the slice of chalk presents a totally different appearance when placed under the microscope. The general mass of it is made up of very minute granules; but, imbedded in this matrix, are innumerable bodies, some smaller and some larger, but, on a rough average, not more than a hundredth of an inch in diameter, having a well-defined shape and structure. A cubic inch of some specimens of chalk may contain hundreds of thousands of these bodies, compacted together with incalculable millions of the granules.

The examination of a transparent slice gives a good notion of the manner in which the components of the chalk are arranged, and of their relative proportions. But, by rubbing up some chalk with a brush in water and then pouring off the milky fluid, so as to obtain sediments of different degrees of fineness, the granules and the minute rounded bodies may be pretty well separated from one another, and submitted to microscopic examination, either as opaque or as transparent objects. By combining the views obtained in these various methods, each of the rounded bodies may be proved to be a beautifully-constructed calcareous fabric, made up of a number of chambers, communicating freely with one another. The chambered bodies are of various forms. One of the commonest is something like a badly-grown raspberry, being formed of a number of nearly globular chambers of different sizes congregated together. It is called *Globigerina,* and some specimens of chalk consist of little else than *Globigerinae* and granules. Let us fix our attention upon the *Globigerina.* It is the spoor of the game we are tracking. If we can learn what it is and what are the conditions of its existence, we shall see our way to the origin and past history of the chalk.

A suggestion which may naturally enough present itself is, that these curious bodies are the result of some process of aggregation which has taken place in the carbonate of lime; that, just as in winter, the rime on our windows simulates the most delicate and elegantly arborescent foliage — proving that the mere mineral water may, under certain conditions, assume the outward form of organic bodies — so this mineral substance, carbonate of lime, hidden away in the bowels of the earth, has taken the shape of these chambered bodies. I am not raising a merely fanciful and unreal objection. Very learned men, in former days, have even entertained the notion that all the formed things found in rocks are of this nature; and if no such conception is at present held to be admissible, it is because long and varied experience has now shown that mineral water never does assume the form and structure we find in fossils. If any one were to try to persuade you that an oyster-shell (which is also chiefly composed of carbonate of lime) had crystallized out of sea-water, I suppose you would laugh at the absurdity. Your laughter would be justified by the fact that all experience tends to show that oyster-shells are formed by the agency of oysters, and in no other way. And if there were no better reasons, we should be justified, on like grounds, in believing that *Globigerina* is not the product of anything but vital activity.

Happily, however, better evidence in proof of the organic nature of the *Globigerinae* than that of analogy is forthcoming. It so happens that calcareous skeletons, exactly similar to the *Globigerinae* of the chalk are being formed, at the present moment, by minute living creatures, which flourish in multitudes, literally more numerous than the sands of the sea-shore, over a large extent of that part of the earth's surface which is covered by the ocean.

The history of the discovery of these living *Globigerinae,* and of the part which they play in rock building, is singular enough. It is a discovery which, like others of no less scientific importance, has arisen, incidentally out of work devoted to very different and exceedingly practical interests. When men first took to the sea, they speedily learned to look out for shoals and rocks; and the more the burthen of their ships increased, the more imperatively necessary it became of sailors to ascertain with precision the depth of the waters they traversed. Out of this necessity grew the use of the lead and sounding line; and, ultimately, marine-surveying, which is the recording of the form of coasts and of the depth of the sea, as ascertained by the sounding-lead upon charts.

At the same time, it became desirable to ascertain and to indicate the nature of the sea-bottom, since this circumstance greatly affects its goodness as holding ground for anchors. Some ingenious tar, whose name deserves a better fate than the oblivion into which it has fallen, attained this object by "arming" the bottom of the lead with a lump of grease, to which more or less of the sand or mud, or broken shells, as the case might be, adhered, and was brought to the surface. But, however well adapted such an apparatus might be for rough nautical purposes, scientific accuracy could not be expected from the armed lead, and to remedy its defects (especially when applied to sounding in great depths) Lieut. Brooke, of the American Navy, some years ago invented a most ingenious machine, by which a considerable portion of the superficial layer of the sea-bottom can be scooped out and brought up from any depth to which the lead descends. In 1853, Lieut. Brooke obtained mud from the bottom of the North Atlantic, between Newfoundland and the Azores, at a depth of more than 10,000 feet, or two miles, by the help of this sounding apparatus. The specimens were sent for examination to Ehrenberg of Berlin, and to Bailey of West Point, and those able microscopists found that this deep-sea mud was almost entirely composed of the skeletons of living organisms — the greater proportion of these being just like the *Globigerinae* already known to occur in the chalk.

Thus far, the work had been carried on simply in the interests of science, but Lieut. Brooke's method of sounding acquired a high commercial value, when the enterprise of laying down the telegraph-cable between this country and the United States was undertaken. For it became a matter of immense importance to know, not only the depth of the sea over the whole line along which the cable was to be laid, but the exact nature of the bottom, so as to guard against chances of cutting or fraying the strands of that costly rope. The Admiralty consequently ordered Captain Dayman, an old friend and shipmate of mine, to ascertain the depth over the whole line of the cable, and to bring back specimens of the bottom. In former days, such a command as this might have sounded very much like one of the impossible things which the young Prince in the Fairy Tales is ordered to do before he can obtain the hand of the Princess. However, in the months of June and July, 1857, my friend performed the task assigned to him with great expedition and precision, without, so far as I know, having met with any reward of that kind. The specimens of Atlantic mud which he procured were sent to me to be examined and reported upon.

The result of all these operations is, that we know the contours and the nature of the surface-soil covered by the North Atlantic for a distance of 1,700 miles from east to west, as well as we know that of any part of the dry land. It is a prodigious plain — one of the widest and most even plains in the world. If the sea were drained off, you might drive a waggon all the way from Valentia, on the west coast of Ireland, to Trinity Bay, in Newfoundland. And, except upon one sharp incline about 200 miles from Valentia, I am not quite sure that it would even be necessary to put the skid on, so gentle are the ascents and descents upon that long route. From Valentia the road would lie down-hill for about 200 miles to the point at which the bottom is now covered by 1,700 fathoms of sea-water. Then would come the central plain, more than a thousand miles wide, the inequalities of the surface of which would be hardly perceptible, though the depth of water upon it now varies from 10,000 to 15,000 feet; and there are places in which Mont Blanc might be sunk without showing its peak above water. Beyond this, the ascent on the American side commences, and gradually leads, for about 300 miles, to the Newfoundland shore.

Almost the whole of the bottom of this central plain (which extends for many hundred miles in a north and south direction) is covered by a fine mud, which, when brought to the surface, dries into a grayish white friable substance. You can write with this on a blackboard, if you are so inclined; and, to the eye, it is quite like very soft, grayish chalk. Examined chemically, it proves to be composed almost wholly of carbonate of lime; and if you make a section of it, in the same way as that of the piece of chalk was made, and view it with the microscope, it presents innumerable *Globigerinae* embedded in a granular matrix. Thus this deep-sea mud is substantially chalk. I say substantially, because there are a good many minor differences; but as these have no bearing on the question immediately before us, — which is the nature of the *Globigerinae* of the chalk — it is unnecessary to speak of them.

Globigerinae of every size, from the smallest to the largest, are associated together in the Atlantic mud, and the chambers of many are filled by a soft animal matter. This soft substance is, in fact, the remains of the creature to which the *Globigerina* shell, or rather skeleton, owes its existence — and which is an animal of the simplest imaginable description. It is, in fact, a mere particle of living jelly, without defined parts of any kind — without a mouth, nerves, muscles, or distinct organs, and only manifesting its vitality to ordinary observation by thrusting out and retracting from all parts of its surface, long filamentous processes, which serve for arms and legs. Yet this amorphous particle, devoid of everything which, in the higher animals, we call organs, is capable of feeding, growing, and multiplying; of separating from the ocean the small proportion of carbonate of lime which is dissolved in sea-water; and of building up that substance into a skeleton for itself, according to a pattern which can be imitated by no other known agency.

The notion that animals can live and flourish in the sea, at the vast depths from which apparently living *Globigerinae* have been brought up, does not agree very well with our usual conceptions respecting the conditions of animal life; and it is not so absolutely impossible as it might at first sight appear to be, that the *Globigerinae* of the Atlantic sea-bottom do not live and die where they are found.

As I have mentioned, the soundings from the great Atlantic plain

are almost entirely made up of *Globigerinae,* with the granules which
have been mentioned, and some few other calcareous shells; but a
small percentage of the chalky mud — perhaps at most some five per
cent. of it — is of a different nature, and consists of shells and skeletons composed of silex, or pure flint. These silicious bodies belong
partly to the lowly vegetable organisms which are called *Diatomaceae,*
and partly to the minute, and extremely simple, animals, termed
Radiolaria. It is quite certain that these creatures do not live at the
bottom of the ocean, but at its surface, where they may be obtained
in prodigious numbers by the use of a properly constructed net.
Hence it follows that these silicious organisms, though they are not
heavier than the lightest dust, must have fallen, in some cases, through
fifteen thousand feet of water, before they reached their final resting-
place on the ocean floor. And considering how large a surface these
bodies expose in proportion to their weight, it is probable that they
occupy a great length of time in making their burial journey from
the surface of the Atlantic to the bottom.

But if the *Radiolaria* and Diatoms are thus rained upon the bottom of the sea, from the superficial layer of its waters in which they
pass their lives, it is obviously possible that the *Globigerinae* may
be similarly derived; and if they were so, it would be much more
easy to understand how they obtain their supply of food than it is
at present. Nevertheless, the positive and negative evidence all points
the other way. The skeletons of the full-grown, deep-sea *Globigerinae*
are so remarkably solid and heavy in proportion to their surface as
to seem little fitted for floating; and, as a matter of fact, they are not
to be found along with the Diatoms and *Radiolaria* in the uppermost
stratum of the open ocean. It has been observed, again, that the
abundance of *Globigerinae,* in proportion to other organisms, of
like kind, increases with the depth of the sea; and that deep-water
Globigerinae are larger than those which live in shallower parts of
the sea; and such facts negative the supposition that these organisms
have been swept by currents from the shallows into the deeps of the
Atlantic. It therefore seems to be hardly doubtful that these wonderful creatures live and die at the depths in which they are found.

However, the important points for us are, that the living *Globigerinae* are exclusively marine animals, the skeletons of which
abound at the bottom of deep seas; and that there is not a shadow
of reason for believing that the habits of *Globigerinae* of the chalk
differed from those of the existing species. But if this be true, there
is no escaping the conclusion that the chalk itself is the dried mud
of an ancient deep sea.

In working over the soundings collected by Captain Dayman, I was surprised to find that many of what I have called "granules" of that mud were not, as one might have been tempted to think at first, the mere powder and waste of *Globigerinae,* but that they had a definite form and size. I termed these bodies *"coccoliths,"* and doubted their organic nature. Dr. Wallich verified my observation, and added the interesting discovery that, not unfrequently, bodies similar to these "coccoliths" were aggregated together into spheroids, which he termed *"coccospheres."* So far as we knew, these bodies, the nature of which is extremely puzzling and problematical, were peculiar to the Atlantic soundings. But, a few years ago, Mr. Sorby, in making a careful examination of the chalk by means of thin sections and otherwise, observed, as Ehrenberg had done before him, that much of its granular basis possesses a definite form. Comparing these formed particles with those in the Atlantic soundings, he found the two to be identical; and thus proved that the. chalk, like the surroundings, contains these mysterious coccoliths and coccospheres. Here was a further and most interesting confirmation, from internal evidence, of the essential identity of the chalk with modern deep-sea mud. *Globigerinae,* coccoliths, and coccospheres are found as the chief constituents of both, and testify to the general similarity of the conditions under which both have been formed.

The evidence furnished by the hewing, facing, and superposition of the stones of the Pyramids, that these structures were built by men, has no greater weight than the evidence that the chalk was built by *Globigerinae;* and the belief that those ancient pyramid-builders were terrestrial and air-breathing creatures like ourselves, is not better based than the conviction that the chalk-makers lived in the sea. But as our belief in the building of the Pyramids by men is not only grounded on the internal evidence afforded by these structures, but gathers strength from multitudinous collateral proofs, and is clinched by the total absence of any reason for a contrary belief; so the evidence drawn from the *Globigerinae* that the chalk is an ancient sea-bottom, is fortified by innumerable independent lines of evidence; and our belief in the truth of the conclusion to which all positive testimony tends, receives the like negative justification from the fact that no other hypothesis has a shadow of foundation.

It may be worth while briefly to consider a few of these collateral proofs that the chalk was deposited at the bottom of the sea. The great mass of the chalk is composed, as we have seen, of the skeletons of *Globigerinae,* and other simple organisms, imbedded in granular matter. Here and there, however, this hardened mud of the ancient sea reveals the remains of higher animals which have lived and died, and left their hard parts in the mud, just as the oysters die and leave their shells behind them, in the mud of the present seas.

There are, at the present day, certain groups of animals which are never found in fresh waters, being unable to live anywhere but in the sea. Such are the corals; those corallines which are called *Polyzoa;* those creatures which fabricate the lamp-shells, and are called *Brachiopoda;* the pearly *Nautilus,* and all animals allied to it; and all the forms of sea-urchins and star-fishes. Not only are all these creatures confined to salt water at the present day; but, so far as our records of the past go, the conditions of their existence have been the same: hence, their occurrence in any deposit is as strong evidence as can be obtained, that that deposit was formed in the sea. Now the remains of animals of all the kinds which have been enumerated, occur in the chalk, in greater or less abundance; while not one of those forms of shellfish which are characteristic of fresh water has yet been observed in it. When we consider that the remains of more than three thousand distinct species of aquatic animals have been discovered among the fossils of the chalk, that the great majority of them are of such forms as are now met with only in the sea, and that there is no reason to believe that any one of them inhabited fresh water — the collateral evidence that the chalk represents an ancient sea-bottom acquires as great force as the proof derived from the nature of the chalk itself. I think you will now allow that I did not overstate my case when I asserted that we have as strong grounds for believing that all the vast area of dry land, at present occupied by the chalk, was once at the bottom of the sea, as we have for any matter of history whatever; while there is no justification for any other belief.

No less certain it is that the time during which the countries we now call south-east England, France, Germany, Poland, Russia, Egypt, Arabia, Syria, were more or less completely covered by a deep sea, was of considerable duration. We have already seen that the chalk is, in places, more than a thousand feet thick. I think you will agree with me, that it must have taken some time for the skeletons of animalcules of a hundredth of an inch in diameter to heap up such a mass as that. I have said that throughout the thickness of the chalk the remains of other animals are scattered. These remains are often in the most exquisite state of preservation. The valves of the shell-fishes are commonly adherent; the long spines of some of the sea-urchins, which would be detached by the smallest jar, often remain in their places. In a word, it is certain that these animals have lived and died when the place which they now occupy was the surface of as much of the chalk as had then been deposited; and that each has been covered up by the layer of *Globigerina* mud, upon which the creatures imbedded a little higher up have, in like manner, lived and died. But some of these remains prove the existence of reptiles of vast size in the chalk sea. These lived their time, and had their ancestors and descendants, which assuredly implies time, reptiles being of slow growth.

There is more curious evidence, again, that the process of covering up, or, in other words, the deposit of *Globigerina* skeletons, did not go on very fast. It is demonstrable that an animal of the cretaceous sea might die, that its skeleton might lie uncovered upon the sea-bottom long enough to lose all its outward coverings and appendages by putrefaction; and that, after this had happened, another animal might attach itself to the dead and naked skeleton, might grow to maturity, and might itself die before the calcareous mud had buried the whole.

Cases of this kind are admirably described by Sir Charles Lyell. He speaks of the frequency with which geologists find in the chalk a fossilized sea-urchin, to which is attached the lower valve of a *Crania*. This is a kind of shell-fish, with a shell composed of two pieces, of which, as in the oyster, one is fixed and the other free.

"The upper valve is almost invariably wanting, though occasionally found in a perfect state of preservation in the white chalk at some distance. In this case, we see clearly that the sea-urchin first lived from youth to age, then died and lost its spines, which were carried away. Then the young *Crania* adhered to the bared shell, grew and perished in its turn; after which, the upper valve was separated from the lower, before the Echinus became enveloped in chalky mud."

A specimen in the Museum of Practical Geology, in London, still further prolongs the period which must have elapsed between the death of the sea-urchin, and its burial by the *Globigerinae*. For the outward face of the valve of a *Crania,* which is attached to a sea-urchin, (*Micraster*), is itself overrun by an incrusting coralline, which spreads thence over more or less of the surface of the sea-urchin. It follows that, after the upper valve of the *Crania* fell off, the surface of the attached valve must have remained exposed long enough to allow of the growth of the whole coralline, since corallines do not live embedded in mud.

The progress of knowledge may, one day, enable us to deduce from such facts as these the maximum rate at which the chalk can have

142

accumulated, and thus to arrive at the minimum duration of the chalk period. Suppose that the valve of the *Crania* upon which a coralline has fixed itself in the way just described, is so attached to the sea-urchin that no part of it is more than an inch above the face upon which the sea-urchin rests. Then, as the coralline could not have fixed itself, if the *Crania* had been covered up with chalk mud, and could not have lived had itself been so covered, it follows, that an inch of chalk mud could not have accumulated within the time between the death and decay of the soft parts of the sea-urchin and the growth of the coralline to the full size which it has attained. If the decay of the soft parts of the sea-urchin; the attachment, growth to maturity, and decay of the *Crania;* and the subsequent attachment and growth of the coralline, took a year (which is a low estimate enough), the accumulation of the inch of chalk must have taken more than a year: and the deposit of a thousand feet of chalk must, consequently, have taken more than twelve thousand years.

The foundation of all this calculation is, of course, a knowledge of the length of time the *Crania* and the coralline needed to attain their full size; and, on this head, precise knowledge is at present wanting. But there are circumstances which tend to show, that nothing like an inch of chalk has accumulated during the life of a *Crania;* and, on any probable estimate of the length of that life, the chalk period must have had a much longer duration than that thus roughly assigned to it.

Thus not only is it certain that the chalk is the mud of an ancient sea-bottom; but it is no less certain, that the chalk sea existed during an extremely long period, though we may not be prepared to give a precise estimate of the length of that period in years. The relative duration is clear, though the absolute duration may not be definable. The attempt to affix any precise date to the period at which the chalk sea began, or ended, its existence, is baffled by difficulties of the same kind. But the relative age of the cretaceous epoch may be determined with as great ease and certainty as the long duration of that epoch.

You will have heard of the interesting discoveries recently made, in various parts of Western Europe, of flint implements, obviously worked into shape by human hands, under circumstances which show conclusively that man is a very ancient denizen of these regions. It has been proved that the whole populations of Europe, whose exist-ence has been revealed to us in this way, consisted of savages, such as the Esquimaux are now; that, in the country which is now France, they hunted the reindeer, and were familiar with the ways of the mammoth and the bison. The physical geography of France was in those days different from what it is now — the river Somme, for instance, having cut its bed a hundred feet deeper between that time and this; and, it is probable, that the climate was more like that of Canada or Siberia, than that of Western Europe.

The existence of these people is forgotten even in the traditions of the oldest historical nations. The name and fame of them had utterly vanished until a few years back; and the amount of physical change which has been effected since their day renders it more than probable that, venerable as are some of the historical nations, the workers of the chipped flints of Hoxne or of Amiens are to them, as they are to us, in antiquity. But, if we assign to these hoar relics of long-vanished generations of men the greatest age that can possibly be claimed for them, they are not older than the drift, or boulder clay, which, in comparison with the chalk, is but a very juvenile deposit. You need go no further than your own sea-board for evidence of this fact. At one of the most charming spots on the coast of Norfolk, Cromer, you will see the boulder clay forming a vast mass, which lies upon the chalk, and must consequently have come into existence after it. Huge boulders of chalk are, in fact, included in the clay, and have evidently been brought to the position they now occupy by the same agency as that which has planted blocks of syenite from Norway side by side with them.

The chalk, then, is certainly older than the boulder clay. If you ask how much, I will again take you no further than the same spot upon your own coasts for evidence. I have spoken of the boulder clay and drift as resting upon the chalk. That is not strictly true. Interposed between the chalk and the drift is a comparatively insignificant layer, containing vegetable matter. But that layer tells a wonderful history. It is full of stumps of trees standing as they grew. Fir-trees are there with their cones, and hazel-bushes with their nuts; there stand the stools of oak and yew trees, beeches and alders. Hence this stratum is appropriately called the "forest-bed."

It is obvious that the chalk must have been upheaved and converted into dry land, before the timber trees could grow upon it. As the bolls of some of these trees are from two to three feet in diameter, it is no less clear that the dry land thus formed remained in the same condition for long ages. And not only do the remains of stately oaks and well-grown firs testify to the duration of this condition of things, but additional evidence to the same effect is afforded by the abundant remains of elephants, rhinoceroses, hippopotamuses, and other great wild beasts, which it has yielded to the zealous search of such men as the Rev. Mr. Gunn. When you look at such a collection as he has formed, and bethink you that these elephantine bones did veritably carry their owners about, and these great grinders crunch, in the dark woods of which the forest-bed is now the only trace, it is impossible not to feel that they are as good evidence of the lapse of time as the annual rings of the tree stumps.

Thus there is a writing upon the wall of cliffs at Cromer, and whoso runs may read it. It tells us, with an authority which cannot be impeached, that the ancient sea bed of the chalk sea was raised up, and remained dry land, until it was covered with forest, stocked with the great game the spoils of which have rejoiced your geologists. How long it remained in that condition cannot be said; but "the whirligig of time brought its revenges" in those days as in these. That dry land, with the bones and teeth of generations of long-lived elephants, hidden away among the gnarled roots and dry leaves of its ancient trees, sank gradually to the bottom of the icy sea, which covered it with huge masses of drift and boulder clay. Sea-beasts, such as the walrus, now restricted to the extreme north, paddled about where birds had twittered among the topmost twigs of the fir-trees. How long this state of things endured we know not, but at length it came to an end. The upheaved glacial mud hardened into the soil of modern Norfolk. Forests grew once more, the wolf and the beaver replaced the reindeer and the elephant; and at length what we call the history of England dawned.

Thus you have, within the limits of your own county, proof that the chalk can justly claim a very much greater antiquity than even the oldest physical traces of mankind. But we may go further and demonstrate, by evidence of the same authority as that which testifies to the existence of the father of men, that the chalk is vastly older than Adam himself. The Book of Genesis informs us that Adam, immediately upon his creation, and before the appearance of Eve, was placed in the Garden of Eden. The problem of the geographic position of Eden has greatly vexed the spirits of the learned in such matters, but there is one point respecting which, so far as I know, no commentator has ever raised a doubt. This is, that of the four rivers which are said to run out of it, Euphrates and Hiddekel are identical with the rivers now known by the names of Euphrates and Tigris. But the whole country in which these mighty rivers take their origin, and through which they run, is composed of rocks which are either of the same age as the chalk, or of later date. So that the chalk must not only have been formed, but, after its formation, the time required for the deposit of these later rocks, and for their upheaval into dry land, must have elapsed, before the smallest brook which feeds the swift stream of "the great river, the river of Babylon," began to flow.

Thus, evidence which cannot be rebutted and which need not be strengthened, though if time permitted I might indefinitely increase its quantity, compels you to believe that the earth, from the time of the chalk to the present day, has been the theatre of a series of changes as vast in their amount, as they were slow in their progress. The area on which we stand has been first sea and then land, for at least four alternations; and has remained in each of these conditions for a period of great length.

Nor have these wonderful metamorphoses of sea into land, and of land into sea, been confined to one corner of England. During the chalk period, or "cretaceous epoch," not one of the present great physical features of the globe was in existence. Our great mountain ranges, Pyrenees, Alps, Himalayas, Andes, have all been upheaved since the chalk was deposited, and the cretaceous sea flowed over the sites of Sinai and Ararat. All this is certain, because rocks of cretaceous, or still later, date have shared in the elevatory movements which gave rise to these mountain chains; and may be found perched up, in some cases, many thousand feet high upon their flanks. And evidence of equal cogency demonstrates that, though, in Norfolk, the forest-bed rests directly upon the chalk, yet it does so, not because the period at which the forest grew immediately followed that at which the chalk was formed, but because an immense lapse of time, represented elsewhere by thousands of feet of rock, is not indicated at Cromer.

I must ask you to believe that there is no less conclusive proof that a still more prolonged succession of similar changes occurred, before the chalk was deposited. Nor have we any reason to think that the first term in the series of these changes is known. The oldest sea-beds preserved to us are sands, and mud, and pebbles, the wear and tear of rocks which were formed in still older oceans.

But, great as is the magnitude of these physical changes of the world, they have been accompanied by a no less striking series of modifications in its living inhabitants. All the great classes of animals, beasts of the field, fowls of the air, creeping things, and things which dwell in the waters, flourished upon the globe long ages before the chalk was deposited. Very few, however, if any, of these ancient forms of animal life were identical with those which now live. Certainly not one of the higher animals was of the same species as any of those now in existence. The beasts of the field, in the days before the chalk, were not our beasts of the field, nor the fowls of the air such as those which the eye of men has seen flying, unless his antiquity dates infinitely further back than we at present surmise. If we could be carried back into those times, we should be as one suddenly set down in Australia before it was colonized. We should see mammals, birds, reptiles, fishes, insects, snails, and the like, clearly recognizable as such, and yet not one of them would be just the same as those with which we are familiar, and many would be extremely different.

From that time to the present, the population of the world has undergone slow and gradual, but incessant, changes. There has been no grand catastrophe — no destroyer has swept away the forms of life of one period, and replaced them by a totally new creation: but one species has vanished and another has taken its place; creatures of one type of structure have diminished, those of another have increased, as time has passed on. And thus, while the differences between the living creatures of the time before the chalk and those of the present day appear startling, if placed side by side, we are led from one to the other by the most gradual progress, if we follow the course of Nature through the whole series of those relics of her operations which she has left behind. It is by the population of the chalk sea that the ancient and the modern inhabitants of the world are most completely connected. The groups which are dying out flourish, side by side, with the groups which are now the dominant forms of life. Thus the chalk contains remains of those strange flying and swimming reptiles, the pterodactyl, the ichthyosaurus, and the plesiosaurus, which are found in no later deposits, but abounded in preceding ages. The chambered shells called ammonites and belemnites, which are so characteristic of the period preceding the cretaceous, in like manner die with it.

But, amongst these fading remainders of a previous state of things, are some very modern forms of life, looking like Yankee pedlers among a tribe of Red Indians. Crocodiles of modern type appear; bony fishes, many of them very similar to existing species, almost supplant the forms of fish which predominate in more ancient seas; and many kinds of living shell-fish first become known to us in the chalk. The vegetation acquires a modern aspect. A few living animals are not even distinguishable as species, from those which existed at that remote epoch. The *Globigerina* of the present day, for example, is not different specifically from that of the chalk; and the same may be said of many other *Foraminifera*. I think it probable that critical and unprejudiced examination will show that more than one species of much higher animals have had a similar longevity; but the only example which I can at present give confidently is the snake's head lamp-shell (*Terebratulina caput serpentis*), which lives in our English seas and abounded (as *Terebratulina striata* of authors) in the chalk.

The longest line of human ancestry must hide its diminished head before the pedigree of this insignificant shell-fish. We Englishmen are proud to have an ancestor who was present at the Battle of Hastings. The ancestors of *Terebratulina caput serpentis* may have been present at a battle of *Ichthyosauria* in that part of the sea which, when the chalk was forming, flowed over the site of Hastings. While all around has changed, this *Terebratulina* has peacefully propagated its species from generation to generation, and stands to this day, as a living testimony to the continuity of the present with the past history of the globe.

Up to this moment I have stated, so far as I know, nothing but well-authenticated facts, and the immediate conclusions which they force upon the mind. But the mind is so constituted that it does not willingly rest in facts and immediate causes, but seeks always after a knowledge of the remoter links in the chain of causation.

Taking the many changes of any given spot of the earth's surface, from sea to land and from land to sea, as an established fact, we cannot refrain from asking ourselves how these changes have occurred. And when we have explained them — as they must be explained — by the alternate slow movements of elevation and depression which have affected the crust of the earth, we go still further back, and ask, Why these movements?

I am not certain that any one can give you a satisfactory answer to that question. Assuredly I cannot. All that can be said, for certain, is, that such movements are part of the ordinary course of nature, inasmuch as they are going on at the present time. Direct proof may be given, that some parts of the land in the northern hemisphere are at this moment insensibly rising and others insensibly sinking; and there is indirect, but perfectly satisfactory, proof, that an enormous area now covered by the Pacific has been deepened thousands of feet, since the present inhabitants of that sea came into existence. Thus there is not a shadow of a reason for believing that the physical changes of the globe, in past times, have been effected by other than natural causes. Is there any more reason for believing that the concomitant modifications in the forms of the living inhabitants of the globe have been brought about in other ways?

Before attempting to answer this question, let us try to form a distinct mental picture of what has happened in some special case. The crocodiles are animals which, as a group, have a very vast antiquity. They abounded ages before the chalk was deposited; they throng the rivers in warm climates, at the present day. There is a difference in the form of the joints of the backbone, and in some minor particulars, between the crocodiles of the present epoch and those which lived before the chalk; but, in the cretaceous epoch, as I have already mentioned, the crocodiles have assumed the modern type of structure. Notwithstanding this, the crocodiles of the chalk are not identically the same as those which lived in the times called "older tertiary," which succeeded the cretaceous epoch; and the crocodiles of the older tertiaries are not identical with those of the newer tertiaries, nor are these identical with existing forms. I leave open the question whether particular species may have lived on from epoch to epoch. But each epoch has had its peculiar crocodiles; though all, since the chalk, have belonged to the modern type, and differ simply in their proportions, and in such structural particulars as are discernible only to trained eyes.

How is the existence of this long succession of different species of crocodiles to be accounted for? Only two suppositions seem to be open to us — Either each species of crocodile has been specially created, or it has arisen out of some pre-existing form by the operation of natural causes. Choose your hypothesis; I have chosen mine. I can find no warranty for believing in the distinct creation of a score of successive species of crocodiles in the course of countless ages of time. Science gives no countenance to such a wild fancy; nor can even the perverse ingenuity of a commentator pretend to discover this sense, in the simple words in which the writer of Genesis records the proceedings of the fifth and sixth days of the Creation.

On the other hand, I see no good reason for doubting the necessary alternative, that all these varied species have been evolved from pre-existing crocodilian forms, by the operation of causes as completely a part of the common order of nature as those which have effected the changes of the inorganic world. Few will venture to affirm that the reasoning which applies to crocodiles loses its force among other animals, or among plants. If one series of species has come into existence by the operation of natural causes, it seems folly to deny that all may have arisen in the same way.

A small beginning has led us to a great ending. If I were to put the bit of chalk with which we started into the hot but obscure flame of burning hydrogen, it would presently shine like the sun. It seems to me that this physical metamorphosis is no false image of what has been the result of our subjecting it to a jet of fervent, though nowise brilliant, thought to-night. It has become luminous, and its clear rays, penetrating the abyss of the remote past, have brought within our ken some stages of the evolution of the earth. And in the shifting "without haste, but without rest" of the land and sea, as in the endless variation of the forms assumed by living beings, we have observed nothing but the natural product of the forces originally possessed by the substance of the universe.

THOMAS BELT (1832–1878)

Driver Ants

From *The Naturalist in Nicaragua.* 1874

[THIS ENGLISH GEOLOGIST and gold-mining expert traveled widely in Australia, Siberia, and the United States. He wrote able technical papers on such diverse subjects as the effects of the glacial period and the origin of whirlwinds. At the age of forty-two, only four years before his death, he published an account, more or less in diary form, of several years spent in Nicaragua. It is the work of a truly observant naturalist interested in every phase of nature, and his English is far above that of the usual type of travelogue. Even today it has lost nothing of accuracy or charm.

One paragraph, although isolated from its context, gives a curiously simple yet vivid idea of a spot where Belt lunched one day, and will serve to point my remarks about his style. Few men could see so much in so little time while engaged in eating, yet in the case of each creature mentioned we visualize something besides the name: "Eight leagues from Matagalpa we reached the small town of Tierrabona, where, as the name implies, the land was very good. Every house had an enclosure around it, planted with maize and beans; and though it was evident that the land was cropped year after year, it still seemed to bear well. We stopped at a small brook just outside the town, and eat some provisions we had brought from Matagalpa. Some speckled tiger-beetles ran about the dusty road; and on wet, muddy places near the stream groups of butterflies collected to suck the moisture. Amongst them were some fine swallow-tails (*Papilio*), quivering their wings as they drank, and lovely blue hairstreaks (*Theclae*). The latter, when they alight, rub their wings together, moving their curious tail-like appendages up and down. Great dragon-flies hawked after flies; and on the surface of still pools 'whirligigs' (*Gyrinidae*) wheeled about in mazy gyrations, just as they do at home."]

THE ECITONS, or foraging ants, are very numerous throughout Central America. Whilst the leaf-cutting ants are entirely vegetable feeders, the foraging ants are hunters, and live solely on insects or other prey; and it is a curious analogy that, like the hunting races of mankind, they have to change their hunting-grounds

151

when one is exhausted, and move on to another. In Nicaragua they are generally called "Army Ants." One of the smaller species (*Eciton predator*) used occasionally to visit our house and swarm over the floors and walls, searching every cranny, and driving out the cockroaches and spiders, many of which were caught, pulled, bitten to pieces and carried off. The individuals of this species were of various sizes; the smallest measuring one and a quarter lines, and the largest three lines, or a quarter of an inch.

I saw many large armies of this, or a closely allied species, in the forest. My attention was generally first called to them by the twittering of some small birds, belonging to several different species, that follow the ants in the woods. On approaching, a dense body of the ants, three or four yards wide, and so numerous as to blacken the ground, would be seen moving rapidly in one direction, examining every cranny, and underneath every fallen leaf. On the flanks, and in advance of the main body, smaller columns would be pushed out. These smaller columns would generally first flush the cockroaches, grasshoppers, and spiders. The pursued insects would rapidly make off, but many, in their confusion and terror, would bound right into the midst of the main body of ants. At first the grasshopper, when it found itself in the midst of its enemies, would give vigorous leaps, with perhaps two or three of the ants clinging to its legs. Then it would stop a moment to rest, and that moment would be fatal, for the tiny foes would swarm over the prey, and after a few more ineffectual struggles it would succumb to its fate, and soon be bitten to pieces and carried off to the rear. The greatest catch of the ants was, however, when they got amongst some fallen brushwood. The cockroaches, spiders, and other insects, instead of running right away, would ascend the fallen branches and remain there, whilst the host of ants were occupying all the ground below. By-and-by up would come some of the ants, following every branch, and driving before them their prey to the ends of the small twigs, when nothing remained for them but to leap, and they would alight in the very throng of their foes, with the result of being certainly caught and pulled to pieces. Many of the spiders would escape by hanging suspended by a thread of silk from the branches, safe from the foes that swarmed both above and below.

I noticed that spiders generally were most intelligent in escaping, and did not, like the cockroaches and other insects, take shelter in the first hiding-place they found, only to be driven out again, or perhaps caught by the advancing army of ants. I have often seen large spiders making off many yards in advance, and apparently determined to put a good distance between themselves and the foe. I once saw one of the false spiders, or harvest-men (*Phalangidae*), standing in the midst of an army of ants, and with the greatest circumspection and coolness lifting, one after the other, its long legs, which supported its body above their reach. Sometimes as many as five out of its eight legs would be lifted at once, and whenever an ant approached one of those on which it stood, there was always a clear space within reach to put down another, so as to be able to hold up the threatened one out of danger.

I was much more surprised with the behaviour of a green, leaf-like locust. This insect stood immovably amongst a host of ants, many of which ran over its legs, without ever discovering there was food within their reach. So fixed was its instinctive knowledge that its safety depended on its immovability, that it allowed me to pick it up and replace it amongst the ants without making a single effort to escape. This species closely resembles a green leaf, and the other senses, which in the Ecitons appear to be more acute than that of sight, must have been completely deceived. It might easily have escaped from the ants by using its wings, but it would only have fallen into as great a danger, for the numerous birds that accompany the army ants are ever on the outlook for any insect that may fly up, and the heavy flying locusts, grasshoppers and cockroaches have no chance of escape. Several species of ant-thrushes always accompany the army ants in the forest. They do not, however, feed on the ants, but on the insects they disturb. Besides the ant-thrushes, trogons, creepers, and a variety of other birds, are often seen on the branches of trees above where an ant army is foraging below, pursuing and catching the insects that fly up.

The insects caught by the ants are dismembered, and their too bulky bodies bitten to pieces and carried off to the rear; and behind the army there are always small columns engaged on this duty. I have followed up these columns often; generally they led to dense masses of impenetrable brushwood, but twice they led me to cracks in the ground, down which the ants dragged their prey. These habitations are only temporary, for in a few days not an ant would be seen in the neighborhood, but all would have moved off to fresh hunting-grounds.

Another much larger species of foraging ant (*Eciton hamata*) hunts sometimes in dense armies, sometimes in columns, according to the prey it may be after. When in columns, I found that it was generally, if not always, in search of the nests of another ant (*Hypoclinea sp.*), which bear their young in holes in rotten trunks of

153

fallen timber, and are very common in cleared places. The Ecitons hunt about in columns, which branch off in various directions. When a fallen log is reached, the column spreads out over it, searching through all the holes and cracks. The workers are of various sizes, and the smallest are here of use, for they squeeze themselves into the narrowest holes, and search out their prey in the furthest ramifications of the nests. When a nest of the *Hypoclinea* is attacked, the ants rush out, carrying the larvae and pupae in their jaws, but are immediately despoiled of them by the Ecitons, which are running about in every direction with great swiftness. Whenever they come across a *Hypoclinea* carrying a larva or pupa, they take it from it so quickly, that I could never ascertain exactly how it was done.

As soon as an Eciton gets hold of its prey, it rushes off back along the advancing column, which is composed of two sets, one hurrying forward, the other returning laden with their booty, but all and always in the greatest haste and apparent hurry. About the nest which they are harrying, all appears in confusion, Ecitons running here and there and everywhere in the greatest haste and disorder; but the result of all this apparent confusion is that scarcely a single *Hypoclinea* gets away with a pupa or larva. I never saw the Ecitons injure the Hypoclineas themselves, they were always contented with despoiling them of their young. The ant that is attacked is a very cowardly species, and never shows fight. I often found it running about sipping at the glands of leaves, or milking aphides, leaf-hoppers, or scale-insects that it found unattended by other ants. On the approach of another, though of a much smaller species, it would immediately run away. Probably this cowardly and unantly disposition has caused it to become the prey of the Eciton. At any rate, I never saw the Ecitons attack the nest of any other species.

The moving columns of Ecitons are composed almost entirely of workers of different sizes, but at intervals of two or three yards there are larger and lighter coloured individuals that often stop, and sometimes run a little backward, stopping and touching some of the ants with their antennae. They look like the officers giving orders and directing the march of the column.

This species is often met with in the forest, not in quest of one particular form of prey, but hunting, like *Eciton predator,* only spread out over a much greater space of ground. Crickets, grasshoppers, scorpions, centipedes, wood-lice, cockroaches, and spiders are driven out from below the fallen leaves and branches. Many of them are caught by the ants, others that get away are picked up by the numerous birds that accompany the ants, as vultures follow the armies of the East. The ants send off exploring parties up the trees, which hunt for nests of wasps, bees, and probably birds. If they find any, they soon communicate the intelligence to the army below, and a column is sent up immediately to take possession of the prize. I have seen them pulling out the larvae and pupae from the cells of a large wasp's nest, whilst the wasps hovered about, powerless, before the multitude of the invaders, to render any protection to their young. '

I have no doubt that many birds have acquired instincts to combat or avoid the great danger to which their young are exposed by the attacks of these and other ants. Trogons, parrots, toucans, motmots, and many other birds build in holes of trees or in the ground, and these, with their heads ever turned to the only entrance, are in the best possible position to pick off the solitary parties when they first approach, and thus prevent them from carrying to the main army intelligence about the nest. Some of these birds, and especially the toucans, have bills beautifully adapted for picking up the ants before they reach the nest. Many of the smaller birds build on the branches of the bull's-horn thorn, which is always thickly covered with small stinging honey-eating ants, that would not allow the Ecitons to ascend these trees.

Amongst the mammalia the opossums can convey their young out of danger in their pouches, and the females of many of the tree-rats and mice have a hard callosity near the teats, to which the young cling with their milk teeth, and can be dragged away by the mother to a place of safety.

The eyes in the Ecitons are very small, in some of the species imperfect, and in others entirely absent; in this they differ greatly from the *Pseudomyrma* ants, which hunt singly and which have the eyes greatly developed. The imperfection of eyesight in the Ecitons is an advantage to the community, and to their particular mode of hunting. It keeps them together, and prevents individual ants from starting off alone after objects that, if their eyesight were better, they might discover at a distance; the Ecitons and most other ants follow each other by scent, and, I believe, they can communicate the presence of danger, of booty, or other intelligence, to a distance by the different intensity or qualities of the odours given off. I one day saw a column of *Eciton hamata* running along the foot of a nearly perpendicular tramway cutting, the side of which was about six feet high. At one point I noticed a sort of assembly of about a dozen individuals that appeared in consultation. Suddenly one ant left the conclave, and ran with great speed up the perpendicular face of the cutting without stopping. It was followed by others, which, however, did not keep straight on like the first, but ran a short way, then returned, then again followed a little further than the first time. They were evidently scenting the trail of the pioneer, and making it permanently recognisable. These ants followed the exact line taken by the first one, although it was far out of sight. Wherever it had made a slight detour they did so likewise. I scraped with my knife a small portion of the clay on the trail, and the ants were completely at fault for a time which way to go. Those ascending and those descending stopped at the scraped portion, and made short circuits until they hit the scented trail again, when all their hesitation vanished, and they ran up and down it with the greatest confidence. On gaining the top of the cutting, the ants entered some brushwood suitable for hunting. In a very short space of time the information was communicated to the ants below, and a dense column rushed up to search for their prey. The Ecitons are singular amongst the ants in this respect, that they have no fixed habitations, but move on from one place to another, as they exhaust the hunting grounds around them. I think *Eciton hamata* does not stay more than four or five days in one place. I have sometimes come across the migratory columns; they may easily be known by all the common workers moving in one direction, many of them carrying the larvae and pupae carefully in their jaws. Here and there one of the light-coloured officers moves backwards and forwards directing the columns. Such a column is of enormous length, and contains many thousands if not millions of individuals. I have sometimes followed them up for two or three hundred yards without getting to the end.

They make their temporary habitations in hollow trees, and sometimes underneath large fallen trunks that offer suitable hollows. A nest that I came across in the latter situation was open at one side. The ants were clustered together in a dense mass, like a great swarm of bees, hanging from the roof, but reaching to the ground below. Their innumerable long legs looked like brown threads binding together the mass, which must have been at least a cubic yard in bulk, and contained hundreds of thousands of individuals, although many columns were outside, some bringing in the pupae of ants, others the legs and dissected bodies of various insects. I was surprised to see in this living nest tubular passages leading down to the centre of the mass, kept open just as if it had been formed of inorganic materials. Down these holes the ants who were bringing in booty passed with their prey. I thrust a long stick down to the centre of the cluster, and brought out clinging to it many ants holding larvae and pupae, which probably were kept warm by the crowding together of the ants. Besides the common dark-coloured workers and light-coloured officers, I saw here many still larger individuals with enormous jaws. These they go about holding wide open in a threatening manner, and I found, contrary to my expectation, that they could give a severe bite with them, and that it was difficult to withdraw the jaws from the skin again.

One day when watching a small column of these ants, I placed a little stone on one of them to secure it. The next that approached, as soon as it discovered its situation, ran backwards in an agitated manner, and soon communicated the intelligence to the others. They rushed to the rescue, some bit at the stone and tried to move it, others seized the prisoner by the legs, and tugged with such force that I thought the legs would be pulled off, but they persevered until they got the captive free. I next covered one up with a piece of clay, leaving only the ends of its antennae projecting. It was soon discovered by its fellows, which set to work immediately, and by biting off pieces of the clay, soon liberated it. Another time I found a very few of them passing along at intervals. I confined one of these under a piece of clay, at a little distance from the line, with his head projecting. Several ants passed it, but at last one discovered it and tried to pull it out, but could not. It immediately set off at a great rate, and I thought it had deserted its comrade, but it had only gone for assistance, for in a short time about a dozen ants came hurrying up, evidently fully informed of the circumstances of the case, for they made directly for their imprisoned comrade, and soon set him free. I do not see how this action could be instinctive. It was sympathetic help, such as man only among the higher mammalia shows. The excitement and ardour with which they carried on their unflagging exertions for the rescue of their comrade could not have been greater if they had been human beings, and this to meet a danger that can be only of the rarest occurrence. Amongst the ants of Central America I place Eciton as the first in intelligence, and as such at the head of the Articulata. Wasps and bees come next, and then others of the Hymenoptera. Between ants and the lower forms of insects there is a greater difference in reasoning powers than there is between man and the lowest mammalian. A recent writer has augured that of all animals ants approach nearest to man in their social condition. Perhaps if we could learn their wonderful language we should find that even in their mental condition they also rank next to humanity.

I shall relate two more instances of the use of a reasoning faculty in these ants. I once saw a wide column trying to pass along a crumbling, nearly perpendicular, slope. They would have got very slowly over it, and many of them would have fallen, but a number having secured their hold, and reaching to each other, remained stationary, and over them the main column passed. Another time they were crossing a water-course along a small branch, not thicker than a goose-quill. They widened this natural bridge to three times its width by a number of ants clinging to it and to each other on each side, over which the column passed three or four deep; whereas excepting for this expedient they would have had to pass over in single file, and treble the time would have been consumed. Can it be contended that such insects are not able to determine by reasoning powers which is the best way of doing a thing, or that their actions are not guided by thought and reflection? This view is much strengthened by the fact that the cerebral ganglia in ants are more developed than in any other insects, and that in all the Hymenoptera, at the head of which they stand, "they are many times larger than in the less intelligent orders, such as beetles" (Darwin).

The Hymenoptera standing at the head of the Articulata, and the Mammalia at the head of the Vertebrata, it is curious to mark how in zoological history the appearance and development of these two orders (culminating in the one in the Ants, and in the other in the Primates) run parallel. The Hymenoptera and the Mammalia both make their first appearance early in the secondary period, and it is not until the commencement of the tertiary epoch that ants and monkeys appear upon the scene. There the parallel ends; no one species of ant has attained any great superiority above all its fellows, whilst man is very far in advance of all the other Primates.

When we see these intelligent insects dwelling together in orderly communities of many thousands of individuals, their social instincts developed to a high degree of perfection, making their marches with the regularity of disciplined troops, showing ingenuity in the crossing of difficult places, assisting each in danger, defending their nests at the risk of their own lives, communicating information rapidly to a great distance, making a regular division of work, the whole community taking charge of the rearing of the young, and all imbued with the stronger sense of industry, each individual labouring not for itself alone but for all its fellows — we may imagine that Sir Thomas More's description of Utopia might have been applied with greater justice to such a community than to any human society. "But in Utopia, where every man has a right to everything, they do all know that if care is taken to keep the public stores full, no private man can want anything; for among them there is no unequal distribution, so that no man is poor, nor in any necessity, and though no man has anything, yet they are all rich; for what can make a man so rich as to lead a serene and cheerful life, free from anxieties, neither apprehending want himself, nor vexed with the endless complaints of his wife? He is not afraid of the misery of his children, nor is he contriving how to raise a portion for his daughters, but is secure in this, that both he and his wife, his children and grandchildren, to as many generations as he can fancy, will all live both plentifully and happily."

W. H. HUDSON (1841–1922)

Patagonian Memories

From *The Naturalist in La Plata,* 1892, and *Idle Days in Patagonia,* 1893

[Although Hudson was born on the pampas in Argentina, his parents were Massachusetts Americans. In 1868, when twenty-seven years old he left for England and lived there until his death. He is more widely known for some of his mystical tales and novels, such as *Green Mansions,* but several other volumes have put him in the forefront of the literature of natural history. The subjects of these books are diverse, but have birds as the dominant note. For many years after he left South America there was little or no writing from his particular field; also, his observations were along the lines of habits, such as migration, the uses of the senses, music and fear in nature. All this has combined to make his work of lasting value, and for long to come many of his notes will be significant, as the work of the field naturalist turns more and more from collecting and taxonomy to the study and understanding of life activities.

Our first selection, taken from *Idle Days in Patagonia,* is almost a mental picture of Hudson, his mind stimulated by the strange country which was his early home. The second choice is from *The Naturalist in La Plata* and shows his observational and descriptive powers at their best.

The latter book was written twenty-five years after Hudson had reached England, yet it has all the accuracy and freshness of day-to-day observation. Few naturalists have identified themselves so completely with the animal life of one particular region as has this author; few have depended so fully and so vividly on their own first-hand observing.]

THE PLAINS OF PATAGONIA

NEAR THE END of Darwin's famous narrative of the voyage of the *Beagle* there is a passage which, for me, has a very special interest and significance. It is as follows, and the italicization is mine: — "In calling up images of the past, I find the plains of Patagonia frequently cross before my eyes; yet these plains are pro-

160

nounced by all to be most wretched and useless. They are characterized only by negative possessions; without habitations, without
water, without trees, without mountains, they support only a few
dwarf plants. *Why, then — and the case is not peculiar to myself —
have these arid wastes taken so firm possession of my mind?* Why
have not the still more level, the greener and more fertile pampas,
which are serviceable to mankind, produced an equal impression? I
can scarcely analyze these feelings, but it must be partly owing to the
free scope given to the imagination. The plains of Patagonia are
boundless, for they are scarcely practicable, and hence unknown;
they bear the stamp of having thus lasted for ages, and there appears
no limit to their duration through future time. If, as the ancients
supposed, the flat earth was surrounded by an impassable breadth
of water, or by deserts heated to an intolerable excess, who would
not look at these last boundaries to man's knowledge with deep but
ill-defined sensations?"

That he did not in this passage hit on the right explanation of
the sensations he experienced in Patagonia, and of the strength of the
impressions it made on his mind, I am quite convinced; for the thing
is just as true of today as of the time, in 1836, when he wrote that
the case was not peculiar to himself. Yet since that date — which now,
thanks to Darwin, seems so remote to the naturalist — those desolate
regions have ceased to be impracticable, and, although still unin-
habited and uninhabitable, except to a few nomads, they are no
longer unknown. During the last twenty years the country has been
crossed in various directions, from the Atlantic to the Andes, and
from the Rio Negro to the Straits of Magellan, and has been found
all barren. The mysterious illusive city, peopled by whites, which
was long believed to exist in the unknown interior, in a valley called
Trapalanda, is to moderns a myth, a mirage of the mind, as little to
the traveller's imagination as the glittering capital of great Manoa,
which Alonzo Pizarro and his false friend Orellana failed to discover. The traveller of today really expects to see nothing more exciting than a solitary huanaco keeping watch on a hilltop, and a few
grey-plumaged rheas flying from him, and, possibly, a band of long-
haired roving savages, with their faces painted black and red. Yet,
in spite of accurate knowledge, the old charm still exists in all its
freshness; and after all the discomforts and sufferings endured in a
desert cursed with eternal barrenness, the returned traveller finds in
after years that it still keeps its hold on him, that it shines brighter in
memory, and is dearer to him than any other region he may have
visited.

161

We know that the more deeply our feelings are moved by any scene the more vivid and lasting will its image be in memory — a fact which accounts for the comparatively unfading character of the images that date back to the period of childhood, when we are most emotional. Judging from my own case, I believe that we have here the secret of the persistence of Patagonian images, and their frequent recurrence in the minds of many who have visited that grey, monotonous, and, in one sense, eminently uninteresting region. It is not the effect of the unknown, it is not imagination; it is that nature in these desolate scenes, for a reason to be guessed at by-and-by, moves us more deeply than in others. In describing his rambles in one of the most desolate spots in Patagonia, Darwin remarks: "Yet, in passing over these scenes, without one bright object near, an ill-defined but strong sense of pleasure is vividly excited." When I recall a Patagonian scene, it comes before me so complete in all its vast extent, with all its details so clearly outlined, that, if I were actually gazing on it, I could scarcely see it more distinctly; yet other scenes, even those that were beautiful and sublime, with forest, and ocean, and mountain, and over all the deep blue sky and brilliant sunshine of the tropics, appear no longer distinct and entire in memory, and only become more broken and clouded if any attempt is made to regard them attentively. Here and there I see a wooded mountain, a grove of palms, a flowery tree, green waves dashing on a rocky shore — nothing but isolated patches of bright colour, the parts of the picture that have not faded on a great blurred canvas, or series of canvases. These last are images of scenes which were looked on with wonder and admiration — feelings which the Patagonian wastes could not inspire — but the grey, monotonous solitude woke other and deeper feelings, and in that mental state the scene was indelibly impressed on the mind.

I spent the greater part of one winter at a point on the Rio Negro, seventy or eighty miles from the sea, where the valley on my side of the water was about five miles wide. The valley alone was habitable, where there was water for man and beast, and a thin soil producing grass and grain; it is perfectly level, and ends abruptly at the foot of the bank or terrace-like formation of the higher barren plateau. It was my custom to go out every morning on horseback with my gun, and, followed by one dog, to ride away from the valley; and no sooner would I climb the terrace and plunge into the grey universal thicket, than I would find myself as completely alone and cut off from all sight and sound of human occupancy as if five hundred instead of only five miles separated me from the hidden green valley and river. So wild and solitary and remote seemed that grey waste, stretching away into infinitude, a waste untrodden by man, and where the wild animals are so few that they have made no discoverable path in the wilderness of thorns. There I might have dropped down and died, and my flesh been devoured by birds, and my bones bleached white in sun and wind, and no person would have found them, and it would have been forgotten that one had ridden forth in the morning and had not returned. Or if, like the few wild animals there — puma, huanaco, and hare-like dolichotis, or Darwin's rhea and the crested tinamou among the birds — I had been able to exist without water, I might have made myself a hermitage of brushwood or dug-out in the side of a cliff, and dwelt there until I had grown grey as the stones and trees around me, and no human foot would have stumbled on my hiding-place.

Not once, nor twice, nor thrice, but day after day I returned to this solitude, going to it in the morning as if to attend a festival, and leaving it only when hunger and thirst and the westering sun compelled me. And yet I had no object in going — no motive which could be put into words; for although I carried a gun, there was nothing to shoot — the shooting was all left behind in the valley. Sometimes a dolichotis, starting up at my approach, flashed for one moment on my sight, to vanish the next moment in the continuous thicket; or a covey of tinamous sprang rocket-like into the air, and fled away with long wailing notes and loud whirr of wings; or on some distant hillside a bright patch of yellow, of a deer that was watching me, appeared and remained motionless for two or three minutes. But the animals were few, and sometimes I would pass an entire day without seeing one animal, and perhaps not more than a dozen birds of any size. The weather at that time was cheerless, generally with a grey film of cloud spread over the sky, and a bleak wind, often cold enough to make my bridle hand feel quite numb. Moreover, it was not possible to enjoy a canter; the bushes grew so close together that it was as much as one could do to pass through at a walk without brushing against them; and at this slow pace, which would have seemed intolerable in other circumstances, I would ride about for hours at a stretch. In the scene itself there was nothing to delight the eye. Everywhere through the light-grey mould, grey as ashes and formed by the ashes of myriads of generations of dead trees, where the wind had blown on it, or the rain had washed it away, the underlying yellow sand appeared, and the old ocean-polished pebbles, dull red, and grey, and green, and yellow. On arriving at a hill, I would slowly ride to its summit, and stand there to survey the prospect. On every side it stretched away in great undulations; but the undulations were wild and irregular; the hills were rounded and cone-shaped, they were solitary and in groups and ranges; some sloped gently, others were ridge-like and stretched away in league-long terraces, with other terraces beyond; and all alike were clothed in the grey everlasting thorny vegetation. How grey it all was! hardly less so near at hand than on the haze-wrapped horizon, where the hills were dim and outline blurred by distance. Sometimes I would see the large eagle-like, white-breasted buzzard, *Buteo erythronotus,* perched on the summit of a bush half a mile away; and so long as it would continue stationed motionless before me my eyes would remain involuntarily fixed on it, just as one keeps his eyes on a bright light shining in the gloom; for the whiteness of the hawk seemed to exercise a fascinating power on the vision, so surpassingly bright was it by contrast in the midst of that universal unrelieved greyness. Descending from my look-out, I would take up my aimless wanderings again, and visit other elevations to gaze on the same landscape from another point; and so on for hours, and at noon I would dismount and sit or lie on my folded poncho for an hour or longer. One day, in these rambles, I discovered a small grove composed of twenty to thirty trees, about eighteen feet high, and taller than the surrounding trees. They were growing at a convenient distance apart, and had evidently been resorted to by a herd of deer or other wild animals for a very long time, for the boles were polished to a glassy smoothness with much rubbing, and the ground beneath was trodden to a floor of clean, loose yellow sand. This grove was on a hill differing in shape from other hills in its neighbourhood, so that it was easy for me to find it on other occasions; and after a time I made a point of finding and using it as a resting-place every day at noon. I did not ask myself why I made choice of that one spot, sometimes going miles out of my way to sit there, instead of sitting down under any one of the millions of trees and bushes covering the country, on any other hillside. I thought nothing at all about it, but acted unconsciously; only afterwards, when revolving the subject it seemed to me that after having rested there once, each time I wished to rest again the wish came associated with the image of that particular clump of trees, with polished stems and clean bed of sand beneath; and in a short time I formed a habit of returning, animal-like, to repose at that same spot.

It was perhaps a mistake to say that I would sit down and rest, since I was never tired: and yet without being tired, that noonday pause, during which I sat for an hour without moving, was strangely grateful. All day the silence seemed grateful, it was very perfect, very profound. There were no insects, and the only bird sound — a feeble chirp of alarm emitted by a small skulking wren-like species — was not heard oftener than two or three times an hour. The only sounds as I rode were the muffled hoof-strokes of my horse, scratching of twigs against my boot or saddle-flap, and the low panting of the dog. And it seemed to be a relief to escape even from these sounds when I dismounted and sat down: for in a few moments the dog would stretch his head out on his paws and go to sleep, and then there would be no sound, not even the rustle of a leaf. For unless the wind blows strong there is no fluttering motion and no whisper in the small stiff undeciduous leaves; and the bushes stand unmoving as if carved out of stone. One day while *listening* to the silence, it occurred to my mind to wonder what the effect would be if I were to shout aloud. This seemed at the time a horrible suggestion of fancy, a "lawless and uncertain thought" which almost made me shudder, and I was anxious to dismiss it quickly from my mind. But during those solitary days it was a rare thing for any thought to cross my mind; animal forms did not cross my vision or bird-voices assail my hearing more rarely. In that novel state of mind I was in, thought had become impossible. Elsewhere I had always been able to think most freely on horseback; and on the pampas, even in the most lonely places, my mind was always most active when I travelled at a swinging gallop. This was doubtless habit; but now, with a horse under me, I had become incapable of reflection: my mind had suddenly transformed itself from a thinking machine into a machine for some other unknown purpose. To think was like setting in motion a noisy engine in my brain; and there was something there which bade me be still, and I was forced to obey. My state was one of *suspense* and *watchfulness:* yet I had no expectation of meeting with an adventure, and felt as free from apprehension as I feel now when sitting in a room in London. The change in me was just as great and wonderful as if I had changed my identity for that of another man or animal; but at the time I was powerless to wonder at or speculate about it; the state seemed familiar rather than strange, and although accompanied by a strong feeling of elation, I did not know it — did not know that something had come between me and my intellect — until I lost it and returned to my former self — to thinking, and the old insipid existence.

Such changes in us, however brief in duration they may be, and in most cases they are very brief, but which so long as they last seem to affect us down to the very roots of our being, and come as a great surprise — a revelation of an unfamiliar and unsuspected nature hidden under the nature ·we are conscious of — can only be attributed to an instantaneous reversion to the primitive and wholly savage mental conditions. Probably not many men exist who would be unable to recall similar cases in their own experience; but it frequently happens that the revived instinct is so purely animal in character and repugnant to our refined or humanitarian feelings, that it is sedulously concealed and its promptings resisted. In the military and seafaring vocations, and in the lives of travel of adventure, these sudden and surprising reversions are most frequently experienced. The excitement affecting men going into battle, which even affects those who are constitutionally timid and will cause them to exhibit a reckless daring and contempt of danger astonishing to themselves, is a familiar instance. This instinctive courage has been compared to intoxication, but it does not, like alcohol, obscure a man's faculties: on the contrary, he is far more keenly active to everything going on around him than the person who keeps perfectly cool. The man who is coolly courageous in fight has his faculties in their ordinary condition: the faculties of the man who goes into battle inflamed with instinctive, joyous excitement are sharpened to a preternatural keenness. When the constitutionally timid man has had an experience of this kind he looks back on the day that brought it to him as the happiest he has known, one that stands out brightly and shines with a strange glory among his days.

When we are suddenly confronted with any terrible danger, the change of nature we undergo is equally great. In some cases fear paralyzes us, and, like animals, we stand still, powerless to move a step in flight, or to lift a hand in defence of our lives; and sometimes we are seized with panic, and, again, act more like the inferior animals than rational beings. On the other hand, frequently in cases of sudden extreme peril, which cannot be escaped by flight, and must be instantly faced, even the most timid men at once, as if by miracle, become possessed of the necessary courage, sharp, quick apprehension, and swift decision. This is a miracle very common in nature; man and the inferior animals alike, when confronted with almost certain death "gather resolution from despair." We are accustomed to call this the "courage of despair"; but there can be no trace of so debilitating a feeling in the person fighting, or prepared to fight, for dear life. At such times the mind is clearer than it has ever been; the nerves are steel; there is nothing felt but a wonderful strength and fury and daring. Looking back at certain perilous moments in my own life, I remember them with a kind of joy; not that there was any joyful excitement then, but because they brought me a new experience — a new nature, as it were — and lifted me for a time above myself. And yet, comparing myself with other men, I find that on ordinary occasions my courage is rather below than above the average. And probably this instinctive courage, which flashes out so brightly on occasions, is inherited by a very large majority of the male children born into the world; only in civilized life the exact conjuncture of circumstances needed to call it into activity rarely occurs.

In hunting, again, instinctive impulses come very much to the surface. Leech caricatured Gallic ignorance of fox-hunting in England when he made his French gentleman gallop over the hounds and dash away to capture the fox himself; but the sketch may be also taken as a comic illustration of a feeling that exists in every one of us. If any sportsman among my readers has ever been confronted with some wild animal — a wild dog, a pig, or cat, let us say, — when he had no firearm or other weapon to kill it in the usual civilized way, and has nevertheless attacked it, driven by a sudden uncontrollable impulse, with a hunting knife, or anything that came to hand, and has succeeded in slaying it, I would ask such a one whether this victory did not give him a greater satisfaction than all his other achievements in the field? After it, all legitimate sport would seem illegitimate, and whole hecatombs of hares and pheasants, and even large animals, fallen before his gun, would only stir in him a feeling of disgust and self-contempt. He would probably hold his tongue about a combat of that brutal kind, but all the same he would gladly remember how in some strange, unaccountable way he suddenly became possessed of the daring, quickness, and certitude necessary to hold his wily, desperate foe in check, to escape its fangs and claws, and finally to overcome it. Above all, he would remember the keen feeling of savage joy experienced in the contest. This would make all ordinary sport seem insipid; to kill a rat in some natural way would seem better to him than to murder elephants scientifically from a safe distance. The feeling occasionally bursts out in the *Story of My Heart:* "To shoot with a gun is nothing . . . give me an iron mace that I may crush the savage beast and hammer him down. A spear to thrust him through with, so that I may feel the long blade enter, and push of the shaft." And more in the same strain, shocking to some, perhaps, but showing that gentle Richard Jefferies had in him some of the elements of a fine barbarian.

But it is in childhood and boyhood, when instincts are nearest to the surface, and ready when occasion serves to spring into activity.

Inherited second nature is weakest then; and habit has not progressed far in weaving its fine network of restraining influences over the primitive nature. The network is continually being strengthened in the individual's life, and, in the end he is cased, like the caterpillar, in an impervious cocoon; only, as we have seen, there are in life miraculous moments when the cocoon suddenly dissolves, or becomes transparent, and he is permitted to see himself in his original nakedness. The delight which children experience on entering the woods and other wild places is very keen; and this feeling, although it diminishes as we advance in life, remains with us to the last. Equally great is their delight at finding wild fruits, honey, and other natural food; and even when not hungry they will devour it with strange zest. They will gladly feast on sour, acrid fruits, which at table, and picked in the garden, would only excite disgust. This instinctive seeking for food, and the delight experienced in finding it, occasionally comes up in very unexpected and surprising ways. "As I came through the wood," says Thoreau, "I caught a glimpse of a woodchuck stealing across my path, and felt a strange thrill of savage delight, and was strongly tempted to seize and devour him raw; not that I was hungry then, except for the wildness which he represented."

In almost all cases — those in which danger is encountered and rage experienced being exceptions — the return to an instinctive or primitive state of mind is accompanied by this feeling of elation, which, in the very young, rises to an intense gladness, and sometimes makes them mad with joy, like animals newly escaped from captivity. And, for a similar reason, the civilized life is one of continual repression, although it may not seem so until a glimpse of nature's wildness, a taste of adventure, an accident, suddenly makes it seem unspeakably irksome; and in that state we feel that our loss in departing from nature exceeds our gain.

It was elation of this kind, the feeling experienced on going back to a mental condition we have outgrown, which I had in the Patagonian solitude; for I had undoubtedly *gone back;* and that state of intense watchfulness, or alertness rather, with suspension of the higher intellectual faculties, represented the mental state of the pure savage. He thinks little, reasons little, having a surer guide in his instinct; he is in perfect harmony with nature, and is nearly on a level, mentally, with the wild animals he preys on, and which in their turn sometimes prey on him. If the plains of Patagonia affect a person in this way, even in a much less degree than in my case, it is not strange that they impress themselves so vividly on the mind, and remain fresh in memory, and return frequently; while other scenery, however grand or beautiful, fades gradually away, and is at last forgotten. To a slight, in most cases probably very slight, extent, all natural sights and sounds affect us in the same way; but the effect is often transitory, and is gone with the first shock of pleasure, to be followed in some cases by a profound and mysterious melancholy. The greenness of earth; forest and river and hill; the blue haze and distant horizon; shadows of clouds sweeping over the sun-flushed landscape — to see it all is like returning to a home, which is more truly our home than any habitation we know. The cry of the wild bird pierces us to the heart; we have never heard that cry before, and it is more familiar to us than our mother's voice. "I heard," says Thoreau, "a robin in the distance, the first I had heard for many a thousand years, methought, whose note I shall not forget for many a thousand more, — the same sweet and powerful song as of yore. O the evening robin!" Hafiz sings: —

O breeze of the morning blow me a memory of the ancient time;
If after a thousand years thy odours should float o'er my dust,
My bones, full of gladness uprising, would dance in the sepulchre!

And we ourselves are the living sepulchres of a dead past — that past which was ours for so many thousands of years before this life of the present began; its old bones are slumbering in us — dead, and yet not dead nor deaf to Nature's voices; the noisy burn, the roar of the waterfall, and thunder of long waves on the shore, and the sound of rain and whispering winds in the multitudinous leaves, bring it a memory of the ancient time; and the bones rejoice and dance in their sepulchre.

Professor W. K. Parker, in his work *On Mammalian Descent*, speaking of the hairy covering almost universal in this class of animals, says: "This has become, as every one knows, a custom among the race of men, and shows, at present, no sign of becoming obsolete. Moreover, that first correlation, namely, milk-glands and a hairy covering, appears to have entered the very soul of creatures of this class, and to have become *psychical* as well as *physical,* for in that type, which is only inferior to the angels, the fondness for this kind of outer covering is a strong and ineradicable passion." I am not sure that this view accords with some facts in our experience, and with some instinctive feelings which we all have. Like Waterton, I have found that the feet take very kindly to the earth, however hot or cold or rough it may be, and that shoes, after being left off for a short time, seem as uncomfortable as a mask. The face is always uncovered;

why does the supposed correlation not apply to this part? The face is always pleasantly warm when the too delicate body shivers with cold under its covering; and pleasantly cool when the sun shines hot on us. When the wind strikes us on a hot day, or during violent exercise, the sensation to the face is extremely agreeable, but far from agreeable to the body where the covering does not allow the moisture to evaporate rapidly. The umbrella has not entered the soul — not yet; but it is miserable to get wet in the rain, yet pleasant to feel the rain on the face. "I am all face," the naked American savage said, to explain why he felt no discomfort from the bleak wind which made his civilized fellow-traveller shiver in his furs. Again, what a relief, what a pleasure, to throw off the clothes when occasion permits. Leigh Hunt wrote an amusing paper on the pleasures of going to bed, when the legs, long separated by unnatural clothing, delightedly rub against and renew their acquaintance with one another. Everyone knows the feeling. If it were convenient, and custom not so tyrannical, many of us would be glad to follow Benjamin Franklin's example, and rise not to dress, but to settle comfortably down to our morning's work, with nothing on. When, for the first time, in some region where nothing but a figleaf has "entered the soul," we see men and women going about naked and unashamed, we experience a slight shock; but it has more pleasure than pain in it, although we are reluctant to admit the pleasure, probably because we mistake the nature of the feeling. If, after seeing them for a few days in their native simplicity, our new friends appear before us clothed, we are shocked again, and this time disagreeably so; it is like seeing those who were free and joyous yesterday now appear with fettered feet and sullen downcast faces.

To leave this question; what has truly entered our soul and become psychical is our environment — that wild nature in which and to which we were born at an inconceivably remote period, and which made us what we are. It is true that we are eminently adaptive, that we have created, and exist in some sort of harmony with new conditions, widely different from those to which we were originally adapted; but the old harmony was infinitely more perfect than the new, and if there be such a thing as historical memory in us, it is not strange that the sweetest moment in any life, pleasant or dreary, should be when Nature draws near to it, and, taking up her neglected instrument, plays a fragment of some ancient melody, long unheard on the earth.

It might be asked: If nature has at times this peculiar effect on us, restoring instantaneously the old vanished harmony between organism and environment, why should it be experienced in a greater degree in the Patagonian desert than in other solitary places, — a desert which is waterless, where animal voices are seldom heard, and vegetation is grey instead of green? I can only suggest a reason for the effect being so much greater in my own case. In sub-tropical woods and thickets, and in wild forests in temperate regions, the cheerful verdure and bright colours of flower and insects, if we have acquired a habit of looking closely at these things, and the melody and noises of bird-life engages the senses; there is movement and brightness; new forms, animal and vegetable, are continually appearing, curiosity and expectation are excited, and the mind is so much occupied with novel objects that the effect of wild nature in its entirety is minimized. In Patagonia the monotony of the plains, or expanse of low hills, the universal unrelieved greyness of everything, and the absence of animal forms and objects new to the eye, leave the mind open and free to receive an impression of visible nature as a whole. One gazes on the prospect as on the sea, for it stretches away sea-like, without change, into infinitude; but without the sparkle of water, the changes of hue which shadows and sunlight and nearness and distance give, and motion of waves and white flash of foam. It has a look of antiquity, of desolation, of eternal peace, of a desert that has been a desert from of old and will continue a desert for ever; and we know that its only human inhabitants are a few wandering savages, who live by hunting as their progenitors have done for thousands of years. Again, in fertile savannahs and pampas there may appear no signs of human occupancy, but the traveller knows that eventually the advancing tide of humanity will come with its flocks and herds, and the ancient silence and desolation will be no more; and this thought is like human companionship, and mitigates the effect of nature's wildness on the spirit. In Patagonia no such thought or dream of the approaching changes to be wrought by human agency can affect the mind. There is no water there, the arid soil is sand and gravel — pebbles rounded by the action of ancient seas, before Europe was; and nothing grows except the barren things that nature loves — thorns, and a few woody herbs, and scattered tufts of wiry bitter grass.

Doubtless we are not all affected in solitude by wild nature in the same degree; even in the Patagonian wastes many would probably experience no such mental change as I have described. Others have their instincts nearer to the surface, and are moved deeply by nature in any solitary place; and I imagine that Thoreau was such a one. At all events, although he was without the Darwinian lights which we have, and these feelings were always to him "strange," "mysterious," "unaccountable," he does not conceal them. This is the "something uncanny in Thoreau" which seems inexplicable and startling to such as have never been startled by nature, nor deeply moved; but which, to others, imparts a peculiarly delightful aromatic flavour to his writings. It is his wish towards a more primitive mode of life, his strange abandonment when he scours the wood like a half-starved hound, and no morsel could be too savage for him; the desire to take a ranker hold on life and live more as the animals do; the sympathy with nature so keen that it takes his breath away; the feeling that all the elements were congenial to him, which made the wildest scenes unaccountably familiar, so that he came and went with a strange liberty in nature. Once only he had doubts, and thought that human companionship might be essential to happiness; but he was at the same time conscious of a slight insanity in the mood; and he soon again became sensible of the sweet beneficent society of nature, of an infinite and unaccountable friendliness all at once like an atmosphere sustaining him.

In the limits of a chapter it is impossible to do more than touch the surface of so large a subject as that of the instincts and remains of instincts existing in us. Dr. Wallace doubts that there are any human instincts, even in the perfect savage; which seems strange in so keen an observer, and one who has lived so much with nature and uncivilized men; but it must be borne in mind that his peculiar theories with regard to man's origin — the acquisition of large brains, naked body, and the upright form not through but in spite of natural selection — would predispose him to take such a view. My own experience and observation have led me to a contrary conclusion, and my belief is that we might learn something by looking more beneath the hardened crust of custom into the still burning core. For instance, that experience I had in Patagonia — the novel state of mind I have described — seemed to furnish an answer to a question frequently asked with regard to men living in a state of nature. When we consider that our intellect, unlike that of the inferior animals, is progressive, how wonderful it seems that communities and tribes of men should exist — "are contented to exist," we often say, just as if they had any choice in the matter — for ages and for thousands of years in a state of pure barbarism, living from hand to mouth, exposed to extremes of temperature, and to frequently-recurring famine even in the midst of the greatest fertility, when a little foresight — "The smallest amount of intelligence possessed by the lowest of mankind," we say — would be sufficient to make their condition immeasurably better. If, in the wild natural life, their normal state is like that into which I temporarily fell, then it no longer appears strange to me that they take no thought for the morrow, and remain stationary, and are only a little removed from other mammalians, their superiority in this respect being only sufficient to counterbalance their physical disadvantages. That instinctive state of the human mind, when the higher faculties appear to be nonexistent, a state of intense alertness and preparedness, which compels the man to watch and listen and go silently and stealthily, must be like that of the lower animals: the brain is then like a highly-polished mirror, in which all visible nature — every hill, tree, leaf — is reflected with miraculous clearness; and we can imagine that if the animal could think and reason, thought would be superfluous and a hindrance, since it would dim that bright perception on which his safety depends.

That is a part, the lesser part, of the lesson I learnt in the Patagonian solitude: the second larger part must be cut very short; for on all sides it leads to other questions, some of which would probably be thought "more curious than edifying." That hidden fiery core is nearer to us than we ordinarily imagine, and its heat still permeates the crust to keep us warm. This is, no doubt, a matter of annoyance and even grief to those who grow impatient at Nature's unconscionable slowness; who wish to be altogether independent of such an underlying brute energy; to live on a cool crust and rapidly grow angelic. But, as things are, it is, perhaps, better to be still, for a while, a little lower than the angels: we are hardly in a position just yet to dispense with the unangelic qualities, even in this exceedingly complex state, in which we appear to be so effectually "hedged in from harm." I recall here an incident witnessed by a friend of mine of an Indian he and his fellow-soldiers were pursuing who might easily have escaped unharmed; but when his one companion was thrown to the ground through his horse falling the first Indian turned deliberately, sprang to the earth, and, standing motionless by the other's side, received the white men's bullets. Not for love — it would be absurd to suppose such a thing — but inspired by that fierce instinctive spirit of defiance which in some cases will actually cause a man to go out of his way to seek death. Why are we, children of light — the light which makes us timid — so strongly stirred by a deed like this, so useless and irrational, and feel an admiration so great that compared with it that which is called forth by the noblest virtue, or the highest achievement of the intellect, seems like a pale dim feeling? It is because in our inmost natures, our deepest feelings, we are still one with the savage. We admire a Gordon less for his godlike qualities — his spirituality, and crystal purity of heart, and justice, and love of his kind — than for that more ancient nobility, the qualities he had in common with the wild man of childish intellect, an old Viking, a fighting Colonel Burnaby, a Captain Webb who madly flings his life away, a vulgar Welsh prizefighter who enters a den full of growling lions, and drives them before him like frightened sheep. It is due to this instinctive savage spirit in us, in spite of our artificial life and all we have done to rid ourselves of an inconvenient heritage, that we are capable of so-called heroic deeds; of cheerfully exposing ourselves to the greatest privations and hardships, suffering them stoically, and facing death without blenching, sacrificing our lives, as we say, in the cause of humanity, or geography, or some other branch of science.

It is related that a late aged prime minister of England on one occasion stood for several hours at his sovereign's side at a reception, in an oppressive atmosphere, and suffering excruciating pains from a gouty foot; yet making no sign and concealing his anguish under a smiling countenance. We have been told that this showed his good blood: that because he came of good stock, and had the training and traditional feelings of a gentleman, he was able to suffer in that calm way. This pretty delusion quickly vanishes in a surgical hospital, or on a field covered with wounded men after a fight. But the savage always endures pain more stoically than the civilized man. He is

> Self-balanced against contingencies,
> As the trees and animals are.

However great the sufferings of the gouty premier may have been, they were less than those which any Indian youth in Guiana and Venezuela voluntarily subjects himself to before he ventures to call himself a man, or to ask for a wife. Small in comparison, yet he did not endure them smilingly because the traditional pride and other feelings of a gentleman made it possible for him to do so, but because that more ancient and nobler pride, the stern instinct of endurance of the savage, came to his aid and sustained him.

These things do not, or at all events should not, surprise us. They can only surprise those who are without the virile instinct, or who have never become conscious of it on account of the circumstances of their lives. The only wonder is that the stern indomitable spirit in us should ever in any circumstances fail a man, that even on the scaffold or with the world against him he should be overcome by despair, and burst into weak tears and lamentations, and faint in the presence of his fellows. In one of the most eloquent passages of his finest work Herman Melville describes as follows that manly spirit or instinct in us, and the effect produced on us by the sight of its failure: "Men may seem detestable as joint-stock companies and nations; knaves, fools, and murderers there may be; men may have mean and meagre faces; but man, in the ideal, is so noble and so sparkling, such a grand and glowing creature, that over any ignominious blemish in him all his fellows should run to throw their costliest robes. That immaculate manliness we feel in ourselves — so far within us that it remains intact though all the outer character seems gone — bleeds with keenest anguish at the spectacle of a valour-ruined man. Nor can piety itself, at such a shameful sight, completely stifle her upbraidings against the permitting stars. But this august dignity I treat of, is not the dignity of kings and robes, but the abounding dignity which has no robed investiture. Thou shalt see it shining in the arm that wields a pick and drives a spike; that democratic dignity which, on all hands, radiates without end from God Himself."

There is then something to be said in favour of this animal and primitive nature in us. Thoreau, albeit so spiritually-minded, could yet "reverence" that lower nature in him which made him brother to the brute. He experienced and fully appreciated its tonic effect. And until we get a better civilization more equal in its ameliorating effect on all classes — if there must be classes — and more likely to endure, it is perhaps a fortunate thing that we have so far failed to eliminate the "savage" in us — the "Old Man" as some might prefer to call it. Not a respectable Old Man, but a very useful one occasionally, when we stand in sore need of his services and he comes promptly and unsummoned to our aid.

SEEN AND LOST

We can imagine what the feelings of a lapidary would be — an enthusiast whose life is given to the study of precious stones, and whose sole delight is in the contemplation of their manifold beauty — if a stranger should come in to him, and, opening his hand, exhibit a new unknown gem, splendid as ruby or as sapphire, yet manifestly no mere variety of any familiar stone, but differing as widely from all others as diamond from opal or cat's eye; and then, just when he is beginning to rejoice in that strange exquisite loveliness, the hand should close and the stranger, with a mocking smile on his lips, go forth and disappear from sight in the crowd. A feeling such as that would be is not unfrequently experienced by the field naturalist whose favoured lot it is to live in a country not yet "thoroughly

175

worked out," with its every wild inhabitant scientifically named, accurately described, and skilfully figured in some colossal monograph. One swift glance of the practised eye, ever eagerly searching for some new thing, and he knows that here at length is a form never previously seen by him; but his joy is perhaps only for a few minutes, and the prize is snatched from sight for ever. The lapidary might have some doubts; he might think that the stranger had, after all, only mocked him with the sight of a wonderful artificial gem, and that a close examination would have proved its worthlessness; but the naturalist can have no doubts; if he is an enthusiast, well acquainted with the fauna of his district, and has good eyesight, he knows that there is no mistake; for there it is, the new strange form, photographed by instantaneous process on his mind, and there it will remain a tantalizing image, its sharp lines and fresh colouring unblurred by time.

Walking in some open forest glade, he may look up just in time to see a great strange butterfly — a blue Morpho, let us say, wandering in some far country where this angel insect is unknown — passing athwart his vision with careless, buoyant flight, the most sylph-like thing in nature, and all blue and pure like its aerial home, but with a more delicate and wonderful brilliance in its cerulean colour, giving such unimaginable glory to its broad airy wings; and then, almost before his soul has had time to feel its joy, it may soar away unloitering over the tall trees, to be seen no more.

But the admiration, the delight, and the desire are equally great, and the loss just as keenly felt, whether the strange species seen happens to be one surpassingly beautiful or not. Its newness is to the naturalist its greatest attraction. How beautiful beyond all others seems a certain small unnamed brown bird to my mind! so many years have passed and its image has not yet grown dim; yet I saw it only for a few moments, when it hopped out from the thick foliage and perched within two or three yards of me, not afraid, but only curious; and after peering at me first with one eye and then the other, and wiping its small dagger on a twig, it flew away and was seen no more. For many days I sought for it, and for years waited its reappearance, and it was more to me than ninety and nine birds which I had always known; yet it was very modest, dressed in a brown suit, very pale on the breast and white on the throat, and for distinction a straw-coloured stripe over the eye — that ribbon which Queen Nature bestows on so many of her feathered subjects, in recognition, I suppose, of some small and common kind of merit. If I should meet with it in a collection I should know it again; only, in that case it would look plain and homely to me — this little bird that for a time made all others seem unbeautiful.

Even a richer prize may come in sight for a brief period — one of the nobler mammalians, which are fewer in number, and bound to earth like ourselves, and therefore so much better known than the wandering children of the air. In some secluded spot, resting amidst luxuriant herbage or forest undergrowth, a slight rustling makes us start, and, lo! looking at us from the clustering leaves, a strange face; the leaf-like ears erect, the dark eyes round with astonishment, and the sharp black nose twitching and sniffing audibly, to take in the unfamiliar flavour of a human presence from the air, like the pursed-up and smacking lips of a wine-drinker tasting a new vintage. No sooner seen than gone, like a dream, a phantom, the quaint furry face to be thereafter only an image in memory.

Sometimes the prize may be a very rich one, and actually within reach of the hand — challenging the hand, as it were, to grasp it, and yet presently slip away to be seen no more, although it may be sought for day after day, with a hungry longing comparable to that of some poor tramp who finds a gold doubloon in the forest, and just when he is beginning to realize all that it means to him drops it in the grass and cannot find it again. There is not the faintest motion in the foliage, no rustle of any dry leaf, and yet we know that something has moved — something has come or has gone; and, gazing fixedly at one spot, we suddenly see that it is still there, close to us, the pointed ophidian head and long neck, not drawn back and threatening, but sloping forward, dark and polished as the green and purple weed-stems springing from marshy soil, and with an irregular chain of spots extending down the side. Motionless, too, as the stems it is; but presently the tongue, crimson and glistening, darts out and flickers, like a small jet of smoke and flame, and is withdrawn; then the smooth serpent head drops down, and the thing is gone.

One of my earliest experiences of seeing and losing relates to a humming-bird — a veritable "jewel of ornithology." I was only a boy at the time, but already pretty well acquainted with the birds of the district I lived in, near La Plata River, and among them were three species of the humming-bird. One spring day I saw a fourth — a wonderful little thing, only half as big as the smallest of the other three — the well-known *Phaithornis splendens* — and scarcely larger than a bumble-bee. I was within three feet of it as it sucked at the flowers, suspended motionless in the air, the wings appearing formless and mist-like from their rapid vibratory motion, but the rest of the upper plumage was seen distinctly as anything can be seen. The head and

neck and upper part of the back were emerald green, with the metallic glitter usually seen in the burnished scale-like feathers of these
small birds; the lower half of the back was velvet-black; the tail and
tail-coverts white as snow. On two other occasions, at intervals of
a few days, I saw this brilliant little stranger, always very near, and
tried without success to capture it, after which it disappeared from
the plantation. Four years later I saw it once again not far from the
same place. It was late in summer, and I was out walking on the level
plain where the ground was carpeted with short grass, and nothing
else grew there except a solitary stunted cardoon thistle-bush with
one flower on its central stem above the grey-green artichoke-like
leaves. The disc of the great thorny blossom was as broad as that of
a sunflower, purple in colour, delicately frosted with white; on this
flat disc several insects were feeding — flies, fireflies, and small wasps
— and I paused for a few minutes in my walk to watch them. Suddenly
a small misty object flew swiftly downwards past my face, and paused
motionless in the air an inch or two above the rim of the flower.
Once more my lost humming-bird, which I remembered so well! The
exquisitely graceful form, half circled by the misty moth-like wings,
the glittering green and velvet-black mantle, and snow-white tail
spread open like a fan — there it hung like a beautiful bird-shaped
gem suspended by an invisible gossamer thread. One — two — three
moments passed, while I gazed, trembling with rapturous excitement,
and then, before I had time to collect my faculties and make a forlorn attempt to capture it with my hat, away it flew, gliding so swiftly
on the air that form and colour were instantly lost, and in appearance it was only an obscure grey line traced rapidly along the low sky
and fading quickly out of sight. And that was the last I ever saw of it.

The case of this small "winged gem," still wandering nameless in
the wilds, reminds me of yet another bird seen and lost, also remarkable for its diminutive size. For years I looked for it, and when the
wished-for opportunity came, and it was in my power to secure it, I
refrained; and Fate punished me by never permitting me to see it
again. On several occasions while riding on the pampas I had caught
glimpses of this minute bird flitting up mothlike, with uncertain
tremulous flight, and again dipping into the woods, tall grass, or
thistles. Its plumage was yellowish in hue, like sere dead herbage, and
its extremely slender body looked longer and slimmer than it was,
owing to the great length of its tail, or of the two middle tail-feathers.
I knew that it was a *Synallaxis* — a genus of small birds of the Wood-
hewer family. Now, as I have said in a former chapter, these are wise
little birds, more interesting — I had almost said more beautiful —

in their wisdom, or wisdom-simulating instincts, than the quetzal in its resplendent green or the cock-of-the-rock in its vivid scarlet and orange mantle. Wrens and mocking birds have melody for their chief attraction, and the name of each kind is, to our minds, also the name of a certain kind of sweet music; we think of swifts and swallows in connection with the mysterious migration instinct; and humming-birds have a glittering display and the miraculous motions necessary to display its ever-changing iridescent beauty. In like manner, the homely Dendrocolaptidae possess the genius for building, and an account of one of these small birds without its nest would be like a biography of Sir Christopher Wren that made no mention of his works. It was not strange then, that when I saw this small bird the question rose to my mind, What kind of nest does it build?

One morning in the month of October, the great breeding-time for birds in the Southern Hemisphere, while cautiously picking my way through a bed of cardoon bushes, the mysterious little creature flitted up and perched among the clustering leaves quite near to me. It uttered a feeble grasshopper-like chirp; and then a second individual, smaller, paler-coloured, and if possible shyer than the first, showed itself for two or three seconds, after which both birds dived once more into concealment. How glad I was to see them! for there they were, male and female, in a suitable spot in my own fields, where they evidently meant to breed. Every day after that I paid them one cautious visit, and by waiting from five to fifteen minutes, standing motionless among the thistles, I always succeeded in getting them to show themselves for a few moments. I could easily have secured them then, but my wish was to discover their nesting habits; and after watching for some days, I was rewarded by finding their nest; then for three days more I watched it slowly progressing towards completion, and each time I approached it one of the small birds would flit out to vanish into the herbage. The structure was about six inches long, and not more than two inches in diameter, and was placed horizontally on a broad stiff cardoon leaf, sheltered by other leaves above. It was made of the finest dry grass loosely woven, and formed a simple perfectly straight tube, open at both ends. The aperture was so small that I could only insert my little finger, and the bird could not, of course, have turned round in so narrow a passage, and so always went in at one end and left by the other. On visiting the spot on the fourth day I found, to my intense chagrin, that the delicate fabric had been broken and thrown down by some animal; also, that the birds had utterly vanished — for I sought them in vain, both there and in every weedy and thistly spot in the neighbourhood. The bird without the nest had seemed a useless thing to possess; now, for all my pains, I had only a wisp of fine dry grass in my hand, and no bird. The shy, modest little creature, dwelling violet-like amidst clustering leaves, and even when showing itself still "half-hidden from the eye," was thereafter to be only a tantalizing image in memory. Still, my case was not so hopeless as that of the imagined lapidary; for however rare a species may be, and near to its final extinction, there must always be many individuals existing, and I was cheered by the thought that I might yet meet with one at some future time. And, even if this particular species was not to gladden my sight again, there were others, scores and hundreds more, and at any moment I might expect to see one shining, a living gem, on Nature's open extended palm.

Sometimes it has happened that an animal would have been overlooked or passed by with scant notice, to be forgotten, perhaps, but for some singular action or habit which has instantly given it a strange importance, and made its possession desirable.

I was once engaged in the arduous and monotonous task of driving a large number of sheep a distance of two hundred and fifty miles, in excessively hot weather, when sheep prefer standing still to travelling. Five or six gauchos were with me, and we were on the southern pampas of Buenos Ayres, near to a long precipitous stony sierra which rose to a height of five or six hundred feet above the plain. Who that has travelled for eighteen days on a dead level in a broiling sun can resist a hill? That sierra was more sublime to us than Conondagua, than Illimani.

Leaving the sheep, I rode to it with three of the men; and after securing our horses on the lower slope, we began our laborious ascent. Now the gaucho when taken from his horse, on which he lives like a kind of parasite, is a very slow-moving creature, and I soon left my friends far behind. Coming to a place where ferns and flowering herbage grew thick, I began to hear all about me sounds of a character utterly unlike any natural sound I was acquainted with — innumerable low clear voices tinkling or pealing like minute sweet-toned, resonant bells — for the sounds were purely metallic and perfectly bell-like. I was completely ringed round with the mysterious music, and as I walked it rose and sank rhythmically, keeping time to my steps. I stood still, and immediately the sounds ceased. I took a step forwards, and again the fairy-bells were set ringing, as if at each step my foot touched a central meeting point of a thousand radiating threads, each thread attached to a peal of little bells hanging concealed among the herbage. I waited for my companions, and called their attention to the phenomenon, and to them also it was a thing strange and perplexing. "It is the bell-snake!" cried one excitedly. This is the rattle-snake; but although at that time I had no experience of this reptile, I knew that he was wrong. Yet how natural the mistake! The Spanish name of "bell-snake" had made him imagine that the whirring sound of the vibrating rattles, resembling muffled cicada music, is really bell-like in character. Eventually we discovered that the sound was made by grasshoppers; but they were seen only to be lost, for I could not capture one, so excessively shy and cunning had the perpetual ringing of their own little tocsins made them. And presently I had to return to my muttons; and afterwards there was no opportunity of revisiting the spot to observe so singular a habit again and collect specimens. It was a very slender grasshopper, about an inch and a half long, of a uniform, tawny, protective colour — the colour of an old dead leaf. It also possessed a protective habit common to most grasshoppers, of embracing a slender vertical stem with its four fine front legs, and moving cunningly round so as to keep the stem always in front of it to screen itself from sight. Only other grasshoppers are silent when alarmed, and the silence and masking action are related, and together prevent the insect from being detected. But this particular species, or race, or colony, living on the sides of the isolated sierra, had acquired a contrary habit, resembling a habit of gregarious birds and mammals. For this informing sound (unless it mimicked some *warning-sound,* as of a rattlesnake, which it didn't) could not possibly be beneficial to individuals living alone, as grasshoppers generally do, but, on the contrary, only detrimental; and such a habit was therefore purely for the public good, and could only have arisen in a species that always lived in communities.

On another occasion, in the middle of the hot season, I was travelling alone across-country in a locality which was new to me, a few leagues east of La Plata River, in its widest part. About eleven o'clock in the morning I came to a low-lying level plain where the close-cropped grass was vivid green, although elsewhere all over the country the vegetation was scorched and dead, and dry as ashes. The ground being so favourable, I crossed this low plain at a swinging gallop, and in about thirty minutes' time. In that half-hour I saw a vast number of snakes, all of one kind, and a species new to me; but my anxiety to reach my destination before the oppressive heat of the afternoon made me hurry on. So numerous were the snakes in that green place that frequently I had as many as a dozen in sight at one time. It looked to me like a *Coronella* — harmless colubrine snakes — but was more than twice as large as either of the two species of that genus I was already familiar with. In size they varied greatly, ranging from two to fully five feet in length, and the colour was dull yellow or tan, slightly lined and mottled with shades of brown. Among dead or partially withered grass and herbage they would have been undistinguishable at even a very short distance, but on the vivid green turf they were strangely conspicuous, some being plainly visible forty or fifty yards away; and not one was seen coiled up. They were all lying motionless, stretched out full length, and looking like dark yellow or tan-coloured ribbons, thrown on to the grass. It was most unusual to see so many snakes together, although not surprising in the circumstances. The December heats had dried up all the watercourses and killed the vegetation, and made the earth hard and harsh as burnt bricks; and at such times snakes, especially the more active non-venomous kinds, will travel long distances, in their slow way, in search of water. Those I saw during my ride had probably been attracted by the moisture from a large area of country; and although there was no water, the soft fresh grass must have been grateful to them. Snakes are seen coiled up when they are at home; when travelling and far afield, they lie as a rule extended full length, even when resting — and they are generally resting. Pausing at length, before quitting this green plain, to give my horse a minute's rest, I got off and approached a large snake; but when I was quite twelve yards from it, it lifted its head, and, turning deliberately round, came rather swiftly at me. I retreated, and it followed, until, springing on to my horse, I left it, greatly surprised at its action, and beginning to think that it must be venomous. As I rode on the feeling of surprise increased, conquering haste; and in the end, seeing more snakes, I dismounted and approached the largest, when exactly the same thing occurred again, the snake rousing itself and coming angrily at me when I was still (considering the dull lethargic character of the deadliest kinds) at an absurd distance from it. Again and again I repeated the experiment, with the same result. And at length I stunned one with a blow of my whip to examine its mouth, but found no poison-fangs in it.

I then resumed my journey, expecting to meet with more snakes of the same kind at my destination; but there were none, and very soon business called me to a distant place, and I never met with this species afterwards. But when I rode away from that green spot, and was once more on the higher, desolate, wind-swept plain surrounding it — a rustling sea of giant thistles, still erect, although dead, and red as rust, and filling the hot blue sky with silvery down — it was with a very strange feeling. The change from the green and living to the dead and dry and dusty was so great! There seemed to be something mysterious, extra-natural, in that low-level plain, so green and fresh and snaky, where my horse's hoofs had made no sound — a place where no man dwelt, and no cattle pastured, and no wild bird folded its wing. And the serpents there were not like others — the mechanical coiled-up thing we know, a mere bone-and-muscle man-trap, set by the elements, to spring and strike when trodden on: but these had a high intelligence, a lofty spirit, and were filled with a noble rage and astonishment that any other kind of creature, even a man, should venture there to disturb their sacred peace. It was a fancy, born of that sense of mystery which the unknown and the unusual in nature wakes in us — an obsolescent feeling that still links us to the savage. But the simple fact was wonderful enough, and that has been set down simply and apart from all fancies. If the reader happens not to be a naturalist, it is right to tell him that a naturalist cannot exaggerate consciously; and if he be capable of unconscious exaggeration, then he is no naturalist. He should hasten "to join the innumerable caravan that moves" to the fantastic realms of romance. Looking at the simple fact scientifically, it was a case of mimicry — the harmless snake mimicking the fierce deadly kind. Only with this difference: the venomous snake, of all the deadly things in nature, is the slowest to resentment, the most reluctant to enter into a quarrel; whereas in this species angry demonstrations were made when the intruder was yet far off, and before he had shown any hostile intentions.

My last case — the last, that is, of the few I have selected — relates to a singular variation in the human species. On this occasion I was again travelling alone in a strange district on the southern frontier of Buenos Ayres. On a bitterly cold mid-winter day, shortly before noon, I arrived, stiff and tired, at one of those pilgrims' rests on the pampas — a wayside *pulperia,* or public house, where the traveller can procure anything he may require or desire, from a tumbler of Brazilian rum to make glad his heart, to a poncho, or cloak of blue cloth with fluffy scarlet lining, to keep him warm o' nights; and, to speed him on his way, a pair of cast-iron spurs weighing six pounds avoirdupois, with rowels eight inches in diameter, manufactured in this island for the use of barbarous men beyond the sea. The wretched mud-and-grass building was surrounded by a foss crossed by a plank draw-bridge; outside of the enclosure twelve or fourteen saddled horses were standing, and from the loud noise of talk and laughter in the bar I conjectured that a goodly company of rough frontiersmen were already making merry at that early hour. It was necessary for me to go in among them to see the proprietor of the place and ask permission to visit his kitchen in order to make myself a "tin of coffee," that being the refreshment I felt inclined for. When I went in and made my salutation, one man wheeled round square before me, stared straight into my eyes, and in an exceedingly high-pitched reedy or screechy voice and a sing-song tone returned my "good morning," and bade me call for the liquid I loved best at his expense. I declined with thanks, and in accordance with gaucho etiquette added that I was prepared to pay for his liquor. It was then for him to say that he had already been served and so let the matter drop, but he did not do so: he screamed out in his wild animal voice that he would take gin. I paid for his drink, and would, I think, have felt greatly surprised at his strange insolent behaviour, so unlike that of the usually courteous gaucho, but this thing affected me not at all, so profoundly had his voice and singular appearance impressed me; and for the rest of the time I remained in the place I continued to watch him narrowly. Professor Huxley has somewhere said, "A variation frequently occurs, but those who make notice of it take no care about noting down the particulars." That is not a failing of mine, and this is what I noted down while the man's appearance was still fresh in memory. He was about five feet eleven inches in height — very tall for a gaucho — straight and athletic, with exceedingly broad shoulders, which made his round head look small; long arms and huge hands. The round flat face, coarse black hair, swarthy reddish colour, and smooth hairless cheeks seemed to show that he had more Indian than Spanish blood in him, while his round black eyes were even more like those of a rapacious animal in expression than in the pure-blooded Indian. He also had the Indian or half-breed's moustache, when that natural ornament is permitted to grow, and which is composed of thick bristles standing out like a cat's whiskers. The mouth was the marvellous feature, for it was twice the size of an average mouth, and the two lips were alike in thickness. This mouth did not smile, but snarled, both when he spoke and when he should have smiled; and when he snarled the whole of his teeth and a part of the gums were displayed. The teeth were not as in other human beings — incisors, canines, and molars: they were all exactly alike, above and below, each tooth a gleaming white triangle, broad at the gum where it touched its companion teeth, and with a point sharp as the sharpest-pointed dagger. They were like the teeth of a shark or crocodile. I noticed that when he showed them, which was very often, they were not set together as in dogs, weasels, and other savage snarling animals, but apart, showing the whole terrible serration in the huge red mouth.

After getting his gin he joined in the boisterous conversation with the others, and this gave me an opportunity of studying his face for several minutes, all the time with a curious feeling that I had put myself into a cage with a savage animal of horrible aspect, whose instincts were utterly unknown to me, and were probably not very pleasant. It was interesting to note that whenever one of the others addressed him directly, or turned to him when speaking, it was with a curious expression, not of fear, but partly amusement and partly something else which I could not fathom. Now, one might think that this was natural enough purely on account of the man's extraordinary appearance. I do not think that a sufficient explanation; for however strange a man's appearance may be, his intimate friends and associates soon lose all sense of wonder at his strangeness, and even forget that he is unlike others. My belief is that this curiosity, or whatever it was they showed in their faces, was due to something in his character — a mental strangeness, showing itself at unexpected times, and which might flash out at any moment to amuse or astonish them. There was certainly a correspondence between the snarling action of the mouth and the dangerous form of the teeth, perfect as that in any snarling animal; and such animals, it should be remembered, snarl not only when angry and threatening, but in their playful moods as well. Other and more important correspondences or correlations might have existed; and the voice certainly was unlike any human voice I have ever heard, whether in white, red, or black man. But the time I had for observation was short, the conversation revealed nothing further, and by-and-by I went away in search of the odorous kitchen, where there would be hot water for coffee, or at all events cold water and a kettle, and materials for making a fire — to wit, bones of dead cattle, "buffalo chips," and rancid fat.

I have never been worried with the wish or ambition to be head-hunter in the Dyak sense, but on this one occasion I did wish that it had been possible, without violating any law, or doing anything to a fellow-creature which I should not like done to myself, to have obtained possession of this man's head, with its set of unique and terrible teeth. For how, in the name of Evolution, did he come by them, and by other physical peculiarities — the snarling habit and that high-pitched animal voice, for instance — which made him a being different from others — one separate and far apart? Was he, so admirably formed, so complete and well-balanced, merely a freak of nature, to use an old-fashioned phrase — a sport, or spontaneou_ individual variation — an experiment for a new human type, imagined by Nature in some past period, inconceivably long ago, but which she had only now, too late, found time to carry out? Or rather was he like that little hairy maiden exhibited not long ago in London, a reproduction of the past, the mystery called reversion — a something in the life of a species like memory in the life of an individual, the memory which suddenly brings back to the old man's mind the image of his childhood? For no dream-monster in human form ever appeared to me with so strange and terrible a face; and this was no dream but sober fact, for I saw and spoke with this man; and unless cold steel has given him his quietus, or his own horse has crushed him, or a mad bull gored him — all natural forms of death in that wild land — he is probably living and in the prime of life, and perhaps at this very moment drinking gin at some astonished traveller's expense at that very bar where I met him. The old Palaeolithic man, judging from the few remains we have of him, must have had an unspeakably savage and, to our way of thinking, repulsive and horrible aspect, with his villainous low receding forehead, broad nose, great projecting upper jaw, and retreating chin; to meet such a man face to face in Piccadilly would frighten a nervous person of the present time. But his teeth were not unlike our own, only very much larger and more powerful, and well adapted to their work of masticating the flesh, underdone and possibly raw, of mammoth and rhinoceros. If, then, this living man recalls a type of the past, it is of a remoter past, a more primitive man, the volume of whose history is missing from the geological records. To speculate on such a subject seems idle and useless; and when I coveted possession of that head it was not because I thought that it might lead to any fresh discovery. A lower motive inspired the feeling. I wished for it only that I might bring it over the sea, to drop it like a new apple of discord, suited to the spirit of the times, among the anthropologists and evolutionists generally of this old and learned world. Inscribed, of course, "To the most learned," but giving no locality and no particulars. I wished to do that for the pleasure — not a very noble kind of pleasure, I allow — of witnessing from some safe hiding-place the stupendous strife that would have ensued — a battle more furious, lasting and fatal to many a brave knight of biology, than was ever yet fought over any bone or bony fragment or fabric ever picked up, including the celebrated cranium of the Neanderthal.

JOHN MUIR (1838–1914)

The Water Ouzel

From *The Mountains of California.* 1894

[FROM HIS OWN PEN we know of Johnnie Muir as a very human little
Scotch boy, as full of humour and devilment as a healthy boy should
be. This phase continued in his new home in Wisconsin and began
to be sublimated with a keen, sober, but uncreative interest, almost
passion, for wild things and wild scenes. He became a hesperian
Thoreau, with less philosophy but infinitely more physical guts
than the Cambridge dreamer, and he could guide his thoughts more
consistently into an essay on a given subject.

Yet with the years Muir's writings have seemed to me to have
grown thinner, while Thoreau's often inconsequential thoughts
have held to their initial force. As with some other very modern
authors there are slightly too many words in Muir's paragraphs.]

THE WATERFALLS of the Sierra are frequented by only one bird, —
the Ouzel or Water Thrush (*Cinclus Mexicanus,* Sw.). He is a
singularly joyous and lovable little fellow, about the size of a
robin, clad in a plain waterproof suit of bluish gray, with a tinge of
chocolate on the head and shoulders. In form he is about as smoothly
plump and compact as a pebble that has been whirled in a pot-hole,
the flowing contour of his body being interrupted only by his strong
feet and bill, the crisp wing-tips, and the upslanted wren-like tail.

Among all the countless waterfalls I have met in the course of
ten years' exploration in the Sierra, whether among the icy peaks,
or warm foot-hills, or in the profound yosemitic cañons of the middle region, not one was found without its Ouzel. No cañon is too
cold for this little bird, none too lonely, provided it be rich in falling
water. Find a fall, or cascade, or rushing rapid, anywhere upon a
clear stream, and there you will surely find its complementary Ouzel,
flitting about in the spray, diving in foaming eddies, whirling like
a leaf among beaten foam-balls; ever vigorous and enthusiastic, yet
self-contained, and neither seeking nor shunning your company.

If disturbed while dipping about in the margin shallows, he either

187

sets off with a rapid whir to some other feeding-ground up or down the stream, or alights on some half-submerged rock or snag out in the current, and immediately begins to nod and courtesy like a wren, turning his head from side to side with many other odd dainty movements that never fail to fix the attention of the observer.

He is the mountain streams' own darling, the humming-bird of blooming waters, loving rocky ripple-slopes and sheets of foam as a bee loves flowers, as a lark loves sunshine and meadows. Among all the mountain birds, none has cheered me so much in my lonely wanderings, — none so unfailingly. For both in winter and summer he sings, sweetly, cheerily, independent alike of sunshine and of love, requiring no other inspiration than the stream on which he dwells. While water sings, so must he, in heat or cold, calm or storm, ever attuning his voice in sure accord; low in the drought of summer and the drought of winter, but never silent.

During the golden days of Indian summer, after most of the snow has been melted, and the mountain streams have become feeble, — a succession of silent pools, linked together by shallow, transparent currents and strips of silvery lacework — then the song of the Ouzel is at its lowest ebb. But as soon as the winter clouds have bloomed, and the mountain treasuries are once more replenished with snow, the voices of the streams and ouzels increase in strength and richness until the flood season of early summer. Then the torrents chant their noblest anthems, and then is the flood-time of our songster's melody. As for weather, dark days and sun days are the same to him. The voices of most song-birds, however joyous, suffer a long winter eclipse; but the Ouzel sings on through all the seasons and every kind of storm. Indeed no storm can be more violent than those of the waterfalls in the midst of which he delights to dwell. However dark and boisterous the weather, snowing, blowing, or cloudy, all the same he sings, and with never a note of sadness. No need of spring sunshine to thaw *his* song, for it never freezes. Never shall you hear anything wintry from *his* warm breast; no pinched cheeping, no wavering notes between sorrow and joy; his mellow, fluty voice is ever tuned to downright gladness, as free from dejection as cock-crowing.

It is pitiful to see wee frost-pinched sparrows on cold mornings in the mountain groves shaking the snow from their feathers, and hopping about as if anxious to be cheery, then hastening back to their hidings out of the wind, puffing out their breast-feathers over their toes, and subsiding among the leaves, cold and breakfastless, while the snow continues to fall, and there is no sign of clearing. But the Ouzel never calls forth a single touch of pity; not because he is strong to endure, but rather because he seems to live a charmed life beyond the reach of every influence that makes endurance necessary.

One wild winter morning, when Yosemite Valley was swept its length from west to east by a cordial snow-storm, I sallied forth to see what I might learn and enjoy. A sort of gray, gloaming-like darkness filled the valley, the huge walls were out of sight, all ordinary sounds were smothered, and even the loudest booming of the falls was at times buried beneath the roar of the heavy-laden blast. The loose snow was already over five feet deep on the meadows, making extended walks impossible without the aid of snow-shoes. I found no great difficulty, however, in making my way to a certain ripple on the river where one of my ouzels lived. He was at home, busily gleaning his breakfast among the pebbles of a shallow portion of the margin, apparently unaware of anything extraordinary in the weather. Presently he flew out to a stone against which the icy current was beating, and turning his back to the wind, sang as delightfully as a lark in the springtime.

After spending an hour or two with my favorite, I made my way across the valley, boring and wallowing through the drifts, to learn as definitely as possible how the other birds were spending their time. The Yosemite birds are easily found during the winter because all of them excepting the Ouzel are restricted to the sunny north side of the valley, the south side being constantly eclipsed by the great frosty shadow of the wall. And because the Indian Cañon groves, from their peculiar exposure, are the warmest, the birds congregate there, more especially in severe weather.

I found most of the robins cowering on the lee side of the larger branches where the snow could not fall upon them, while two or three of the more enterprising were making desperate efforts to reach the mistletoe berries by clinging nervously to the under side of the snow-crowned masses, back downward, like woodpeckers. Every now and then they would dislodge some of the loose fringes of the snow-crown, which would come sifting down on them and send them screaming back to camp, where they would subside among their companions with a shiver, muttering in low, querulous chatter like hungry children.

Some of the sparrows were busy at the feet of the larger trees gleaning seeds and benumbed insects, joined now and then by a robin weary of his unsuccessful attempts upon the snow-covered berries. The brave woodpeckers were clinging to the snowless sides of the larger boles and overarching branches of the camp trees, making short flights from side to side of the grove, pecking now and then at the acorns they had stored in the bark, and chattering aimlessly as if unable to keep still, yet evidently putting in the time in a very dull way, like storm-bound travelers at a country tavern. The hardy nut-hatches were threading the open furrows of the trunks in their usual industrious manner, and uttering their quaint notes, evidently less distressed than their neighbors. The Steller jays were of course making more noisy stir than all the other birds combined; ever coming and going with loud bluster, screaming as if each had a lump of melting sludge in his throat, and taking good care to improve the favorable opportunity afforded by the storm to steal from the acorn stores of the woodpeckers. I also noticed one solitary gray eagle braving the storm on the top of a tall pine-stump just outside the main grove. He was standing bolt upright with his back to the wind, a tuft of snow piled on his square shoulders, a monument of passive endurance. Thus every snow-bound bird seemed more or less uncomfortable if not in positive distress. The storm was reflected in every gesture, and not one cheerful note, not to say song, came from a single bill; their cowering, joyless endurance offering a striking contrast to the spontaneous, irrepressible gladness of the Ouzel, who could no more help exhaling sweet song than a rose sweet fragrance. He *must* sing though the heavens fall. I remember noticing the distress of a pair of robins during the violent earthquake of the year 1872, when the pines of the Valley, with strange movements, flapped and waved their branches, and beetling rock-brows came thundering down to the meadows in tremendous avalanches. It did not occur to me in the midst of the excitement of other observations to look for the ouzels, but I doubt not they were singing straight on through it all, regarding the terrible rock-thunder as fearlessly as they do the booming of the waterfalls.

What may be regarded as the separate songs of the Ouzel are exceedingly difficult of description, because they are so variable and at the same time so confluent. Though I have been acquainted with my favorite ten years, and during most of this time have heard him sing nearly every day, I still detect notes and strains that seem new to me. Nearly all of his music is sweet and tender, lapsing from his round breast like water over the smooth lip of a pool, then breaking farther on into a sparkling foam of melodious notes, which glow with subdued enthusiasm, yet without expressing much of the strong, gushing ecstasy of the bobolink or skylark.

The more striking strains are perfect arabesques of melody, composed of a few full, round, mellow notes, embroidered with delicate trills which fade and melt in long slender cadences. In a general way his music is that of the streams refined and spiritualized. The deep booming notes of the falls are in it, the trills of rapids, the gurgling of margin eddies, the low whispering of level reaches, and the sweet tinkle of separate drops oozing from the ends of mosses and falling into tranquil pools.

The Ouzel never sings in chorus with other birds, nor with his kind, but only with the streams. And like flowers that bloom beneath the surface of the ground, some of our favorite's best song-blossoms never rise above the surface of the heavier music of the water. I have often observed him singing in the midst of beaten spray, his music completely buried beneath the water's roar; yet I knew he was surely singing by his gestures and the movements of his bill.

His food, as far as I have noticed, consists of all kinds of water insects, which in summer are chiefly procured along shallow margins. Here he wades about ducking his head under water and deftly turning over pebbles and fallen leaves with his bill, seldom choosing to go into deep water where he has to use his wings in diving.

He seems to be especially fond of the larvae of mosquitos, found in abundance attached to the bottom of smooth rock channels where the current is shallow. When feeding in such places he wades upstream, and often while his head is under water the swift current is deflected upward along the glossy curves of his neck and shoulders, in the form of a clear, crystalline shell, which fairly incloses him like a bell-glass, the shell being broken and re-formed as he lifts and dips his head; while ever and anon he sidles out to where the too powerful current carries him off his feet; then he dexterously rises on the wing and goes gleaning again in shallower places.

But during the winter, when the stream-banks are embossed in snow, and the streams themselves are chilled nearly to the freezing-point, so that the snow falling into them in stormy weather is not wholly dissolved, but forms a thin, blue sludge, thus rendering the current opaque — then he seeks the deeper portions of the main rivers, where he may dive to clear water beneath the sludge. Or he repairs to some open lake or millpond, at the bottom of which he feeds in safety. . . .

The Ouzel seldom swims more than a few yards on the surface, for, not being web-footed, he makes rather slow progress, but by means of his strong, crisp wings he swims, or rather flies, with celerity under the surface, often to considerable distances. But it is in withstanding the force of heavy rapids that his strength of wing in this respect is most strikingly manifested. The following may be regarded as a fair illustration of his power of sub-aquatic flight. One stormy morning in winter when the Merced River was blue and green with unmelted snow, I observed one of my ouzels perched on a snag out in the midst of a swift-rushing rapid, singing cheerily, as if everything was just to his mind; and while I stood on the bank admiring him, he suddenly plunged into the sludgy current, leaving his song abruptly broken off. After feeding a minute or two at the bottom, and when one would suppose that he must inevitably be swept far down-stream, he emerged just where he went down, alighted on the same snag, showered the water-beads from his feathers, and continued his unfinished song, seemingly in tranquil ease as if it had suffered no interruption.

The Ouzel alone of all birds dares to enter a white torrent. And though strictly terrestrial in structure, no other is so inseparably related to water, not even the duck or the bold ocean albatross, or the stormy-petrel. For ducks go ashore as soon as they finish feeding in undisturbed places, and very often make long flights overland from lake to lake or field to field. The same is true of most other aquatic birds. But the Ouzel, born on the brink of a stream, or on a snag or boulder in the midst of it, seldom leaves it for a single moment. For, notwithstanding he is often on the wing, he never flies overland, but whirs with rapid, quail-like beat above the stream, tracing all its windings. Even when the stream is quite small, say from five to ten feet wide, he seldom shortens his flight by crossing a bend, however abrupt it may be; and even when disturbed by meeting some one on the bank, he prefers to fly over one's head, to dodging out over the ground. When, therefore, his flight along a crooked stream is viewed endwise, it appears most strikingly wavered — a description on the air of every curve with lightning-like rapidity.

The vertical curves and angles of the most precipitous torrents he traces with the same rigid fidelity, swooping down the inclines of cascades, dropping sheer over dizzy falls amid the spray, and ascending with the same fearlessness and ease, seldom seeking to lessen the steepness of the aclivity by beginning to ascend before reaching the base of the fall. No matter though it may be several hundred feet in height he holds straight on, as if about to dash headlong into the throng of booming rockets, then darts abruptly upward, and, after alighting at the top of the precipice to rest a moment, proceeds to feed and sing. His flight is solid and impetuous, without any intermission of wing-beats, — one homogeneous buzz like that of a laden bee on its way home. And while thus buzzing freely from fall to fall, he is frequently heard giving utterance to a long outdrawn train of unmodulated notes, in no way connected with his song, but corresponding closely with his flight in sustained vigor.

Were the flights of all the ouzels in the Sierra traced on a chart, they would indicate the direction of the flow of the entire system of ancient glaciers, from about the period of the breaking up of the ice-sheet until near the close of the glacial winter; because the streams which the ouzels so rigidly follow are, with the unimportant exceptions of a few side tributaries, all flowing in channels eroded for them out of the solid flank of the range by the vanished glaciers, — the streams tracing the ancient glaciers, the ouzels tracing the streams. Nor do we find so complete compliance to glacial conditions in the life of any other mountain bird, or animal of any kind. Bears frequently accept the pathways laid down by glaciers as the easiest to travel; but they often leave them and cross over from cañon to cañon. So also, most of the birds trace the moraines to some extent, because the forests are growing on them. But they wander far, crossing the cañons from grove to grove, and draw exceedingly angular and complicated courses.

The Ouzel's nest is one of the most extraordinary pieces of bird architecture I ever saw, odd and novel in design, perfectly fresh and beautiful, and in every way worthy of the genius of the little builder. It is about a foot in diameter, round and bossy in outline, with a neatly arched opening near the bottom, somewhat like an old-fashioned brick oven, or Hottentot's hut. It is built almost exclusively of green and yellow mosses, chiefly the beautiful fronded hypnum that covers the. rocks and old drift-logs in the vicinity of waterfalls. These are deftly interwoven, and felted together into a charming little hut; and so situated that many of the outer mosses continue to flourish as if they had not been plucked. A few fine, silky-stemmed grasses are occasionally found interwoven with the mosses, but, with the exception of a thin layer lining the floor, their presence seems accidental, as they are of a species found growing with the mosses and are probably plucked with them. The site chosen for this curious mansion is usually some little rock-shelf within reach of the lighter particles of the spray of a waterfall, so that its walls are kept green and growing, at least during the time of high water.

No harsh lines are presented by any portion of the nest as seen in place, but when removed from its shelf, the back and bottom, and sometimes a portion of the top, is found quite sharply angular, because it is made to conform to the surface of the rock upon which and against which it is built, the little architect always taking advantage of slight crevices and protuberances that may chance to offer, to render his structure stable by means of a kind of gripping and dovetailing.

In choosing a building-spot, concealment does not seem to be taken into consideration; yet notwithstanding the nest is large and guilelessly exposed to view, it is far from being easily detected, chiefly because it swells forward like any other bulging moss-cushions growing naturally in such situations. This is more especially the case where the nest is kept fresh by being well sprinkled. Sometimes these romantic huts have their beauty enhanced by rock-ferns and grasses that spring up around the mossy walls, or in front of the door-sill, dripping with crystal beads.

Furthermore, at certain hours of the day, when the sunshine is poured down at the required angle, the whole mass of the spray enveloping the fairy establishment is brilliantly irised; and it is through so glorious a rainbow atmosphere as this that some of our blessed ouzels obtain their first peep at the world.

Ouzels seem so completely part and parcel of the streams they inhabit, they scarce suggest any other origin than the streams themselves; and one might almost be pardoned in fancying they come direct from the living waters, like flowers from the ground. At least, from whatever cause, it never occurred to me to look for their nests until more than a year after I had made the acquaintance of the birds themselves, although I found one the very day on which I began the search. In making my way from Yosemite to the glaciers at the heads of the Merced and Tuolumne rivers, I camped in a particularly wild and romantic portion of the Nevada canon where in previous excursions I had never failed to enjoy the company of my favorites, who were attracted here, no doubt, by the safe nesting-places in the shelving rocks, and by the abundance of food and falling water. The river, for miles above and below, consists of a succession of small falls from ten to sixty feet in height, connected by flat, plume-like cascades that go flashing from fall to fall, free and almost channelless, over waving folds of glacier-polished granite.

On the south side of one of the falls, that portion of the precipice which is bathed by the spray presents a series of little shelves and tablets caused by the development of planes of cleavage in the granite, and by the consequent fall of masses through the action of the water. "Now here," said I, "of all places, is the most charming spot for an Ouzel's nest." Then carefully scanning the fretted face of the precipice through the spray, I at length noticed a yellowish moss-cushion, growing on the edge of a level tablet within five or six feet of the outer folds of the fall. But apart from the fact of its being situated where one acquainted with the lives of ouzels would fancy an Ouzel's nest ought to be, there was nothing in its appearance visible at first sight, to distinguish it from other bosses of rock-moss similarly situated with reference to perennial spray; and it was not until I had scrutinized it again and again, and had removed my shoes and stockings and crept along the face of the rock within eight or ten feet of it, that I could decide certainly whether it was a nest or a natural growth.

In these moss huts three or four eggs are laid, white like foam-bubbles; and well may the little birds hatched from them sing water songs, for they hear them all their lives, and even before they are born.

I have often observed the young just out of the nest making their odd gestures, and seeming in every way as much at home as their experienced parents, like young bees on their first excursion to the flower fields. No amount of familiarity with people and their ways seems to change them in the least. To all appearance their behavior is just the same on seeing a man for the first time, as when they have seen him frequently.

On the lower reaches of the rivers where mills are built, they sing on through the din of the machinery, and all the noisy confusion of dogs, cattle, and workmen. On one occasion, while a woodchopper was at work on the river-bank, I observed one cheerily singing within reach of the flying chips. Nor does any kind of unwonted disturbance put him in bad humor, or frighten him out of calm self-possession. In passing through a narrow gorge, I once drove one ahead of me from rapid to rapid, disturbing him four times in quick succession where he could not very well fly past me on account of the narrowness of the channel. Most birds under similar circumstances fancy themselves pursued, and become suspiciously uneasy; but, instead of growing nervous about it, he made his usual dippings, and sang one of his most tranquil strains. When observed within a few yards their eyes are seen to express remarkable gentleness and intelligence; but they seldom allow so near a view unless one wears clothing of about the same color as the rocks and trees, and knows how to sit still. On one occasion, while rambling along the shore of a mountain lake, where the birds, at least those born that season, had never seen a man, I sat down to rest on a large stone close to the water's edge, upon which it seemed the ouzels and sandpipers were in the habit of alighting when they came to feed on that part of the shore, and some of the other birds also, when they came down to wash or drink. In a few minutes, along came a whirring Ouzel and alighted on the stone beside me, within reach of my hand. Then suddenly observing me, he stopped nervously as if about to fly on the instant, but as I remained as motionless as the stone, he gained confidence, and looked me steadily in the face for about a minute, then flew quietly to the outlet and began to sing. Next came a sandpiper and gazed at me with much the same guileless expression of eye as the Ouzel. Lastly, down with a swoop came a Stellar's jay out of a fir-tree, probably with the intention of moistening his noisy throat. But instead of sitting confidingly as my other visitors had done, he rushed off at once, nearly tumbling heels over head into the lake in his suspicious confusion, and with loud screams roused the neighborhood. . . .

Such, then, is our little cinclus, beloved of every one who is so fortunate as to know him. Tracing on strong wing every curve of the most precipitous torrents from one extremity of the Sierra to the other; not fearing to follow them through their darkest gorges and coldest snow-tunnels; acquainted with every waterfall, echoing their divine music; and throughout the whole of their beautiful lives interpreting all that we in our unbelief call terrible in the utterances of torrents and storms, as only varied expressions of God's eternal love.

MAURICE MAETERLINCK (1862–   )

The Swarm

From *The Life of the Bee.* 1901

[HERE IS A Belgian-French dramatist and poet, who when almost forty years of age suddenly produced the *Life of the Bee*. To be sure, there is a modicum of fancy, and unentomological philosophy is not wholly lacking in this narrative, but the proportion of good natural history is so preponderant that the exquisite style is sufficiently leavened for inclusion in our volume.

For play-by-play history of the world of bees read Edwin Way Teale's *Golden Throng*, an account doubly reinforced by a series of most excellent photographs, and unrolling in excellent English. But for sheer drama and superb simile turn back, now and again, to Maeterlinck.]

WE WILL NOW, so as to draw more closely to nature, consider the different episodes of the swarm as they come to pass in an ordinary hive. . . .

Here, then, they have shaken off the torpor of winter. The queen started laying again in the very first days of February, and the workers have flocked to the willows and nut-trees, gorse and violets, anemones and lungworts. Then spring invades the earth, and cellar and stream with honey and pollen, while each day beholds the birth of thousands of bees. The overgrown males now all sally forth from their cells, and disport themselves on the combs; and so crowded does the too prosperous city become that hundreds of belated workers, coming back from the flowers towards evening, will vainly seek shelter within, and will be forced to spend the night on the threshold, where they will be decimated by the cold.

Restlessness seizes the people, and the old queen begins to stir. She feels that a new destiny is being prepared. She has religiously fulfilled her duty as a good creatress; and from this duty done there result only tribulation and sorrow. An invincible power menaces her tranquillity; she will soon be forced to quit this city of hers, where she has reigned. But this city is her work, it is she, herself. She is not its queen in the sense in which men use the word. She issues

197

no orders; she obeys, as meekly as the humblest of her subjects, the masked power, sovereignly wise, that for the present, and till we attempt to locate it, we will term the "spirit of the hive." But she is the unique organ of love; she is the mother of the city. She founded it amid uncertainty and poverty. She has peopled it with her own substance; and all who move within its walls — workers, males, larvae, nymphs, and the young princesses whose approaching birth will hasten her own departure, one of them being already designed as her successor by the "spirit of the hive" — all these have issued from her flanks.

What is this "spirit of the hive" — where does it reside? It is not like the special instinct that teaches the bird to construct its well planned nest, and then seek other skies when the day for migration returns. Nor is it a kind of mechanical habit of the race, or blind craving for life, that will fling the bees upon any wild hazard the moment an unforeseen event shall derange the accustomed order of phenomena. On the contrary, be the event never so masterful, the "spirit of the hive" still will follow it, step by step, like an alert and quickwitted slave, who is able to derive advantage even from his master's most dangerous orders.

It disposes pitilessly of the wealth and the happiness, the liberty and life, of all this winged people; and yet with discretion, as though governed itself by some great duty. It regulates day by day the number of births, and contrives that these shall strictly accord with the number of flowers that brighten the country-side. It decrees the queen's deposition or warns her that she must depart; it compels her to bring her own rivals into the world, and rears them royally, protecting them from their mother's political hatred. So, too, in accordance with the generosity of the flowers, the age of the spring, and the probable dangers of the nuptial flight, will it permit or forbid the first-born of the virgin princesses to slay in their cradles her younger sisters, who are singing the song of the queens. At other times, when the season wanes, and flowery hours grow shorter, it will command the workers themselves to slaughter the whole imperial brood, that the era of revolutions may close, and work become the sole object of all. The "spirit of the hive" is prudent and thrifty, but by no means parsimonious. And thus, aware, it would seem, that nature's laws are somewhat wild and extravagant in all that pertains to love, it tolerates, during summer days of abundance, the embarrassing presence in the hive of three or four hundred males, from whose ranks the queen about to be born shall select her lover; three or four hundred foolish, clumsy, useless, noisy creatures, who are pretentious, gluttonous, dirty, coarse, totally and scandalously idle, insatiable, and enormous.

But after the queen's impregnation, when flowers begin to close sooner, and open later, the spirit one morning will coldly decree the simultaneous and general massacre of every male. It regulates the workers' labours, with due regard to their age; it allots their task to the nurses who tend the nymphs and the larvae, the ladies of honour who wait on the queen and never allow her out of their sight; the house-bees who air, refresh, or heat the hive by fanning their wings, and hasten the evaporation of the honey that may be too highly charged with water; the architects, masons, wax-workers, and sculptors who form the chain and construct the combs; the foragers who sally forth to the flowers in search of the nectar that turns into honey, of the pollen that feeds the nymphs and the larvae, the propolis that welds and strengthens the buildings of the city, or the water and salt required by the youth of the nation. Its orders have gone to the chemists who ensure the preservation of the honey by letting a drop of formic acid fall in from the end of their sting; to the capsule-makers who seal down the cells when the treasure is ripe, to the sweepers who maintain public places and streets most irreproachably clean, to the bearers whose duty it is to remove the corpses; and to the amazons of the guard who keep watch on the threshold by night and by day, question comers and goers, recognise the novices who return from their very first flight, scare away vagabonds, marauders and loiterers, expel all intruders, attack redoubtable foes in a body, and, if need be, barricade the entrance.

Finally, it is the "spirit of the hive" that fixes the hour of the great annual sacrifice to the genius of the race: the hour, that is, of the swarm; when we find a whole people, who have attained the topmost pinnacle of prosperity and power, suddenly abandoning to the generation to come their wealth and their palaces, their homes and the fruits of their labour; themselves content to encounter the hardships and perils of a new and distant country. This act, be it conscious or not, undoubtedly passes the limits of human morality. Its result will sometimes be ruin, but poverty always; and the thrice-happy city is scattered abroad in obedience to a law superior to its own happiness. Where has this law been decreed, which, as we soon shall find, is by no means as blind and inevitable as one might believe? Where, in what assembly, what council, what intellectual and moral sphere, does this spirit reside to whom all must submit, itself being vassal to an heroic duty, to an intelligence whose eyes are persistently fixed on the future?

It comes to pass with the bees as with most of the things of the world; we remark some few of their habits; we say they do this, they work in such and such fashion, their queens are born thus, their workers are virgin, they swarm at a certain time. And then we imagine we know them, and ask nothing more. We watch them hasten from flower to flower, we see the constant agitation within the hive; their life seems very simple to us, and bounded, like every life, by the instinctive cares of reproduction and nourishment. But let the eye draw near, and endeavour to see; and at once the least phenomenon of all becomes overpoweringly complex; we are confronted by the enigma of intellect, of destiny, will, aim, means, causes; the incomprehensible organisation of the most insignificant act of life.

Our hive, then, is preparing to swarm; making ready for the great immolation to the exacting gods of the race. In obedience to the order of the spirit — an order that to us may well seem incomprehensible, for it is entirely opposed to all our own instincts and feelings — 60,000 or 70,000 bees out of the 80,000 or 90,000 that form the whole population, will abandon the maternal city at the prescribed hour. They will not leave at a moment of despair; or desert, with sudden and wild resolve, a home laid waste by famine, disease, or war. No, the exile has long been planned, and the favourable hour patiently awaited. Were the hive poor, had it suffered from pillage or storm, had misfortune befallen the royal family, the bees would not forsake it. They leave it only when it has attained the apogee of its prosperity; at a time when, after the arduous labours of the spring, the immense palace of wax has its 120,000 well-arranged cells overflowing with new honey, and with the many-coloured flour, known as "bees' bread," on which nymphs and larvae are fed.

Never is the hive more beautiful than on the eve of its heroic renouncement, in its unrivalled hour of fullest abundance and joy; serene for all its apparent excitement and feverishness.

Let us endeavour to picture it to ourselves, not as it appears to the bees, — for we cannot tell in what magical, formidable fashion things may be reflected in the 6,000 or 7,000 facets of their lateral eyes and the triple cyclopean eye on their brow, — but as it would seem to us, were we of their stature. From the height of a dome more colossal than that of St. Peter's at Rome waxen walls descend to the ground, balanced in the void and the darkness; gigantic and manifold, vertical and parallel geometric constructions, to which, for relative precision, audacity, and vastness, no human structure is comparable. Each of these walls, whose substance still is immaculate and fragrant, of virginal, silvery freshness, contains thousands of cells, that are stored with provisions sufficient to feed the whole people for several weeks. Here, lodged in transparent cells, are the pollens, love-ferment of every flower of spring, making brilliant splashes of red and yellow, of black and mauve. Close by, in twenty thousand reservoirs, sealed with a seal that shall only be broken on days of supreme distress, the honey of April is stored, most limpid and perfumed of all, wrapped round with long and magnificent embroidery of gold, whose borders hang stiff and rigid. Still lower the honey of May matures, in great open vats, by whose side watchful cohorts maintain an incessant current of air. In the centre, and far from the light whose diamond rays steal in through the only opening, in the warmest part of the hive, there stands the abode of the future; here does it sleep, and wake. For this is the royal domain of the brood-cells, set apart for the queen and her acolytes; about 10,000 cells wherein the eggs repose, 15,000 or 16,000 chambers tenanted by larvae, 40,000 dwellings inhabited by white nymphs to whom thousands of nurses minister. And finally, in the holy of holies of these parts, are the three, four, six, or twelve sealed palaces, vast in size compared with the others, where the adolescent princesses lie who await their hour, wrapped in a kind of shroud, all of them motionless and pale, and fed in the darkness.

On the day, then, that the "spirit of the hive" has ordained, a certain part of the population will go forth, selected in accordance with sure and immovable laws, and make way for hopes that as yet are formless. In the sleeping city there remain the males, from whose ranks the royal lover shall come, the very young bees that tend the brood-cells, and some thousands of workers who continue to forage abroad, to guard the accumulated treasure, and preserve the moral traditions of the hive. For each hive has its own code of morals. There are some that are very virtuous and some that are very perverse; and a careless bee-keeper will often corrupt his people, destroy their respect for the property of others, incite them to pillage, and induce in them habits of conquest and idleness which will render them sources of danger to all the little republics around. These things result from the bee's discovery that work among distant flowers, whereof many hundreds must be visited to form one drop of honey, is not the only or promptest method of acquiring wealth, but that it is easier to enter ill-guarded cities by stratagem, or force her way into others too weak for self-defence. Nor is it easy to restore to the paths of duty a hive that has become thus depraved.

All things go to prove that it is not the queen, but the "spirit of the hive," that decides on the swarm. With this queen of ours it happens as with many a chief among men, who though he appear to give orders, is himself obliged to obey commands far more mysterious, far more inexplicable, than those he issues to his subordinates. The hour once fixed, the spirit will probably let it be known at break of dawn, or the previous night, if indeed not two nights before; for scarcely has the sun drunk in the first drops of dew when a most unaccustomed stir, whose meaning the bee-keeper rarely will fail to grasp, is to be noticed within and around the buzzing city. At times one would almost appear to detect a sign of dispute, hesitation, recoil. It will happen even that for day after day a strange emotion apparently without cause, will appear and vanish in this transparent, golden throng. Has a cloud that we cannot see crept across the sky that the bees are watching; or is their intellect battling with a new regret? Does a winged council debate the necessity of the departure? Of this we know nothing; as we know nothing of the manner in which the spirit conveys its resolution to the crowd. Certain as it may seem that the bees communicate with each other, we know not whether this be done in human fashion. It is possible even that their own refrain may be inaudible to them: the murmur that comes to us heavily laden with perfume of honey, the ecstatic whisper of fairest summer days that the bee-keeper loves so well, the festival song of labour that rises and falls around the hive in the crystal of the hour, and might almost be the chant of the eager flowers, hymn of their gladness and echo of their soft fragrance, the voice of the white carnations, the marjoram, and the thyme. They have, however, a whole gamut of sounds that we can distinguish, ranging from profound delight to menace, distress, and anger; they have the ode of the queen, the song of abundance, the psalms of grief, and, lastly, the long and mysterious war-cries the adolescent princesses send forth during the combats and massacres that precede the nuptial flight. May this be a fortuitous music that fails to attain their inward silence? In any event they seem not the least disturbed at the noises we make near the hive; but they regard these perhaps as not of their world, and possessed of no interest for them. It is possible that we on our side hear only a fractional part of the sounds that the bees produce, and that they have many harmonies to which our ears are not attuned. We soon shall see with what startling rapidity they are able to understand each other, and adopt concerted measures, when, for instance, the great honey thief, the huge sphinx atropos, the sinister butterfly that bears a death's head on its back, penetrates into the hive, humming its own strange note, which acts as a kind of irresistible incantation; the news spreads quickly from group to group, and from the guards at the threshold to the workers on the furthest combs, the whole population quivers.

It was for a long time believed that when these wise bees, generally so prudent, so far-sighted and economical, abandoned the treasures of their kingdom and flung themselves upon the uncertainties of life, they were yielding to a kind of irresistible folly, a mechanical impulse, a law of the species, a decree of nature, or to the force that for all creatures lies hidden in the revolution of time. It is our habit, in the case of the bees no less than our own, to regard as fatality all that we do not as yet understand. But now that the hive has surrendered two or three of its material secrets, we have discovered that this exodus is neither instinctive nor inevitable. It is not a blind emigration, but apparently the well-considered sacrifice of the present generation in favour of the generation to come. The bee-keeper has only to destroy in their cells the young queens that still are inert, and, at the same time, if nymphs and larvae abound, to enlarge the store-houses and dormitories of the nation, for this unprofitable tumult instantaneously to subside, for work to be at once resumed, and the flowers revisited; while the old queen, who now is essential again, with no successor to hope for, or perhaps to fear, will renounce for this year her desire for the light of the sun. Reassured as to the future of the activity that will soon spring into life, she will tranquilly resume her maternal labours, which consist in the laying of two or three thousand eggs a day, as she passes, in a methodical spiral, from cell to cell, omitting none, and never pausing to rest.

Where is the fatality here, save in the love of the race of today for the race of tomorrow? This fatality exists in the human species also, but its extent and power seem infinitely less. Among men it never gives rise to sacrifices as great, as unanimous, or as complete. What farseeing fatality, taking the place of this one, do we ourselves obey? We know not; as we know not the being who watches us as we watch the bees.

But the hive that we have selected is disturbed in its history by no interference of man; and as the beautiful day advances with radiant and tranquil steps beneath the trees, its ardour, still bathed in dew, makes the appointed hour seem laggard. Over the whole surface of the golden corridors that divide the parallel walls the workers are busily making preparation for the journey. And each one will first of all burden herself with provision of honey sufficient for five or six days. From this honey that they bear within them they will distil, by a chemical process still unexplained, the wax required for the immediate construction of buildings. They will provide themselves also with a certain amount of propolis, a kind of resin with which they will seal all the crevices in the new dwelling, strengthen weak places, varnish the walls, and exclude the light; for the bees love to work in almost total obscurity, guiding themselves with their many-faceted eyes, or with their antennae perhaps, the seat, it would seem, of an unknown sense that fathoms and measures the darkness.

They are not without prescience, therefore, of what is to befall them on this the most dangerous day of all their existence. Absorbed by the cares, the prodigious perils of this mighty adventure, they will have no time now to visit the gardens and meadows; and tomorrow, and after tomorrow, it may happen that rain may fall, or there may be wind; that their wings may be frozen or the flowers refuse to open. Famine and death would await them were it not for this foresight of theirs. None would come to their help, nor would they seek help of any. For one city knows not the other, and assistance never is given. And even though the bee-keeper deposit the hive, in which he has gathered the old queen and her attendant cluster of bees, by the side of the abode they have but this moment quitted, they would seem, by the disaster never so great that shall now have befallen them, to have wholly forgotten the peace and the happy activity that once they had known there, the abundant wealth and the safety that had then been their portion; and all, one by one, and down to the last of them, will perish of hunger and cold around their unfortunate queen rather than return to the home of their birth, whose sweet odour of plenty, the fragrance, indeed, of their own past assiduous labour, reaches them even in their distress.

That is a thing, some will say, that men would not do, — a proof that the bee, notwithstanding the marvels of its organisation, still is lacking in intellect and veritable consciousness. Is this so certain? Other beings, surely, may possess an intellect that differs from ours, and produces different results, without therefore being inferior. And besides, are we, even in this little human parish of ours, such infallible judges of matters that pertain to the spirit? Can we so readily divine the thoughts that may govern the two or three people we may chance to see moving and talking behind a closed window, when their words do not reach us? Or let us suppose that an inhabitant of Venus or Mars were to contemplate us from the height of a mountain, and watch the little black specks that we form in space, as we come and go in the streets and squares of our towns. Would the mere sight of our movements, our buildings, machines, and canals, convey to him any precise idea of our morality, intellect, our manner of thinking, and loving, and hoping, — in a word, of our real and intimate self? All he could do, like ourselves when we gaze at the hive, would be to take note of some facts that seem very surprising; and from these facts to deduce conclusions probably no less erroneous, no less uncertain, than those that we choose to form concerning the bee.

This much at least is certain; our "little black specks" would not reveal the vast moral direction, the wonderful unity, that are so apparent in the hive. "Whither do they tend, and what is it they do?" he would ask, after years and centuries of patient watching. "What is the aim of their life, or its pivot? Do they obey some God? I can see nothing that governs their actions. The little things that one day they appear to collect and build up, the next they destroy and scatter. They come and go, they meet and disperse, but no one knows what it is they seek. In numberless cases the spectacle they present is altogether inexplicable. There are some, for instance, who as it were, seem scarcely to stir from their place. They are to be distinguished by their glossier coat, and often too by their more considerable bulk. They occupy buildings ten or twenty times larger than ordinary dwellings, and richer, and more ingeniously fashioned. Every day they spend many hours at their meals, which sometimes indeed are prolonged far into the night. They appear to be held in extraordinary honour by those who approach them; men come from the neighbouring houses, bringing provisions and even from the depths of the country, laden with presents. One can only assume that these persons must be indispensable to the race, to which they render essential service, although our means of investigation have not yet enabled us to discover what the precise nature of this service may be. There are others, again, who are incessantly engaged in the most wearisome labour, whether it be in great sheds full of wheels that forever turn round and round, or close by the shipping, or in obscure hovels, or on small plots of earth that from sunrise to sunset they are constantly delving and digging. We are led to believe that this labour must be an offence, and punishable. For the persons guilty of it are housed in filthy, ruinous, squalid cabins. They are clothed in some colourless hide. So great does their ardour appear for this noxious, or at any rate useless activity, that they scarcely allow themselves time to eat or sleep. In numbers they are to the others as a thousand to one. It is remarkable that the species should have been able to survive to this day under conditions so unfavourable to its development. It should be mentioned, however, that apart from this characteristic devotion to their wearisome toil, they appear inoffensive and docile; and satisfied with the leavings of those who evidently are the guardians, if not the saviours, of the race." . . .

But we are forgetting the hive wherein the swarming bees have begun to lose patience, the hive whose black and vibrating waves are bubbling and overflowing, like a brazen cup beneath an ardent sun. It is noon; and the heat so great that the assembled trees would seem almost to hold back their leaves, as a man holds his breath before something very tender but very grave. The bees give their honey and sweet-smelling wax to the man who attends them; but more precious gift still is their summoning him to the gladness of June, to the joy of the beautiful months; for events in which bees take part happen only when skies are pure, at the winsome hours of the year when flowers keep holiday. They are the soul of the summer, the clock whose dial records the moments of plenty; they are the untiring wing on which delicate perfumes float; the guide of the quivering light-ray, the song of the slumberous, languid air; and their flight is the token, the sure and melodious note, of all the myriad fragile joys that are born in the heat and dwell in the sunshine. They teach us to tune our ear to the softest, most intimate whisper of these good, natural hours. To him who has known them and loved them, a summer where there are no bees becomes as sad and as empty as one without flowers or birds.

The man who never before has beheld the swarm of a populous hive must regard this riotous, bewildering spectacle with some apprehension and diffidence. He will be almost afraid to draw near; he will wonder can these be the earnest, the peace-loving, hard-working bees whose movements he has hitherto followed? It was but a few moments before he had seen them troop in from all parts of the country, as pre-occupied, seemingly, as little housewives might be, with no thoughts beyond household cares. He had watched them stream into the hive, imperceptibly almost, out of breath, eager, exhausted, full of discreet agitation; and had seen all the young amazons stationed at the gate salute them, as they passed by, with the slightest wave of antennae. And then, the inner court reached, they had hurriedly given their harvest of honey to the adolescent portresses always stationed within, exchanging with these at most the three or four probably indispensable words; or perhaps they would hasten themselves to the vast magazines that encircle the brood-cells, and deposit the two heavy baskets of pollen that depend from their thighs, thereupon at once going forth once more, without giving a thought to what might be passing in the royal palace, the work-rooms, or the dormitory where the nymphs lie asleep; without for one instant joining in the babel of the public place in front of the gate, where it is the wont of the cleaners, at time of great heat, to congregate and to gossip.

Today this is all changed. A certain number of workers, it is true, will peacefully go to the fields, as though nothing were happening; will come back, clean the hive, attend to the brood-cells, and hold altogether aloof from the general ecstasy. These are the ones that will not accompany the queen; they will remain to guard the old home, feed the nine or ten thousand eggs, the eighteen thousand larvae, the thirty-six thousand nymphs and seven or eight royal princesses, that today shall all be abandoned. Why they have been singled out for this austere duty, by what law, or by whom, it is not in our power to divine. To this mission of theirs they remain inflexibly, tranquilly faithful; and though I have many times tried the experiment of sprinkling a colouring matter over one of these resigned Cinderellas, that are moreover easily to be distinguished in the midst of the rejoicing crowds by their serious and somewhat ponderous gait, it is rarely indeed that I have found one of them in the delirious throng of the swarm.

And yet, the attraction must seem irresistible. It is the ecstasy of the perhaps unconscious sacrifice the god has ordained; it is the festival of honey, the triumph of the race, the victory of the future: the one day of joy, of forgetfulness and folly; the only Sunday known to the bees. It would appear to be also the solitary day upon which all eat their fill, and revel, to heart's content, in the delights of the treasure they themselves have amassed. It is as though they were prisoners to whom freedom at last had been given, who had suddenly been led to a land of refreshment and plenty. They exult, they cannot contain the joy that is in them. They come and go aimlessly, — they whose every movement has always its precise and useful purpose — they depart and return, sally forth once again to see if the queen be ready, to excite their sisters, to beguile the tedium of waiting. They fly much higher than is their wont, and the leaves of the mighty trees round about all quiver responsive. They have left trouble behind, and care. They no longer are meddling and fierce, aggressive, suspicious, untamable, angry. Man — the unknown master whose sway they never acknowledge, who can subdue them only by conforming to their every law, to their habits of labour, and following step by step the path that is traced in their life by an intellect nothing can thwart or turn from its purpose, by a spirit whose aim is always the good of the morrow — on this day man can approach them, can divide the glittering curtain they form as they fly round and round in songful circles; he can take them up in his hand, and gather them as he would a bunch of grapes; for today, in their gladness, possessing nothing, but full of faith in the future, they will submit to everything and injure no one, provided only they be not separated from the queen who bears that future within her.

But the veritable signal has not yet been given. In the hive there is indescribable confusion; and a disorder whose meaning escapes us. At ordinary times each bee, once returned to her home, would appear to forget her possession of wings; and will pursue her active labours, making scarcely a movement, on that particular spot in the hive that her special duties assign. But today they all seem bewitched; they fly in dense circles round and round the polished walls, like a living jelly stirred by an invisible hand. The temperature within rises rapidly, — to such a degree, at times, that the wax of the buildings will soften, and twist out of shape. The queen, who ordinarily never will stir from the centre of the comb, now rushes wildly, in breathless excitement, over the surface of the vehement crowd that turn and turn on themselves. Is she hastening their departure, or trying to delay it? Does she command or haply implore? Does this prodigious emotion issue from her, or is she its victim? Such knowledge as we possess of the general psychology of the bee warrants the belief that the swarming always takes place against the old sovereign's will. For indeed the ascetic workers, her daughters, regard the queen above all as the organ of love, indispensable, certainly, and sacred, but in herself somewhat unconscious, and often of feeble mind. They treat her like a mother in her dotage. Their respect for her, their tenderness, is heroic and boundless. The purest honey, specially distilled and almost entirely assimilable, is reserved for her use alone. She has an escort that watches over her by day and by night, that facilitates her maternal duties and gets ready the cells wherein the eggs shall be laid; she has loving attendants who pet and caress her, feed her and clean her, and even absorb her excrement. Should the least accident befall her the news will spread quickly from group to group, and the whole population will rush to and fro in loud lamentation. Seize her, imprison her, take her away from the hive at a time when the bees shall have no hope of filling her place, owing, it may be, to her having left no predestined descendants, or to there being no larvae less than three days old (for a special nourishment is capable of transforming these into royal nymphs, such being the grand democratic principle of the hive, and a counterpoise to the prerogatives of maternal predestination) , and then, her loss once known, after two or three hours, perhaps, for the city is vast, work will cease in almost every direction. The young will no longer be cared for; part of the inhabitants will wander in every direction, seeking their mother, in quest of whom others will sally forth from the hive; the workers engaged in constructing the comb will fall asunder and scatter, the foragers no longer will visit the flowers, the guard at the entrance will abandon their post; and foreign marauders, all the parasites of honey, forever on the watch for opportunities of plunder, will freely enter and leave without any one giving a thought to the defence of the treasure that has been so laboriously gathered. And poverty, little by little, will steal into the city; the population will dwindle; and the wretched inhabitants soon will perish of distress and despair, though every flower of summer burst into bloom before them.

But let the queen be restored before her loss has become an accomplished, irremediable fact, before the bees have grown too profoundly demoralised, — for in this they resemble men: a prolonged regret, or misfortune, will impair their intellect and degrade their character, — let her be restored but a few hours later, and they will receive her with extraordinary, pathetic welcome. They will flock eagerly round her; excited groups will climb over each other in their anxiety to draw near; as she passes among them they will caress her with the long antennae that contain so many organs as yet unexplained; they will present her with honey, and escort her tumultuously back to the royal chamber. And order at once is restored, work resumed, from the central comb of the brood-cells to the furthest annex where the surplus honey is stored; the foragers go forth, in long black files, to return in less than three minutes sometimes laden with nectar and pollen; streets are swept, parasites and marauders killed or expelled; and the hive soon resounds with the gentle, monotonous cadence of the strange hymn of rejoicing, which is, it would seem, the hymn of the royal presence.

There are numberless instances of the absolute attachment and devotion that the workers display towards their queen. Should disaster befall the little republic; should the hive or the comb collapse, should man prove ignorant, or brutal; should they suffer from famine, from cold or disease, and perish by thousands, it will still be almost invariably found that the queen will be safe and alive, beneath the corpses of her faithful daughters. For they will protect her, help her to escape; their bodies will provide both rampart and shelter; for her will be the last drop of honey, the wholesomest food. And be the disaster never so great, the city of virgins will not lose heart so long as the queen be alive. Break their comb twenty times in succession, take twenty times from them their young and their food, you still shall never succeed in making them doubt of the future; and though they be starving, and their number so small that it scarcely suffices to shield their mother from the enemy's gaze, they will set about to reorganize the laws of the colony, and to provide for what is most pressing; they will distribute the work in accordance with the new necessities of this disastrous moment, and thereupon will immediately re-assume their labours with an ardour, a patience, a tenacity and intelligence not often to be found existing to such a degree in nature, true though it be that most of its creatures display more confidence and courage than man.

But the presence of the queen is not even essential for their discouragement to vanish and their love to endure. It is enough that she should have left, at the moment of her death or departure, the very slenderest hope of descendants. "We have seen a colony," says Langstroth, one of the fathers of modern apiculture, "that had not bees sufficient to cover a comb of three inches square, and yet endeavoured to rear a queen. For two whole weeks did they cherish this hope; finally, when their number was reduced by one-half, their queen was born, but her wings were imperfect, and she was unable to fly. Impotent as she was, her bees did not treat her with the less respect. A week more, and there remained hardly a dozen bees; yet a few days, and the queen had vanished, leaving a few wretched, inconsolable insects upon the combs." . . .

And now to return to our swarming hive, where the bees have already given the signal for departure, without waiting for these reflections of ours to come to an end. At the moment this signal is given, it is as though one sudden mad impulse had simultaneously flung open wide every single gate in the city; and the black throng issues, or rather pours forth in a double, or treble, or quadruple jet, as the number of exits may be; in a tense, direct, vibrating, uninterrupted stream that at once dissolves and melts into space, where the myriad transparent, furious wings weave a tissue throbbing with sound. And this for some moments will quiver right over the hive, with prodigious rustle of gossamer silks that countless electrified hands might be ceaselessly rending and stitching; it floats undulating, it trembles and flutters like a veil of gladness invisible fingers support in the sky, and wave to and fro, from the flowers to the blue, expecting sublime advent or departure. And at last one angle declines, another is lifted; the radiant mantle unites its four sunlit corners; and like the wonderful carpet the fairy-tale speaks of, that flits across space to obey its master's command, it steers its straight course, bending forward a little as though to hide in its folds the sacred presence of the future, towards the willow, the pear-tree, or lime whereon the queen has alighted; and round her each rhythmical wave comes to rest, as though on a nail of gold, and suspends its fabric of pearls and of luminous wings.

And then there is silence once more; and, in an instant, this mighty tumult, this awful curtain apparently laden with unspeakable menace and anger, this bewildering golden hail that streamed upon every object near — all these become merely a great, inoffensive, peaceful cluster of bees, composed of thousands of little motionless groups, that patiently wait, as they hang from the branch of a tree, for the scouts to return who have gone in search of a place of shelter.

This is the first stage of what is known as the "primary swarm" at whose head the old queen is always to be found. They will settle as a rule on the shrub or the tree that is nearest the hive; for the queen, besides being weighed down by her eggs, has dwelt in constant darkness ever since her marriage-flight, or the swarm of the previous year; and is naturally reluctant to venture far into space, having indeed almost forgotten the use of her wings.

The bee-keeper waits till the mass be completely gathered together; then, having covered his head with a large straw hat, (for the most inoffensive bee will conceive itself caught in a trap if entangled in hair, and will infallibly use its sting), but, if he be experienced, wearing neither mask nor veil; having taken the precaution only of plunging his arms in cold water up to the elbow, he proceeds to gather the swarm by vigorously shaking the bough from which the bees depend over an inverted hive. Into this hive the cluster will fall as heavily as an over-ripe fruit. Or, if the branch be too stout, he can plunge a spoon into the mass; and deposit where he will the living spoonfuls, as though he were ladling out corn. He need have no fear of the bees that are buzzing around him, settling on his face and hands. The air resounds with their song of ecstasy, which is different far from their chant of anger. He need have no fear that the swarm will divide, or grow fierce, will scatter, or try to escape. This is a day, I repeat, when a spirit of holiday would seem to animate these mysterious workers, a spirit of confidence, that apparently nothing can trouble. They have detached themselves from the wealth they had to defend, and they no longer recognise their enemies. They become inoffensive because of their happiness, though why they are happy we know not, except it be because they are obeying their law. A moment of such blind happiness is accorded to every living thing by nature at times, when she seeks to accomplish her end. Nor need we feel any surprise that here the bees are her dupes; we ourselves, who have studied her movements these centuries past, and with a brain more perfect than that of the bee, we too are her dupes, and know not even yet whether she be benevolent or indifferent, or only basely cruel.

There where the queen has alighted the swarm will remain; and had she ascended alone into the hive, the bees would have followed, in long black files, as soon as intelligence had reached them of the maternal retreat. The majority will hasten to her, with utmost eagerness; but large numbers will pause for an instant on the threshold of the unknown abode, and there will describe the circles of solemn rejoicing with which it is their habit to celebrate happy events. "They are beating to arms," say the French peasants. And then the strange home will at once be accepted, and its remotest corners explored; its position in the apiary, its form, its colour, are grasped and retained in these thousands of prudent and faithful little memories. Careful note is taken of the neighbouring landmarks, the new city is founded, and its place established in the mind and the heart of all its inhabitants; the walls resound with the love-hymn of the royal presence, and work begins.

But if the swarm be not gathered by man, its history will not end here. It will remain suspended on the branch until the return of the workers, who, acting as scouts, winged quartermasters, as it were, have at the very first moment of swarming sallied forth in all directions in search of a lodging. They return one by one, and render account of their mission; and as it is manifestly impossible for us to fathom the thought of the bees, we can only interpret in human fashion the spectacle that they present. We may regard it as probable, therefore, that most careful attention is given to the reports of the various scouts. One of them it may be, dwells on the advantage of some hollow tree it has seen; another is in favour of a crevice in a ruinous wall, of a cavity in a grotto, or an abandoned burrow. The assembly often will pause and deliberate until the following morning. Then at last the choice is made, and approved by all. At a given moment the entire mass stirs, disunites, sets in motion, and then, in one sustained and impetuous flight, that this time knows no obstacle, it will steer its straight course, over hedges and cornfields, over haystack and lake, over river and village, to its determined and always distant goal. It is rarely indeed that this second stage can be followed by man. The swarm returns to nature; and we lose the track of its destiny.

JEAN HENRI FABRE (1823–1915)

Tribulations of a Naturalist *and* The Courtship of the Scorpion

From *Souvenirs Entomologiques*. 1869–1907

[FABRE was one of the half-forgotten authors who more than measured up to my boyhood's superlative estimate, when it came time to review his work for this Anthology. The only difficulty was what to omit of this plethora of riches. Like Gilbert White, exactly a century before and across the English Channel, Fabre spent almost all his active life close to his home. In this limited area of France he concentrated on the lives and the doings of some of the lesser folk about him. In patience and in painstaking devotion to intensive watching and interpretation he far surpassed all his predecessors and many entomologists who were to succeed him.

He belittled the knowledge to be found in books, he confined his interests chiefly to wasps, bees, beetles, grasshoppers, and spiders, and he was led to disbelieve in evolution. But all this affected in no way the superb accuracy of his records, nor did his facile and delightful style of expression diminish their value to entomological science.

Fabre's ten volumes teem with these exciting tales of insect life, presented in such vivid and intimate style that each individual wasp or glow-worm or scarab becomes, for the reader, a definite personality.

For the main account I have chosen one of the less-known essays, on the courtship of the scorpion, and, as a prelude to this, a poignant account of the pitiful poverty which often handicapped his work. Also included is a brief exposure of the shy, almost Mr. Milquetoast attitude that aroused his fears at the approach of strangers when he was concentrating on the burrow of some wasp on the public roadside or in the field. Probably few of us have enjoyed the spectacle of a scorpion in the turmoil of passion, but every one of us naturalists has had his stomach turn over at the threatened approach of some fellow human, whose reaction to our intensive observation we will label as unsympathetic for lack of a shorter and uglier word. With Fabre as with ourselves, it is not the personal viewpoint, but the ghastly danger of our particular "wasp nest" being disturbed or destroyed at a critical moment that makes us terrified of intrusion.]

TRIBULATIONS OF A NATURALIST

THE MAGNIFICENT Bee herself, with her dark-violet wings and black-velvet raiment, her rustic edifices on the sun-blistered pebbles amid the thyme, her honey, providing a diversion from the severities of the compass and the square, all made a great impression on my mind; and I wanted to know more than I had learned from the schoolboys, which was just how to rob the cells of their honey with a straw. As it happened, my bookseller had a gorgeous work on insects for sale. It was called *Histoire naturelle des animaux articulés,* by de Castelnau, E. Blanchard and Lucas, and boasted a multitude of most attractive illustrations; but the price of it, the price of it! No matter; was not my splendid income supposed to cover everything, food for the mind as well as food for the body? Anything extra that I gave to the one I could save upon the other: a method of balancing painfully familiar to those who look to science for their livelihood. The purchase was effected. That day my professional emoluments were severely strained: I devoted a month's salary to the acquisition of the book. I had to resort to miracles of economy for some time to come before making up the enormous deficit.

The book was devoured; there is no other word for it. In it, I learned the name of my black Bee; I read for the first time various details of the habits of insects; I found, surrounded in my eyes with a sort of halo, the revered names of Réaumur, Huber, and Léon Dufour; and, while I turned over the pages for the hundredth time, a voice within me seemed to whisper:

"You also shall be of their company!"

'Ah, fond illusions, what has come of you? . . .

When the chemist has fully prepared his plan of research, he mixes his reagent at the most convenient moment and lights a flame under his retort. He is the master of time, place and circumstances. He chooses his hour, shuts himself up in his laboratory, where nothing can come to disturb the business in hand; he produces at will this or that condition which reflection suggests to him: he is in quest of the secrets of inorganic matter, whose chemical activities science can awaken whenever it thinks fit.

The secrets of living matter — not those of anatomical structure, but really those of life in action, especially of instinct — present much more difficult and delicate conditions to the observer. Far from being able to choose his own time, he is the slave of the season, of the day, of the hour, of the very moment. When the opportunity offers, he must seize it as it comes, without hesitation, for it may be long before it presents itself again. And, as it usually arrives at the moment when he is least expecting it, nothing is in readiness for making the most of it. He must then and there improvise his little stock of experimenting-material, contrive his plans, evolve his tactics, devise his tricks; and he can think himself lucky if inspiration comes fast enough to allow him to profit by the chance offered. This chance, moreover, hardly ever comes except to those who look for it. You must watch for it patiently for days and days, now on sandy slopes exposed to the full glare of the sun, now on some path walled in by high banks, where the heat is like that of an oven, or again on some sandstone ledge which is none too steady. If it is in your power to set up your observatory under a meagre olive-tree that pretends to protect you from the rays of a pitiless sun, you may bless the fate that treats you as a sybarite: your lot is an Eden. Above all, keep your eyes open. The spot is a good one; and — who knows? — the opportunity may come at any moment. . . .

Ah, if you could now observe at your ease, in the quiet of your study, with nothing to distract your mind from your subject, far from the profane wayfarer who, seeing you so busily occupied at a spot where he sees nothing, will stop, overwhelm you with queries, take you for some water-diviner, or — a graver suspicion this — regard you as some questionable character searching for buried treasure and discovering by means of incantations where the old pots full of coin lie hidden! Should you still wear a Christian aspect in his eyes, he will approach you, look to see what you are looking at and smile in a manner that leaves no doubt as to his poor opinion of people who spend their time in watching Flies. You will be lucky indeed if the troublesome visitor, with his tongue in his cheek, walks off at last without disturbing things and without repeating in his innocence the disaster brought about by my two conscripts' boots.

Should your inexplicable doings not puzzle the passer-by, they will be sure to puzzle the village keeper, that uncompromising representative of the law in the ploughed acres. He has long had his eye on you. He has so often seen you wandering about, like a lost soul, for no appreciable reason; he has so often caught you rooting in the ground, or, with infinite precautions, knocking down some strip of wall in a sunken road, that in the end he has come to look upon you with dark suspicion. You are nothing to him but a gipsy, a tramp, a poultry-thief, a shady person or, at the best, a madman. Should you be carrying your botanizing-case, it will represent to him the poacher's ferret-cage; and you would never get it out of his head that, regardless of the game-laws and the rights of landlords, you are clearing all the neighbouring warrens of their rabbits. Take care. However thirsty you may be, do not lay a finger on the nearest bunch of grapes: the man with the municipal badge will be there, delighted to have a case at last and so to receive an explanation of your highly perplexing behaviour.

I have never, I can safely say, committed any such misdemeanor; and yet, one day, lying on the sand, absorbed in the details of a Bembex' household, I suddenly heard beside me;

"In the name of the law, I arrest you! You come along with me!"

It was the keeper of Les Angles, who, after vainly waiting for an opportunity to catch me at fault and being daily more anxious for an answer to the riddle that was worrying him, at last resolved upon the brutal expedient of a summons. I had to explain things. The poor man seemed anything but convinced:

"Pooh!" he said. "Pooh! You will never make me believe that you come here and roast in the sun just to watch Flies. I shall keep an eye on you, mark you! And, the first time I . . . ! However, that'll do for the present."

And he went off. I have always believed that my red ribbon had a good deal to do with his departure. And I also put down to that red ribbon certain other little services by which I benefitted during my entomological and botanical excursions. It seemed to me — or was I dreaming? — it seemed to me that, on my botanizing-expeditions up Mont Ventoux, the guide was more tractable and the donkey less obstinate.

The aforesaid bit of scarlet ribbon did not always spare me the tribulations which the entomologist must expect when experimenting on the public way. Here is a characteristic example. Ever since daybreak I have been ambushed, sitting on a stone, at the bottom of a ravine. The subject of my matutinal visit is the Languedocian Sphex. Three women, vine-pickers, pass in a group, on the way to their work. They give a glance at the man seated, apparently absorbed in reflection. At sunset, the same pickers pass again, carrying their full baskets on their heads. The man is still there, sitting on the same stone, with his eyes fixed on the same place. My motionless attitude, my long persistency in remaining at that deserted spot, must have impressed them deeply. As they passed by me, I saw one of them tap her forehead and heard her whisper to the others:

"Un paouré inoucènt, pécaïre!"

And all three made the sign of the Cross.

An innocent, she had said, *un inoucènt,* an idiot, a poor creature, quite harmless, but half-witted; and they had all made the sign of the Cross, an idiot being to them one with God's seal stamped upon him.

"How now!" thought I. "What a cruel mockery of fate! You, who are so laboriously seeking to discover what *is* instinct in the animal and what is reason, you yourself do not even possess your reason in these good women's eyes! What a humiliating reflection!"

THE COURTSHIP OF THE SCORPION

In April, when the Swallow returns to us and the Cuckoo sounds his first note, a revolution takes place among my hitherto peaceable Scorpions. Several whom I have established in the colony in the enclosure, leave their shelter at nightfall, go wandering about and do not return to their homes. A more serious business: often, under the same stone, are two Scorpions of whom one is in the act of devouring the other. Is this a case of brigandage among creatures of the same order, who, falling into vagabond ways when the fine weather sets in thoughtlessly enter their neighbours' houses and there meet with their undoing unless they be the stronger? One would almost think it, so quickly is the intruder eaten up, for days at a time and in small mouthfuls, even as the usual game would be.

Now here is something to give us a hint. The Scorpions devoured are invariably of middling size. Their lighter colouring, their less protuberant bellies, mark them as males, always males. The others, larger, more paunchy and a little darker in shade, do not end in this unhappy fashion. So these are probably not brawls between neighbours who, jealous of their solitude, would soon settle the hash of any visitor and eat him afterwards, a drastic method of putting a stop to further indiscretions; they are rather nuptial rites, tragically performed by the matron after pairing. To determine how much ground there is for this suspicion is beyond my powers until next year: I am still too badly equipped.

Spring returns once more. I have prepared the large glass cage in advance and stocked it with twenty-five inhabitants, each with his bit of crockery. From mid-April onwards, every evening, when it grows dark, between seven and nine o'clock, great animation reigns in the crystal palace. That which seemed deserted by day now becomes a scene of festivity. As soon as supper is finished, the whole household runs out to look on. A lantern hung outside the panes allows us to follow events.

It is our distraction after the worries of the day; it is our playhouse. In this theatre for simple folk, the performances are so highly interesting that, the moment the lantern is lighted, all of us, great and small alike, come and take our places in the stalls; all, down to Tom, the House-dog. Tom, it is true, indifferent to Scorpion affairs, like the true philosopher that he is, lies at our feet and dozes, but only with one eye, keeping the other always open on his friends the children.

Let me try to give the reader an idea of what happens. A numerous assembly soon gathers near the glass panes in the region discreetly lit by the lanterns. Every elsewhere, here, there, single Scorpions walk about and, attracted by the light, leave the shade and hasten to the illuminated festival. The very Moths betray no greater eagerness to flutter to the rays of our lamps. The newcomers mingle with the crowd, while others, tired of their pastimes, withdraw into the shade, snatch a few moments' rest and then impetuously return upon the scene.

These hideous devotees of gaiety provide a dance that is not wholly devoid of charm. Some come from afar; solemnly they emerge from the shadow; then, suddenly, with a rush as swift and easy as a slide, they join the crowd, in the light. Their agility reminds one of mice scurrying along with their tiny steps. They seek one another and fly precipitately the moment they touch, as though they had mutually burnt their fingers. Others, after tumbling about a little with their play-fellows, make off hurriedly, wildly. They take fresh courage in the dark and return. At times, there is a violent tumult: a confused mass of swarming legs, snapping claws, tails curving and clashing, threatening or fondling, it is hard to say which. In this affray, under favourable conditions, twin specks of light flare and shine like carbuncles. One would take them for eyes that emit flashing glances; in reality they are two polished, reflecting facets, which occupy the front of the head. All, large and small alike, take part in the brawl; it might be a battle to the death, a general massacre; and it is just a wanton frolic. Even so do kittens bemaul each other. Soon, the group disperses; all make off in all sorts of directions, without a scratch, without a sprain.

Behold the fugitives collecting once more beneath the lantern. They pass and pass again; they come and go, often meeting front to front. He who is in the greatest hurry walks over the back of the other, who lets him have his way without any protest but a movement of the body. It is no time for blows; at most, two Scorpions meeting will exchange a cuff, that is to say, a rap of the caudal staff. In their community, this friendly thump, in which the point of the sting plays no part, is a sort of a fisticuff in frequent use. There are better things than entangled legs and brandished tails; there are sometimes poses of the highest originality. Face to face, with claws drawn back, two wrestlers proceed to stand on their heads like acrobats, that is to say, resting only on the fore-quarters, they raise the whole hinder portion of the body, so much so that the chest displays the four little lung pockets uncovered. Then the tails, held vertically erect in a straight line, exchange mutual rubs, gliding one over the other, while their extremities are hooked together and repeatedly fastened and unfastened. Suddenly, the friendly pyramid falls to pieces and each runs off hurriedly, without ceremony.

What were these two wrestlers trying to do, in their eccentric posture? Was it a set-to between two rivals? It would seem not, so peaceful is the encounter. My subsequent observations were to tell me that this was the mutual teasing of a betrothed couple. To declare his flame, the Scorpion stands on his head.

To continue as I have begun and give a homogeneous picture of the thousand tiny particulars gathered day by day would have its advantages; the story would sooner be told; but, at the same time deprived of its details, which vary greatly between one observation and the next and are difficult to piece together, it would be less interesting. Nothing must be neglected in the relation of manners so strange and as yet so little known. At the risk of repeating one's self here and there, it is preferable to adhere to chronological order and to tell the story by fragments, as one's observations reveal fresh facts. Order will emerge from this disorder; for each of the more remarkable evenings supplies some feature that corroborates and completes those which go before. I will therefore continue my narration in the form of a diary.

25th April, 1904. — Hullo! What is this, something I have not yet seen? My eyes, ever on the watch, look upon the affair for the first time. Two Scorpions face each other, with claws outstretched and fingers clasped. It is a question of a friendly grasp of the hand and not the prelude to a battle, for the two partners are behaving to each other in the most peaceful way. There is one of either sex. One is paunchy and browner than the other: this is the female; the other is comparatively slim and pale: this is the male. With their tails prettily curled, the couple stroll with measured steps along the pane. The male is ahead and walks backwards, without jolt or jerk, without any resistance to overcome. The female follows obediently, clasped by her finger-tips and face to face with her leader.

The stroll is interrupted by halts that do not affect the method of conjunction; it is resumed, now here, now there, from end to end of the enclosure. Nothing shows the object which the strollers have in view. They loiter, they dawdle, they most certainly exchange ogling glances. Even so in my village, on Sundays, after vespers, do the youth of both sexes saunter along the hedges, every Jack with his Jill.

Often they tack about. It is always the male who decides which fresh direction the pair shall take. Without releasing her hands, he turns gracefully to the left or right about and places himself side by side with companion. Then, for a moment, with tail laid flat, he strokes her spine. The other stands motionless, impassive.

For over an hour, without tiring, I watch these interminable comings and goings. A part of the household lends me its eyes in the presence of the strange sight which no one in the world has yet seen, at least with a vision capable of observing. In spite of the lateness of the hour, which upsets all our habits, our attention is concentrated and no essential thing escapes us.

At last, about ten o'clock, something happens. The male has hit upon a potsherd whose shelter seems to suit him. He releases his companion with one hand, with one alone, and continuing to hold her with the other, he scratches with his legs and sweeps with his tail. A grotto opens. He enters and, slowly, without violence, drags the patient Scorpioness after him. Soon both have disappeared. A plug of sand closes the dwelling. The couple are at home.

To disturb them would be a blunder: I should be interfering too soon, at an inopportune moment, if I tried at once to see what was happening below. The preliminary stages may last for the best part of the night; and it does not do for me, who have turned eighty, to sit up so late. I feel my legs giving way; and my eyes seem full of sand.

All night long I dream of Scorpions. They crawl under my bedclothes, they pass over my face; and I am not particularly excited, so many curious things do I see in my imagination. The next morning, at daybreak, I lift the stoneware. The female is alone. Of the male there is no trace, either in the home or in the neighbourhood. First disappointment, to be followed by many others.

10th May. — It is nearly seven o'clock in the evening; the sky is overcast with signs of an approaching shower. Under one of the potsherds is a motionless couple, face to face, with linked fingers. Cautiously I raise the potsherd and leave the occupants uncovered, so as to study the consequences of the interview at my ease. The darkness of the night falls and nothing, it seems to me, will disturb the calm of the home deprived of its roof. A sharp shower compels me to retire. They, under the lid of the cage, have no need to take shelter against the rain. What will they do, left to their business as they are but deprived of a canopy to their alcove?

An hour later, the rain ceases and I return to my Scorpions. They are gone. They have taken up their abode under a neighbouring tile. Still with their fingers linked, the female is outside and the male indoors, preparing the home. At intervals of ten minutes, the members of my family relieve one another so as not to lose the exact moment of the pairing, which appears to be imminent. Wasted pains: at eight o'clock, it being now quite dark, the couple, dissatisfied with the spot, set out on a fresh ramble, hand in hand, and go prospecting elsewhere. The male, walking backwards, leads the way, chooses the dwelling as he pleases; the female follows with docility. It is an exact repetition of what I saw on the 25th of April.

At last a tile is found to suit them. The male goes in first but this time neither hand releases his companion for a moment. The nuptial chamber is prepared with a few sweeps of the tail. Gently drawn towards him, the Scorpioness enters in the wake of her guide.

I visit them a couple of hours later, thinking that I've given them time enough to finish their preparations. I lift the potsherd. They are there in the same posture, face to face and hand in hand. I shall see no more today.

The next day, nothing new either. Each sits confronting the other, meditatively. Without stirring a limb, the gossips, holding each other by the finger-tips, continue their endless interview under the tile. In the evening, at sunset, after sitting linked together for four-and-twenty hours, the couple separate. He goes away from the tile, she remains; and matters have not advanced by an inch.

This observation gives us two facts to remember. After the stroll to celebrate the betrothal, the couple need the mystery and quiet of a shelter. Never would the nuptials be consummated in the open air, amid the bustling crowd, in sight of all. Remove the roof of the house, by night or day, with all possible discretion; and the husband and wife, who seem absorbed in meditation, march off in search of another spot. Also, the sojourn under the cover of a stone is a long one; we have just seen it spun out to twenty-four hours and even then without a decisive result.

12th May. — What will this evening's sitting teach us? The weather is calm and hot, favourable to nocturnal pastimes. A couple has been formed; how things began I do not know. This time the male is greatly inferior to his corpulent mate. Nevertheless, the skinny wight performs his duty gallantly. Walking backwards, according to rule, with his tail rolled trumpetwise, he marches the fat Scorpioness around the glass ramparts. After one circuit follows another, sometimes in the same, sometimes in the opposite direction.

Pauses are frequent. Then the foreheads touch, bend a little to left and right, as if the two were whispering in each other's ears. The little fore-legs flutter in feverish caresses. What are they saying to each other? How shall we translate their silent epithalamium into words?

The whole household turns out to see this curious team, which our presence in no way disturbs. The pair are pronounced to be "pretty"; and the expression is not exaggerated. Semi-translucent and shining in the light of the lantern, they seem carved out of a block of amber. Their arms outstretched, their tails rolled into graceful spirals, they wander on with a slow movement and with measured tread.

Nothing puts them out. Should some vagabond, taking the evening air and keeping to the wall like themselves, meet them on their way, he stands aside — for he understands these delicate matters — and leaves them a free passage. Lastly, the shelter of a tile receives the strolling pair, the male entering first and backwards: that goes without saying. It is nine o'clock.

The idyll of the evening is followed, during the night, by a hideous tragedy. Next morning, we find the Scorpioness under the potsherd of the previous day. The little male is by her side, but slain, and more or less devoured. He lacks the head, a claw, a pair of legs. I place the corpse in the open, on the threshold of the home. All day long, the recluse does not touch it. When night returns, she goes out and, meeting the deceased on her passage, carries him off to a distance to give him a decent funeral, that is to finish eating him.

This act of cannibalism agrees with what the open-air colony showed me last year. From time to time, I would find, under the stones, a pot-bellied female making a comfortable ritual meal off her companion of the night. I suspected that the male, if he did not break loose in time, once his functions were fulfilled, was devoured, wholly or partly, according to the matron's appetite. I now have the certain proof before my eyes. Yesterday, I saw the couple enter their home after their usual preliminary, the stroll; and, this morning, under the same tile, at the moment of my visit, the bride is consuming her mate.

Well, one supposes that the poor wretch has attained his ends. Were he still necessary to the race, he would not be eaten yet. The couple before us have therefore been quick about the business, whereas, I see that others fail to finish after provocations and contemplations exceeding in duration the time which it takes the hour-hand to go twice around the clock. Circumstances impossible to state with precision — the condition of the atmosphere perhaps, the electric tension, the temperature, the individual ardour of the couple — to a large extent accelerate or delay the finale of the pairing; and this constitutes a serious difficulty for the observer anxious to seize the exact moment whereat the as yet uncertain function of the combs might be revealed.

14th May. — It is certainly not hunger that stirs up my animals night after night. The quest of food has nothing to say to their evening rounds. I have served to the busy crowd a varied bill of fare, selected from that which they appear to like best. It includes tender morsels in the shape of young Locusts; small Grasshoppers, fleshier than the Acridians; moths minus their wings. At a later season, I add Dragon-flies, a highly-appreciated dish, as is proved by their equivalent, the full-grown Ant-lion, of whom I used to find the remnants, the wings, in the Scorpion's cave.

This luxurious game leaves them indifferent; they pay no attention to it. Amid the hubbub, the Locusts hop, the Moths beat the ground with the stumps of their wings, the Dragon-flies quiver; and the Scorpions pass. They tread them underfoot, they topple them over, they push them aside with a stroke of the tail; in short, they absolutely refuse to look at them. They have other business in hand.

Almost all of them skirt the glass wall. Some of them obstinately attempt to scale it: they hoist themselves on their tails, fall down, try again elsewhere. With their outstretched fists they knock against the pane; they want to get away at all costs. And yet the grounds are large enough, there is room for all; the walks lend themselves to long strolls. . . . No matter: they want to roam afar. If they were free, they would disperse in every direction. Last year, at the same time, the colonists of the enclosure left the village and I never saw them again.

The spring pairing-season forces them to set forth exploring. The shy hermits of yesterday now leave their cells, and go on love's pilgrimage; heedless of food, they go in quest of their kind. Among the stones of their domain there must be choice spots at which meetings take place, at which assemblies are held. If I were not afraid of breaking my legs, at night, over the rocky obstacles of their hills, I should love to assist at their matrimonial festivals, amid the delights of liberty. What do they do up there, on their bare slopes? Much the same, apparently, as in the glass enclosure. Having picked a bride, they take her about, for a long stretch of time, hand in hand, through the tufts of lavender. If they miss the attractions of my lantern, they have the moon, that incomparable lamp, to light them.

20th May. — The sight of the first invitation to a stroll is not an event upon which we can count every evening. Several emerge from under their stones already linked in couples. In this concatenation of clasped fingers, they have passed the whole day, motionless, face to face, meditating. When night comes, without separating for a moment, they resume the walk around the glass begun on the evening before, or even earlier. No one knows when or how the junction was effected. Others meet unexpectedly in sequestered passages, difficult of inspection. By the time that I see them, it is too late: the team is on the way.

Today, chance favours me. The acquaintance is made before my eyes, in the full light of the lantern. A frisky, sprightly male, in his hurried rush through the crowd, suddenly finds himself confronting a fair passer-by who takes his fancy. She does not gainsay him; and things move quickly.

The foreheads touch, the claws engage; the tails swing with a spacious gesture: they stand up vertically, hook together at the tips and softly stroke each other with a slow caress. The two animals stand on their heads in the manner already described. Soon, the raised bodies sink to the ground; fingers are clasped and the couple start on their stroll without more ado. The pyramidal pose, therefore, is really the prelude to the harnessing. The pose, it is true, is not rare between two individuals of the same sex on the meeting; but it is then less correct and above all, less marked by ceremony. At such times, we find movements of impatience, instead of friendly excitations; the tails strike in lieu of fondling each other.

Let us watch the male, who hurries away backwards, very proud of his conquest. Other females are met, who stand around and look on inquisitively, perhaps enviously. One of them flings herself upon the ravished bride, clasps her with her legs and makes an effort to stop the team. The male exhausts himself in attempts to overcome this resistance; in vain he shakes, in vain he pulls: things won't move. Undistressed by the accident, he throws up the game. A neighbour is there, close by. Cutting parley short, this time without any further declaration, he takes her hands and invites her to a stroll. She protests, releases herself and runs away.

From among the group of onlookers, a second is solicited, in the same free and easy manner. She accepts, but there is nothing to tell us that she will not escape from her seducer on the way. But what does the coxcomb care? There are more where she came from! And what does he want, when all is said? The first that comes along!

This first-comer he soon finds, for here he is, leading his conquest by the hand. He passes into the belt of light. Exerting all his strength, he tugs and jerks at the other if she refuses to come, but is gentle in his manner when he obtains a docile obedience. Pauses, sometimes rather prolonged, are frequent.

Then the male indulges in some curious exercises. Bringing his claws, or let us say, his arms towards him and then stretching them out again, he compels the female to make a like alternation of movements. The two of them form a system of jointed rods, like a lazy-tongs, opening and closing their quadrilateral by turns. After this gymnastic exercise, the mechanism contracts and remains stationary.

The foreheads now touch; the two mouths come together with tender effusions. The word "kisses" comes to one's mind to express these caresses. It is not applicable; for head, face, lips, cheeks, all are missing. The animal, lopped off short, as though with the shears, has not even a muzzle. Where we look for a face we are confronted with a dead wall of hideous jaws.

And to the Scorpion this represents the supremely beautiful! With his fore-legs, more delicate, more agile than the others, he pats the horrible mask, which in his eyes is an exquisite little face; voluptuously he nibbles and tickles with his jaws the equally hideous mouth opposite. It is all superb in its tenderness and simplicity. The dove is said to have invented the kiss. But I know that he had a forerunner in the Scorpion.

Dulcinea lets her admirer have his way and remains passive, not without a secret longing to slip off. But how is she to set about it? It is quite easy. The Scorpioness makes a cudgel of her tail and brings it down with a bang upon the wrists of her too-ardent wooer, who there and then lets go. The match is broken off, for the time being. Tomorrow, the sulking-fit will be over and things will resume their course. . . .

June sets in. For fear of a disturbance caused by too brilliant an illumination, I have hitherto kept the lantern hung outside, at some distance from the pane. The insufficient light does not allow me to observe certain details of the manner in which the couple are linked when strolling. Do they both play an active part in the scheme of the clasped hands? Are their fingers mutually interlinked? Or is only one of the pair active; and, if so, which? Let us ascertain exactly; the thing is not without importance.

I place the lantern inside, in the centre of the cage. There is good light everywhere. Far from being scared, the Scorpions are gayer than ever. They come hurrying round the beacon; some even try to climb up, so as to be nearer the flame. They succeed in doing so by means of the framework containing the glass panes. They hang on to the edges of the tin strips and stubbornly, heedless of slipping, end by reaching the top. There, motionless, lying partly on the glass, partly on the support of the metal casing, they gaze the whole evening long, fascinated by the burning wick. They remind me of the Great Peacock Moths that used to hang in ecstasy under the reflector of my lamp.

At the foot of the beacon, in the full light, a couple lose no time in standing on their heads. The two fence prettily with their tails and then go a-strolling. The male alone acts. With the two fingers of each claw, he has seized the two fingers of the corresponding claw of the Scorpioness bundled together. He alone exerts himself and squeezes; he alone is at liberty to break the team when he likes: he has but to open his pincers. The female cannot do this; she is a prisoner, handcuffed by her ravisher.

In rather infrequent cases, one may see even more remarkable things. I have caught the Scorpion dragging his sweetheart along by the two fore-arms; I have seen him pull her by one leg and the tail. She had resisted the advances of the outstretched hand; and the bully, forgetful of all reserve, had thrown her on her side and clawed hold of her at random. The thing is quite clear: we have to do with a regular rape, abduction with violence. Even so did Romulus' youths rape the Sabine women.

The brutal ravisher is singularly persistent in his feats of prowess, when we remember that things end tragically sooner or later. The ritual demands that he shall be eaten after the wedding. What a strange world, in which the victim drags the sacrificer by main force to the altar!

From one evening to the next, I become aware that the more corpulent females in my menagerie hardly ever take part in the sport of the linked team; it is nearly always the young, slim-waisted ones to whom the ardent strollers pay their addresses. They must have sprightly flappers. True, there are moments when they have interviews with the others, accompanied by strokes of the tail and attempts at harnessing; but these are brief displays, devoid of any great fervour. No sooner is she seized by the fingers than the portly temp-tress, with a blow of her tail, rebukes the untimely familiarity. The rejected suitor retires from the contest without insisting further. They go their several ways.

The big-bellied ones are therefore elderly matrons, indifferent nowadays to the effusive manners of the pairing-season. This time last year and perhaps even before, they had their own good spell; and that is enough for them henceforth. The female Scorpion's period of gestation is consequently extraordinarily long, longer than will be often found even among animals of a higher order. It takes her a year or more to mature her germs.

Let us return to the couple whom we have just seen forming up beneath the lantern. I inspect them at six o'clock the next morning. They are under the tile linked precisely as though for a stroll, that is to say, face to face and with clasped fingers. While I watch them, a second pair forms and begins to wander to and fro. The early hour of the expedition surprises me: I had never seen such an incident in broad daylight and was seldom to see it again. As a rule it is at nightfall that the Scorpions go strolling in couples. Whence this hurry today?

I seem to catch a glimpse of the reason. It is stormy weather; in the afternoon, there is incessant, very mild thunder. St. Medard, whose feast fell yesterday, is opening his floodgates wide; it pours all night. The great electric tension and the smell of ozone have stirred up the sleepy hermits, who, nervously irritated, for the most part come to the threshold of their cells, stretching their questioning claws out-side and enquiring into the condition of things. Two, more violently excited than the others, have come out, influenced by the intoxica-tion of the pairing which is enhanced by the intoxication of the storm; they suited each other; and here they are solemnly marching to the sound of the thunder-claps.

They pass before open huts and try to go in. The owner objects. He appears in the doorway, shaking his fists, and his action seems to say:

"Go somewhere else; this place is taken."

They go away. They meet with the same refusal at other doors, the same threats from the occupant. At last, for want of anything better, they make their way under the tile where the first couple have been lodging since the day before. The cohabitation entails no quarrelling; the first settlers and the newcomers, side by side, keep very quiet, each couple absorbed in meditation, completely motion-less, with fingers still clasped. And this goes on all day. At five o'clock in the evening, the couples separate. Anxious apparently to take part in the usual twilight rejoicings, the males leave the shelter; the females, on the other hand, remain under the tile. Nothing, so far as I know, has happened during the long interview, nothing despite the stimulating effects of the thunderstorm.

This fourfold occupation of one dwelling is not an isolated instance: groups, regardless of sex, are not infrequent under the potsherds in the glass cage. I have already said that, in their original homes, I have never found two Scorpions under one stone. We must not infer from this that unsociable habits prohibit all intercourse among neighbours; we should be making a mistake: the glazed enclosure tells us so. There are cabins in more than sufficient numbers; each Scorpion would be able to choose himself a dwelling and thenceforth to occupy it as the jealous owner. Nothing of the kind takes place. Once the nocturnal excitement sets in, there is no such thing as a home respected by others. Everything is common property. Whoever wishes to slip under the first tile that offers does so without protest from the occupant. The Scorpions go abroad, walk about and enter any house they may chance upon. In this way, when the twilight diversions are over, groups of three, four, or sometimes more are formed without distinction of sex and, packed pretty closely in the narrow home, spend the rest of the night and the whole of the following day together. For that matter, theirs is only a temporary shanty, which is exchanged next evening for another, according to the strollers' fancy. And these roving gipsies live quite peaceably. There is never any serious strife between them, even when they are five or six in the same messroom.

Now this tolerance prevails only in the adults, due, no doubt, to some degree, to the fear of reprisals. There is another and more imperative reason for peaceful relations: concord is a necessity in assemblies at which the future is being prepared. The Scorpions' characters therefore become assuaged, but not entirely: there are always perverse appetites among the females who are about to enter upon the period of gestation.

I have always present in my mind the memory of the following odious spectacle. A heedless male, who has attained hardly a third or a fourth of his final size, is passing, unthinking of evil, before the door of a dwelling. The fat matron comes out, accosts the poor wretch, picks him up in her claws, kills him with her sting and then quietly eats him.

Scorpion lads and lasses, the one sooner, the other later, perish in the same manner in the glass cage. I scruple to replace the deceased:

it would be providing fresh food for the slaughter. There were a dozen of them; and in a few days I have not one left. Without the excuse of hunger, for the regular victuals are plentiful, the females have devoured them all. Youth is certainly a beautiful thing, but it has terrible drawbacks in the society of these ogresses.

I would gladly ascribe these massacres to the peculiar cravings often provoked by pregnancy. The future mother is suspicious and intolerant; to her everything is an enemy, to be got rid of by eating it, when strength permits. And indeed, when the quickly emancipated family is born, in the middle of August, a profound peace reigns in the menagerie. My vigilance is unable to surprise a single case of these outbreaks of cannibalism which used to occur so often.

On the other hand, the males, indifferent to the safety of the family, know nothing of these tragic frenzies. They are peaceful creatures, blunt in their manners, but in any event incapable of ripping up their fellows. We never see two rivals disputing in mortal combat, for the possession of the coveted bride. Things happen, if not mildly, at least without blows of the dagger.

Two suitors come upon the same Scorpioness. Which of the two will propose to her and take her for a walk? The point will be decided by strength of wrist.

Each takes the beauty by the hand nearest to him with the fingers of one claw. One standing on the right, the other on the left, they pull with all their might in opposite directions. The legs, braced backwards, exert a powerful leverage; the flanks quiver; the tails sway to and fro and suddenly dart forward. Now for it! They tug at the Scorpioness by fits and starts with sudden backward runs; it is as though they meant to pull her in two and each to carry off a piece. A declaration of love implies a threat to rend her asunder.

On the other hand, there is no direct exchange of fisticuffs between them, not even a back-hander with the tail. Only the victim is ill-treated and roughly at that. To see these lunatics struggling, you would think that their arms would be torn out. Nevertheless, there are no dislocations.

Weary of an ineffectual contest, the two competitors at last take each other by the hands that remain at liberty: they form a chain of three and resume the process of jerking and tugging more violently than ever. Each of them bustles to and fro, advances, recoils and pulls his hardest till he is exhausted. Suddenly, the more fatigued of the two throws up the sponge and runs away, leaving his adversary in possession of the object of their passions so vehemently disputed. Then, with his free claw, the victor completes the team and the stroll begins. As for the vanquished, we will not trouble about him: he will soon have found something in the crowd to make amends for his confusion.

I will give you another instance of these meek encounters between rivals. A couple are walking along. The male is of medium size, but nevertheless very eager at the game. When his companion refuses to advance, he pulls at her with jerks which send shudders along his spine. A second male, larger than the first, appears upon the scene. The lady takes his fancy; he desires her. Will he abuse his strength, fling himself on the little chap, beat him, perhaps stab him? By no means. Among Scorpions these delicate matters are not decided by force of arms.

The burly fellow leaves the dwarf alone. He goes straight to the coveted fair and seizes her by the tail. Then the two vie with each other in pulling, one in front, the other behind. A brief contest follows, leaving each of them the master of a claw. With frantic violence, one works on the right, the other on the left, as though they wished to pull the dame to pieces. At length the smaller realizes that he is beaten; he lets go and makes off. The big one lays hold of the abandoned prey; and the team takes the road without further incident.

Thus, evening after evening, for four months, from the end of April to the beginning of September, the preludes to the pairing are indefatigably repeated. The scorching dog-days do not calm these unruly lovers; on the contrary, they inflame them with new ardour. In the spring, I used to surprise the pilgrims' tandems singly, at long intervals; in July I observe them by threes and fours at a time, on the same evening.

I take the opportunity, with not much success, to enquire what goes on under the tiles where the strolling couples take refuge; my wish is to see the details of the tender interview from start to finish. It does me no good to turn over the potsherd, even during the quiet hours of the night. I have tried often and in vain. When deprived of their roof, the linked couples resume their ramble and make for another shelter, where the impossibility of prolonged observation obtains once more. Special circumstances, independent of any intervention on our part, are needed to make the delicate undertaking succeed.

Today these circumstances are present. At seven o'clock in the morning, on the third of July, a couple attracts my attention, a couple whom I saw forming, walking about and selecting a home on the previous evening. The male is under the tile, quite invisible save for the tips of his claws. The cabin was too small to shelter the two. He went in; she, with her mighty paunch, remained outside, clutched by the fingers by her companion.

The tail, curved into a wide arc, is bent slackly to one side, with the point of the sting resting on the ground. The eight legs, firmly planted, are drawn backwards, marking a tendency to escape. The whole body is completely motionless. I inspect the fat Scorpioness twenty times in the course of the day, without perceiving the least movement of the hinder part, the least change in the attitude, the least flexion in the curve of the tail. The animal could be no more lifeless if turned to stone.

The male, on his side, is no more active. Though I cannot see him, I at least observe his fingers, which would tell me of any change of posture. And this petrified condition, which has lasted for the best part of the night, persists all day, until eight o'clock in the evening. What do they feel, facing each other thus? What are they doing, motionless with clasped fingers? If the expression were allowable, I should say that they are meditating profoundly. It is the only term that more or less represents what I see. But no human language could have words to fit and convey the bliss, the ecstasy of the Scorpions thus coupled by the finger-tips. Let us remain silent upon that which we cannot understand.

A little before eight o'clock, when the animation outside the house is already approaching its height, the female suddenly moves; she struggles and, with an effort, contrives to release herself. She flees, with one of the pincers bent back towards her and the other stretched out. To break her seductive bonds, she pulled with such violence that she put one of her shoulders out of joint. She flees, feeling her way with the uninjured claw. The male runs off too. All is over for this evening.

These rambles in pairs, which are customary in the evening all through the summer, are evidently the preliminaries to more serious affairs. The strollers inspect each other, display their graces, show off their qualities before coming to conclusions. But when does the decisive moment arrive? My patience is exhausted in waiting for it; I vainly prolong my vigils and turn over potsherd after potsherd, in my anxiety at last to know the exact part played by the combs; my hopes remain unfulfilled.

It is at a very late hour in the night that the marriage is consummated: of that I have no doubt whatever. If I had any chance of arriving at the right moment, I would struggle against sleep till break of day: my old eyelids are still capable of doing so when the acquisition of an idea is at stake. But how hazardous my perseverance would be!

I am very well aware, having seen it over and over again, that, in the vast majority of cases, we find the couple next morning, under the tile, harnessed together just as they were on the evening before. To succeed, I should have to upset the habits of a lifetime and lie in wait every night for three or four months on end. The plan is beyond my strength; and I give it up.

Once only did I obtain an inkling of the solution of the problem. At the moment when I lift the stone, the male is turning over without releasing the clasp of his hands; with his belly upturned, he slowly slides backwards under his mate. Even so does the Cricket behave when his pleadings at last obtain a hearing. In this posture, the couple would only have to steady themselves, probably with the teeth of their combs, to achieve their ends. But, startled by the violation of their home, the superimposed twain separate then and there. From the little that I have seen, it seems likely, therefore, that the Scorpions end their mating in an attitude similar to that of the Crickets. In addition they have their hands clasped and their combs interlocked.

I am better informed of subsequent events within the cell. Let us mark the tiles under which the couples take refuge in the evening after their stroll. What do we find next morning? As a rule, precisely the same linked couple as the day before, face to face, with fingers united.

Sometimes the female is alone. The male, having finished his business, has found means to release himself and go away. He had grave reasons for cutting short the transports of the alcove. Especially in May, the time of the most ardent enjoyment, I often indeed find the female nibbling and relishing her deceased mate.

Who committed the murder? The Scorpioness, evidently. These are the atrocious customs of the Praying Mantis: the lover is stabbed and then eaten, if he does not retire in time. By the exercise of nimbleness and decision, he can do so sometimes, not always. He is able to release his hands, for it is his that squeeze; by lifting his thumbs, he unclasps them. But there remains the diabolical little mechanism of the combs, an apparatus of sensual pleasure, now a trap. On both sides the long teeth of this interlocking gear, closely fitting and perhaps spasmodically contracted, refuse to come apart as promptly as could be wished. The poor fellow is lost.

He has a poisoned dagger similar to that which threatens him: can he, does he know how to defend himself? It seems as though he cannot, for he is always the victim. It is possible that his reversed posture hinders him in wielding his tail, which he must curve over his back if he wishes to bring it into play. Perhaps also an insuperable instinct prevents him from putting the future mother to death. He allows himself to be pinked by the terrible bride; he perishes without defence.

The widow forthwith begins to eat him. It is a part of the ritual, as with the Spiders, who, deprived of the Scorpion's fatal engine, at least leave the males time to escape if they are prompt enough in forming a decision.

The funeral repast, though frequent, is not indispensable; whether the male is devoured depends a little on the condition of the female's stomach. I have seen some who, despising the nuptial morsel, frugally swallowed the head of the deceased and then flung the corpse outside, without touching it again. I have seen these furies carry their dead husband at arm's length, dragging him about the whole morning, in sight of all, like a trophy, and then, without further ceremony, leaving him untouched and abandoning him to those eager dissectors, the Ants.

THEODORE ROOSEVELT (1858–1919)

Foreword to A *Book-Lover's Holidays in the Open*, and On an East African Ranch

From *A Book-Lover's Holidays in the Open*, 1916, and *African Game Trails*, 1909

[IF ROOSEVELT had not been a statesman, politician, soldier, or explorer and had never held public office of any kind, he would still have confident claim to being one of our best field naturalists. From years of association with him at home and in the field I know this to be true, and it is plainly evident from his contributions to conservation and to the meaning of pattern and color in animals and birds. An appreciative listener and stimulating conversationalist, he held his own with any naturalist or scientist of his day. In his writings valuable original observations on animals are scattered through many travel volumes.

May I be permitted one anecdote which illustrates one of many less-known phases of his mind? It deals with a game that Colonel Roosevelt and I used to play at Sagamore Hill. After an evening of talk, perhaps about the fringes of knowledge, or some new possibility of climbing inside the minds and senses of animals, we would go out on the lawn, where we took turns at an amusing little astronomical rite. We searched until we found, with or without glasses, the faint, heavenly spot of light-mist beyond the lower left-hand corner of the Great Square of Pegasus, when one or the other of us would then recite:

> That is the Spiral Galaxy in Andromeda.
> It is as large as our Milky Way.
> It is one of a hundred million galaxies.
> It is 750,000 light-years away.
> It consists of one hundred billion suns,
>   each larger than our sun.

After an interval Colonel Roosevelt would grin at me and say: "Now I think we are small enough! Let's go to bed."

We must have repeated this salutary ceremony forty or fifty times in the course of years, and it never palled.]

234

FOREWORD to *A Book-Lover's Holidays in the Open*

THE MAN should have youth and strength who seeks adventure in the wide, waste spaces of the earth, in the marshes, and among the vast mountain masses, in the northern forests, amid the steaming jungles of the tropics, or on the deserts of sand or of snow. He must long greatly for the lonely winds that blow across the wilderness, and for sunrise and sunset over the rim of the empty world. His heart must thrill for the saddle and not for the hearthstone. He must be helmsman and chief, the cragsman, the rifleman, the boat steerer. He must be the wielder ot axe and of paddle, the rider of fiery horses, the master of the craft that leaps through white water. His eye must be true and quick, his hand steady and strong. His heart must never fail nor his head grow bewildered, whether he face brute and human foes, or the frowning strength of hostile nature, or the awful fear that grips those who are lost in trackless lands. Wearing toil and hardship shall be his; thirst and famine he shall face, and burning fever. Death shall come to greet him with poison-fang or poison-arrow, in shape of charging beast or of scaly things that lurk in lake and river; it shall lie in wait for him among untrodden forests, in the swirl of wild waters, and in the blast of snow blizzard or thunder-shattered hurricane.

Not many men with wisdom make such a life their permanent and serious occupation. Those whose tasks lie along other lines can lead it for but a few years. For them it must normally come in the hardy vigor of their youth, before the beat of the blood has grown sluggish in their veins.

Nevertheless, older men also can find joy in such a life, although in their case it must be led only on the outskirts of adventure, and although the part they play therein must be that of the onlooker rather than that of the doer. The feats of prowess are for others. It is for other men to face the peril of unknown lands, to master unbroken horses, and to hold their own among their fellows with bodies of supple strength. But much, very much, remains for the man who has "warmed both hands before the fire of life," and who, although he loves the great cities, loves even more the fenceless grass-land and the forest-clad hills.

The grandest scenery of the world is his to look at if he chooses; and he can witness the strange ways of tribes who have survived into an alien age from an immemorial past, tribes whose priests dance in honor of the serpent and worship the spirits of the wolf and the bear. Far and wide, all the continents are open to him as they never were to any of his forefathers; the Nile and the Paraguay are easy of access, and the border-land between savagery and civilization; and the veil of the past has been lifted so that he can dimly see how, in time immeasurably remote, his ancestors — no less remote — led furtive lives among uncouth and terrible beasts, whose kind has perished utterly from the face of the earth. He will take books with him as he journeys; for the keenest enjoyment of the wilderness is reserved for him who enjoys also the garnered wisdom of the present and the past. He will take pleasure in the companionship of the men of the open; in South America, the daring and reckless horsemen who guard the herds of the grazing country, and the dark-skinned paddlers who guide their clumsy dugouts down the dangerous equatorial rivers; the white and red and half-breed hunters of the Rockies, and of the Canadian woodland; and in Africa the faithful black gun-bearers who have stood steadily at his elbow when the lion came on with coughing grunts, or when the huge mass of the charging elephant burst asunder the vine-tangled branches.

The beauty and charm of the wilderness are his for the asking, for the edges of the wilderness lie close beside the beaten roads of present travel. He can see the red splendor of desert sunsets, and the unearthly glory of the afterglow on the battlements of desolate mountains. In sapphire gulfs of ocean he can visit islets, above which the wings of myriads of sea-fowl make a kind of shifting cuneiform script in the air. He can ride along the brink of the stupendous cliff-walled canyon, where eagles soar below him, and cougars make their lairs on the ledges and harry the big-horned sheep. He can journey through the northern forests, the home of the giant moose, the forests of fragrant and murmuring life in summer, the iron-bound and melancholy forests of winter.

The joy of living is his who has the heart to demand it.

ON AN EAST AFRICAN RANCH

The house at which we were staying stood on the beautiful Kitanga hills. They were so named after an Englishman, to whom the natives had given the name of Kitanga; some years ago, as we were told, he had been killed by a lion near where the ranch-house now stood; and we were shown his grave in the little Machakos graveyard. The house was one story high, clean and comfortable, with a veranda running round three sides; and on the veranda were lion-skins and the skull of a rhinoceros. From the house we looked over hills and wide lonely plains; the green valley below, with its flat-topped acacias, was very lovely; and in the evening we could see, scores of miles away, the snowy summit of mighty Kilimanjaro turn crimson in the setting sun. The twilights were not long; and when night fell, stars new to Northern eyes flashed glorious in the sky. Above the horizon hung the Southern Cross, and directly opposite in the heavens was our old familiar friend the Wain, the Great Bear, upside down and pointing to a North Star so low behind a hill that we could not see it. It is a dry country, and we saw it in the second year of a drought; yet I believe it to be a country of high promise for settlers of white race. In many ways it reminds one rather curiously of the great plains of the West, where they slope upward to the foot-hills of the Rockies. It is a white man's country. Although under the equator, the altitude is so high that the nights are cool, and the region as a whole is very healthy. . . .

Game was in sight from the veranda of the house almost every hour of the day. Early one morning, in the mist, three hartbeests came right up to the wire fence, twoscore yards from the house itself; and the black-and-white striped zebra, and ruddy hartbeest, grazed or rested through the long afternoons in plain view, on the hillsides opposite.

It is hard for one who has not himself seen it to realize the immense quantities of game to be found on the Kapiti Plains and Athi Plains and the hills that bound them. The common game of the plains, the animals of which I saw most while at Kitanga and in the neighborhood, were the zebra, wildebeest, hartbeest, Grant's gazelle, and "tommies" or Thomson's gazelle; the zebra and the hartbeest, usually known by the Swahili name of kongoni, being by far the most plentiful. Then there were impala, mountain reedbuck, duiker, steinbuck, and diminutive dikdik. As we travelled and hunted we were hardly ever out of sight of game; and on Pease's farm itself there were many thousand head; and so there were on Slatter's. If wealthy men who desire sport of the most varied and interesting kind would purchase farms like these, they could get, for much less money, many times the interest and enjoyment a deer-forest or grouse-moor can afford.

The wildebeest or gnu were the shyest and least plentiful, but in some ways the most interesting, because of the queer streak of ferocious eccentricity evident in all their actions. They were of all the animals those that were most exclusively dwellers in the open, where there was neither hill nor bush. Their size and their dark-bluish hides, sometimes showing white in the sunlight, but more often black, rendered them more easily seen than any of their companions. But hardly any plains animal of any size makes any effort to escape its enemies by eluding their observation. Very much of what is commonly said about "protective coloration" has no basis whatever in fact. Black and white are normally the most conspicuous colors in nature (and yet are borne by numerous creatures who have succeeded well in the struggle for life) ; but almost any tint, or combination of tints, among the grays, browns and duns, harmonizes fairly well with at least some surroundings, in most landscapes; and in but a few instances among the larger mammals, and in almost none among those frequenting the open plains, is there the slightest reason for supposing that the creature gains any benefit whatever from what is loosely called its "protective coloration." Giraffes, leopards, and zebras, for instance, have actually been held up as instances of creatures that are "protectingly" colored and are benefitted thereby. The giraffe is one of the most conspicuous objects in nature, and never makes the slightest effort to hide; near by its mottled hide is very noticeable, but, as a matter of fact, under any ordinary circumstance any possible foe trusting to eyesight would discover the giraffe so far away that its coloring would seem uniform, that is, would because of the distance be indistinguishable from a general tint which really might have a slight protective value. In other words while it is possible that the giraffe's beautifully waved coloring may under certain circumstances, and in an infinitesimally small number of cases, put it at a slight disadvantage in the struggle for life, in the enormous majority of cases — a majority so great as to make the remaining cases negligible — it has no effect whatever, one way or the other; and it is safe to say that under no conditions is its coloring of the slightest value to it as affording it "protection" from foes trusting to their eyesight. So it is with the leopard; it is undoubtedly much less conspicuous than if it were black — and yet the black leopards, the melanistic individuals, thrive as well as their spotted brothers; while on the whole it is probably slightly more conspicuous than if it were nearly unicolor, like the American cougar. As compared with the cougar's tawny hide the leopard's coloration represents a very slight disadvantage, and not an advantage, to the beast; but its life is led under conditions which make either the advantage or the disadvantage so slight as to be negligible; its peculiar coloration is probably in actual fact of hardly the slightest service to it from the "protective" standpoint whether as regards escaping from its enemies or approaching its prey. It has extraordinary facility in hiding, it is a master of the art of stealthy approach; but it is normally nocturnal, and by night the color of its hide is of no consquence whatever; while by day, as I have already said, its varied coloration renders it slightly more easy to detect than is the case with the cougar.

All of this applies with peculiar force to the zebra, which it has also been somewhat the fashion of recent years to hold up as an example of "protective coloration." As a matter of fact the zebra's coloration is not protective at all; on the contrary it is exceedingly conspicuous, and under the actual conditions of the zebra's life probably never hides it from its foes; the instances to the contrary being due to conditions so exceptional that they may be disregarded. If any man seriously regards the zebra's coloration as "protective," let him try the experiment of wearing a hunting suit of the zebra pattern; he will speedily be undeceived. The zebra is peculiarly a beast of the open plains, and makes no effort to hide from the observation of its foes. It is occasionally found in open forest; and may there now and then escape observation simply as any animal of any color — a dun hartbeest or a nearly black bushbuck — may escape observation. At a distance of over a few hundred yards the zebra's coloration ceases to be conspicuous simply because the distance has caused it to lose all its distinctive character — that is, all the quality which could possibly make it protective. Near by it is always very conspicuous, and if the conditions are such that any animal can be seen at all, a zebra will catch the eye much more quickly than a Grant's gazelle, for instance. These gazelles, by the way, although much less conspicuously colored than the zebra, bear when young, and the females even when adult, the dark side stripe which characterizes all sexes and ages of the smaller gazelle, the "tommy"; it is a very conspicuous marking, quite inexplicable on any theory of protective coloration. The truth is that no game of the plains is helped in any way by its coloration in evading its foes and none seeks to escape the vision of its foes. The larger game animals of the plains are always walking and standing in conspicuous places, and never seek to hide or take advantage of cover; while, on the contrary, the little grass and bush antelopes, like the duiker and steinbuck, trust very much to their power of hiding, and endeavor to escape the sight of their foes by lying absolutely still, in the hope of not being made out against their background. On the plains one sees the wildebeest farthest off and with most ease; the zebra and hartbeest next; the gazelles last. . . .

Zebra share with hartbeest the distinction of being the most abundant game animal on the plains, throughout the whole Athi region. The two creatures are fond of associating together, usually in mixed herds; but sometimes there will merely be one or two individuals of one species in a big herd of the other. They are sometimes, though less frequently than the hartbeest, found in open bush country; but they live in the open plains by choice.

I could not find out that they had fixed times for resting, feeding, and going to water. They and the hartbeests formed the favorite prey of the numerous lions of the neighborhood; and I believe that the nights, even the moonlight nights, were passed by both animals under a nervous strain of apprehension, ever dreading the attack of their arch-enemy, and stampeding from it. Their stampedes cause the utmost exasperation to the settlers, for when in terror of the real or imaginary attack of a lion, their mad, heedless rush takes them through a wire fence as if it were made of twine and pasteboard. But a few months before my arrival a mixed herd of zebra and hartbeest, stampeded either by lions or wild dogs, rushed through the streets of Nairobi, several being killed by the inhabitants, and one of the victims falling just outside the Episcopal church. The zebras are nearly powerless when seized by lions; but they are bold creatures against less formidable foes, trusting in their hoofs and their strong jaws; they will, when in a herd, drive off hyenas or wild dogs, and will turn on hounds, if the hunter is not near. If the lion is abroad in the daytime, they, as well as the other game, seem to realize that he cannot run them down; and though they follow his movements with great alertness, and keep at a respectful distance, they show no panic. Ordinarily, as I saw them, they did not seem very shy of men; but in this respect all the game displayed the widest differences, from time to time, without any real cause, that I could discern, for the difference. At one hour, or on one day, the zebra and hartbeest would flee from our approach when half a mile off; and again they would permit us to come within a couple of hundred yards, before moving slowly away. On two or three occasions at lunch herds of zebra remained for half an hour watching us with much curiosity not over a hundred yards off. Once, when we had been vainly beating for lions at the foot of the Elukania ridge, at least a thousand zebras stood, in herds, on every side of us, throughout lunch; they were from two to four hundred yards distant, and I was especially struck by the fact that those which were to leeward and had our wind were no more alarmed than the others. I have seen them water at dawn and sunset, and also in the middle of the day; and I have seen them grazing at every hour of the day, although I believe most freely in the morning and evening. At noon and until the late afternoon those I saw were quite apt to be resting, either standing or lying down. They are noisy. Hartbeests merely snort or sneeze now and then; but the shrill, querulous barking of the "bonte quaha," as the Boers call the zebra, is one of the common sounds of the African plains, both by day and night. It is usually represented in books by the syllables "qua-ha-ha"; but of course our letters and syllables were not made to represent, and can only in arbitrary and conventional fashion represent, the calls of birds and mammals; the bark of the bonte quagga or common zebra could just as well be represented by the syllables "ba-wa-wa," and as a matter of fact it can readily be mistaken for the bark of a shrill-voiced dog. After one of a herd has been killed by a lion or a hunter his companions are particularly apt to keep uttering their cry. Zebras are very beautiful creatures, and it was an unending pleasure to watch them. I never molested them save to procure specimens for the museums, or food for the porters, who like their rather rank flesh. They were covered with ticks like the other game; on the groin, and many of the tenderest spots, the odious creatures were in solid clusters; yet the zebras were all in high condition, with masses of oily yellow fat. One stallion weighed 650 pounds.

The hartbeest — Coke's hartbeest, known locally by the Swahili name of kongoni — were at least as plentiful, and almost as tame, as the zebras. As with the other game of equatorial Africa, we found the young of all ages; there seems to be no especial breeding-time, and no one period among the males corresponding to the rutting-season among Northern animals. The hartbeests were usually inseparable companions of the zebra; but though they were by preference beasts of the bare plain, they were rather more often found in open bush than were their striped friends. There are in the country numerous ant-hills, which one sees in every stage of development, from a patch of bare earth with a few funnel-like towers to a hillock a dozen feet high and as many yards in circumference. On these big ant-hills one or two kongoni will often post themselves as lookouts, and are then almost impossible to approach. The bulls sometimes fight hard among themselves, and although their horns are not very formidable weapons, yet I knew of one case in which a bull was killed in such a duel, his chest being ripped open by his adversary's horn; and now and then a bull will kneel and grind its face and horns into the dust or mud. Often a whole herd will gather around and on an ant-hill, or even a small patch of level ground, and make it a regular stamping-ground, treading it into dust with their sharp hoofs. They have another habit which I have not seen touched on in the books. Ordinarily their droppings are scattered anywhere on the plain; but again and again I found where hartbeests — and more rarely Grant's gazelles — had in large numbers deposited their droppings for some time in one spot. Hartbeests are homely creatures, with long faces, high withers, showing when first in motion a rather ungainly gait, but they are among the swiftest and most enduring of antelope, and when at speed their action is easy and regular. When pursued by a dog they will often play before him — just as a tommy will — taking great leaps, with all four legs inclined backward, evidently in a spirit of fun and derision. . . .

The days I enjoyed most were those spent alone with my horse and gun-bearers. We might be off by dawn, and see the tropic sun flame splendid over the brink of the world; strange creatures rustled through the bush or fled dimly through the long grass, before the light grew bright; and the air was fresh and sweet as it blew in our faces. When the still heat of noon drew near I would stop under a tree, with my water canteen and my lunch. The men lay in the shade, and the hobbled pony grazed close by, while I either dozed or else watched through my telescope the herds of game lying down or standing drowsily in the distance. As the shadows lengthened I would again mount, and finally ride homeward as the red sunset paled to amber and opal, and all the vast, mysterious African landscape grew to wonderful beauty in the dying twilight.

J. ARTHUR THOMSON (1861–   )

About Tadpoles

From *The Biology of the Seasons*. 1911

[THE NAME OF this distinguished Scottish scientist brings to mind his outstanding works on Darwinism, Herbert Spencer, heredity, the evolution of sex, and especially, in conjunction with Dr. Geddes, the great two-volume work entitled *Life: Outlines of General Biology*. This, more than any other presentation of its kind, shows us life in all its manifestations on earth in a truly encyclopedic yet readable fashion. As illustrative of his versatility and ability to combine sound science and graceful writing, I have chosen to present what he has to say on tadpoles, from *The Biology of the Seasons*.]

THE FROGS are among the earliest heralds of the Spring, for although their croaking (in March or earlier) may not be particularly attractive in our ears, it has the same deep *motif* as the nightingale's song. It is a "love"-call. Awakening after a winter's lethargy and fasting, the frogs creep out of the mud of the pond and call to one another. They unite in couples, and the eggs laid by the female in the water are fertilised by the male just as they are laid. These eggs form the familiar masses of "frog-spawn" that we see in the ditches and ponds — often, it must be allowed, in places which a little more intelligence would have avoided.

It is profitable to pause to take a good look at this frog-spawn, for it illustrates a number of biological ideas, and perhaps we may be fortunate enough to see with a pocket-lens the eggs dividing into two, four, eight, and more cells, as if they were being cut by an invisible knife. Each egg in our common British frog (*Rana temporaria*) is about a tenth of an inch in diameter; it is almost entirely black, all but a small white lower pole; it is surrounded by a large sphere of non-living jelly, corresponding to the white of egg in a hen's egg; and there is no egg-shell. The whole mass, often of 2000 eggs, sinks at first, but afterwards floats freely.

Let us consider the biological significance of these spheres of jelly around the eggs, for it is very interesting to notice how they are justi-

243

fied on count after count, though they are non-living extrinsic invest-
ments. The spheres buoy up the eggs and at the same time obviate
overcrowding. In the little chinks between the spheres there are often
groups of green unicellular plants which liberate oxygen in the sunlight and use up the carbonic acid gas which the developing eggs
produce — a most profitable association, a miniature illustration of
the balance of Nature. But there is a fauna as well as a flora of frog-
spawn, and the chinks are tenanted by small fry — such as water-fleas
and rotifers — some of which eventually loosen the gelatinous envelopes, helping the larval-frogs to escape. Others, it must be admitted,
seem to wait to devour. Once again, the envelopes of jelly are useful
in lessening the risks of jostling — which might be fatal to the delicate
embryos — when the wind raises waves in the pond, or when a water-
hen or coot splashes in among the spawn. Moreover, the jelly seems
to be unpalatable to most water animals, and it is so slippery that
few birds can make anything of it. Finally, it may be that the clear
spheres serve as so many greenhouses, enabling the ova to make the
most of the sun's rays. All this illustrates the scientific view of Nature
as an arena where efficiency in any form always counts.

About a fortnight or three weeks after the individual life began,
that is, after fertilisation, the minute larvae are hatched from the
delicate envelope of the ovum, and begin to wriggle about in the
dissolving jelly. They are somewhat awkward-looking, half-made
creatures at first, and when they emerge from the jelly they are mouth-
less, limbless, eyeless, and gill-less. They attach themselves, often in
long rows, to water-weed, the adhesion being effected by a paired
cement-gland below the position of the future mouth. A bulging on
the ventral surface of the body indicates the position of the still unused remains of the legacy of yolk.

Soon after hatching three pairs of external gills grow out, the first
much the largest, one upon each of the first three branchial arches.
These are not comparable to the external gills of a young shark or
skate, which are really elongated internal gills projecting through
the gill-clefts. They are comparable to the true external gills seen
in the young of some very archaic fishes, still living to-day, the *Polyp-
terus* of the Gambia and some other African rivers, and two of the
mud-fishes, *Protopterus* from Africa, and *Lepidosiren* from the Amazons. One or two days after hatching (in our common *Rana temporaria*) another important structure appears on the larva — the mouth
is formed in the centre of a groove in front of the adhesive organs,
and hundreds of small horny teeth are developed.

When the food-canal becomes open, four pairs of gill-clefts break out from the pharynx, and a gill-cover overlaps the first set of gills. These dwindle and are absorbed, their place being taken by a second set of gills supported by the hinder margin of the lower halves of four gill-arches. As these are enclosed in a gill-chamber and as they form a second set, it is natural to compare them to typical fish-gills. But they are really in the strict sense external, and they are certainly skin-covered. Each gill-chamber has at first its opening, but that of the right side joins with that of the left.

About a month after hatching the larval frog is in many ways fish-like: for instance, it has a two-chambered heart which drives impure blood to the gills, which are enclosed by a gill-cover. It swims by its laterally compressed tail, which shows a well developed unpaired fin, without fin-rays, however, which support the unpaired fins of fishes. *In a very general way* it may be said that the developing frog visibly climbs up its own genealogical tree. It is this general idea, indeed, of recapitulation that makes the study of the frog's life-history perennially interesting. It re-enacts the epoch-making colonisation of the dry land, and in many of its internal changes, e.g. in the making of the three-chambered heart, it probably re-enacts what took place very long ago — before the Coal-Measures were laid down in Britain — when Amphibians evolved from a Piscine stock. But what took the race long ages to accomplish is achieved by the individual in a few days, — a fact so familiar that we are apt to forget its marvellousness — the mystery of cumulative and condensed inheritance.

With the acquisition of a mouth the larva begins to feed eagerly, nibbling at plants in the water, and also eating animal food. As a consequence it grows, and the food-canal, in particular, becomes very long and coiled like a watch-spring. It is interesting to notice the relatively great length of the intestine during the predominantly vegetarian period, for it is usual in the animal kingdom to find a diet of vegetable food — which is somewhat slowly digested — associated with length of food-canal or with some equivalent of length. As the tail becomes stronger and the power of locomotion increases, the horse-shoe shaped adhesive organ is converted into two small discs which gradually disappear.

A new stage is marked by the appearance of the hind-limbs as minute projecting buds at the boundary between trunk and tail. Why should hind-limbs appear earlier than the fore-limbs, which are, moreover, much shorter? Investigation shows that they begin to develop at the same time — a fact which gives additional point to the question. The fore-limbs are delayed by the gill-cover, which does not impede the hind-limbs, and they eventually emerge, the left one through the "spiracle," the right one by a rupture. Perhaps we get some insight into the orderliness of developmental processes when we notice that the microscopic lashes or cilia which have hitherto covered the skin of the larva now disappear. In most cases, except as regard reproduction, what we may call "Animate Nature" (for shortness) is conspicuously *economical*.

After the appearance of the hind-legs, the larvae come often to the surface to breathe. They are learning to use their lungs, which have been slowly developing for some time as pockets projecting into the body-cavity from the under side of the gullet. The tadpoles are now about two months old, and in having lungs as well as gills they may be compared to the double-breathing Mud-Fishes or Dipnoi. As the lungs become established and functional, the gills dwindle, and an intricate series of internal changes leads from an essentially fish-like heart and circulation to the characteristic Amphibian arrangements.

After a period of hearty feeding, with consequent increase in size and strength, the tadpole begins to show signs of approaching metamorphosis. It loses its appetite, it becomes much less energetic. The tail begins to break up internally, its muscles and other structures become disintegrated and dissolved, and most of the material is swept away in the blood stream to help in building up a better head. Wandering amoeboid cells, which are present in almost all animals except threadworms and lancelets, seem to play an important part in the extraordinary process of absorbing the tail, working like sappers and miners among the debris, dissolving some of the material, carrying some away. In certain respects what occurs is comparable to violent inflammation. It is like a pathological process which has become normal, and thus from watching tadpoles we get a glimpse of a deep-reaching theory of disease as "a perturbation which contains no elements essentially different from those of health, but elements presented in a different and less useful order." Often, at least, a disease implies a series of metabolic changes which are not in themselves in any way extraordinary — only they are out of place, out of time, and out of order.

One of the many careful observers of the annual wonder — the metamorphosis of the tadpole — gives the following terse statement of some of the more obvious changes: "The horny jaws are thrown off; the large frilled lips shrink up; the mouth loses its rounded suctorial form and becomes much wider; the tongue, previously small, increases considerably in size; the eyes, which as yet have been beneath the skin become exposed."

As the tail shortens more and more, the tadpole, rapidly ceasing to be a tadpole, recovers its appetite and feeds greedily on animal matter, sometimes on its younger fellows. The abdomen shrinks, the stomach and liver enlarge, the intestine becomes relatively narrower and shorter. The tail is reduced to a short projecting stump, and, apart from this, the adult shape has been reached. Disinclination for a purely aquatic life becomes marked, and the young frogs clamber ashore. As they have lost all trace of gills, they are apt to drown in aquaria unless they have floating rafts to climb on to, or some other means of breathing dry air.

It is difficult to say which aspect of the development of tadpoles is most interesting. As we have seen, it is interesting in its main features as a modified recapitulation of that transition from aquatic to aerial respiration, from water to terra firma, which must have marked one of the most important epochs in the evolution of Vertebrates.

But it is equally interesting to go into minute detail and notice the young tadpole's small tongue has not much muscularity about it; that as the tongue increases in size the muscles also increase, but yet are quite unable to move the tongue, though perhaps of some service in compressing glands; and that, as the metamorphosis is accomplished and the frogling hops ashore, the muscles of the tongue are at length strong enough to shoot out the tongue on the daydreaming fly. The peculiar interest of this is that Amphibians were the first animals to have a movable tongue, that of fishes being even worse than flabby, entirely non-muscular.

It is very interesting to consider in the same way the other momentous acquisitions made by the race of Amphibians — such as fingers and toes, and the power of gripping things, vocal cords, and the power of speech — though how much they have to say in their extraordinary jabber no one knows.

Another interesting consideration is the variety of solutions that this one animal, the frog, offers to the problems of its life. Even in mathematics, we believe, there may be more than one solution to a problem, and every one knows that this is true of the practical problems of human life. There is considerable variety in the solutions of the problems of *Brodwissenschaft,* though in strictness, we suppose, the fact of the matter is that the *conditions* of the typical problems are diverse, and therefore the solutions are diverse. But our point is this, that, to the two great problems of nutrition and respiration (if they are really *two,* for is not oxygen a kind of food?), the frog offers in the course of its life-history an unusual diversity of answer.

It will feed on its legacy of yolk, on unicellular Algae, on the epidermis of aquatic plants, on the vegetable debris in the water, on animal matter by the way, on its own tail (of course in a sort of surreptitious phagocytic fashion), on its own brethren, on dead things in the water (tadpoles clean delicate skeletons beautifully), and, by and by, when it comes to its own, after a remarkable gustatory curriculum, it will feed on living insects and little else. Yet the way it feeds as an adult, *e.g.* on beetles much too large for it, is often far from saying much for its varied gastronomic education.

And again, as regards the fundamental problem of breathing, we find the newly hatched tadpole breathing through its skin in the old-fashioned manner of earthworm and leech; then follow in succession, the first set of external gills, the gill-clefts, the second set of external gills, which are usually called internal; then follows a period with gills and lungs together; then there is the transition to terrestrial life with pulmonary and (retained) cutaneous respiration; finally, in winter, the hibernating frog, retiring into the mud-fortresses of its remote ancestors, breathes by its skin only.

Several experimenters have found that the numerical proportion of the sexes in frogs *appears as if it were* modifiable by changes in the nutrition. Thus Professor Yung, of Geneva, fed tadpoles with minced beef, and found that the percentage of females was 78, instead of 54 as in the control set in natural conditions. He fed another set with fish flesh, and the percentage rose to 81, as against 61 in the control set. In a third set, to which the flesh of frogs was supplied, the percentage rose from 56 to 92! It has to be noted, however, that subsequent experimenters have not confirmed these results, which are also open to the fatal objection that the sex of those tadpoles that died was not determined. It may have been that the change of diet affected the males prejudicially and favoured the survival of females. While the results of the experiments cannot, without further inquiry, bear the interpretation originally put upon them, they are still interesting, even if due to differential mortality. It is a well-known fact that in some places, in natural conditions, the percentage of females is very high, though 57 seems to be the average.

Before leaving the tadpoles, interesting in so many ways, let us think over the year's life of the frog. Throughout the winter months the frogs lie near the pond, buried in the mud, mouth shut, nose shut, eyes shut, with the heart beating feebly, breathing through their skin, and eating nothing. The awakening in Spring is followed immediately by pairing and egg-laying, and the aquatic juvenile life of the tadpoles occupies about three months. In summer there is a remarkable migration to the fields and meadows, and many hundreds of froglings, about the size of a first-finger nail, are seen on the march from the pond. The adults also migrate, and the meaning in both cases is the same — that they seek out places where insects abound. Of the many that go forth, only a remnant returns, for there is great mortality in the fields, where there are many physical risks and many alert enemies. The grass snake alone accounts for a good many in some parts of England. Those that escape — whether youngsters or old experienced hands — return to the pond in the autumn and go into winter-quarters in the mud.

WILLIAM MORTON WHEELER (1865–1937)

The Termitodoxa, or Biology and Society

Read at the Symposium of the American Society of Naturalists, Princeton, December 30, 1919, and first published in the *Scientific Monthly*, February 1920. Reprinted in *Foibles of Insects and Men*, 1928

[LITTLE need be added to my eulogy of William Morton Wheeler in the Introduction. In his conversation and written word, in association with him in such fertile fields as British Guiana and the Galápagos Islands, he remains in my memory as the ideal naturalist and scientist, to whose life and works all living aspirants to those titles should look for inspiration.

Out of a host of memories I tell one, on the occasion of his first visit to a tropical jungle. Soon after our arrival at my station at Kartabo, British Guiana, we were working on some piece of collecting equipment, when he said, apropos of some earlier talk: "Beebe, in regard to mimicry, be sure to hold back; don't accept new instances without complete evidence."

At that moment his son came in from a short walk along the Cuyuni trail and dumped a tobacco tin of half-dead insects in front of his father, saying: "Here are some things I caught; perhaps there's something of interest among them."

Wheeler picked one up, looked at it carefully, and instantly flung it down on the table and examined it again. He then leaned back, grinned at me, and said: "Beebe, believe anything you damned please about mimicry."

He had picked up a dead, yellow-banded, small-waisted wasp, which suddenly came to life and tried to sting him. On more careful scrutiny the insect proved to be a perfectly good moth, which in shape, proportions, antennæ, pattern, color, and movement was an amazingly exact imitation of a wasp.

Lest I be thought to be too prejudiced, let me quote from other appreciations of Dr. Wheeler. "Like Aristotle a stranger to no aspect of life; a master of English, his professional works remain as models of form and' substance, and his addresses as masterpieces of wit, humor and intelligence pungent with wise ridicule."

"His writing was the expression of his sensitive feeling for style and of his ideal of good workmanship. At its best, for instance, in his occasional satirical pieces, like the letter from the king of the termites, and in 'The Dry-Rot of our Academic Biology,' it has a force and a polish, not to mention other qualities, that recall Voltaire."

From among the riches of his prose I have chosen "The Termitodoxa."]

THE TERMITODOXA, OR BIOLOGY AND SOCIETY

JUST BEFORE the World War we seemed to be on the verge of startling revelations in animal behavior. "Rolf," the Ayrdale terrier of Mannheim, was writing affectionate letters to Professor William Mackensie of Genoa, and the Elberfeld stallions were easily solving such problems in mental arithmetic as extracting the cube root of 12,167, to the discomfiture of certain German professors, who had never been able to detect similar signs of intelligence in their students. The possibilities of animal correspondence struck me as so promising that I longed to dispatch letters and questionnaires to all the unusual insects of my acquaintance. But dismayed at the thought of the quantity of mail that might reach me, especially from the many insects that have been misrepresented by the taxonomists or maltreated by the economic entomologists, I decided to proceed with caution and to confine myself at first to a single letter to the most wonderful of all insects, the queen of the West African *Termes bellicosus*. During the autumn of 1915 my friend Mr. George Schwab, missionary to the Kamerun, kindly undertook to deliver my communication to a populous termitarium of this species in his back yard in the village of Okani Olinga. He subsequently wrote me that my constant occupation with the ants must have blinded me to the fact that the termitarium, unlike the formicarium, contains a king as well as a queen, but that the *bellicosus* king was so accustomed to being overlooked, even by his own offspring, that he not only pardoned my discourtesy, but condescended to answer my letter. Mr. Schwab embarked for Boston in 1917. Off the coast of Sierra Leone his steamer was shelled by a German submarine camouflaged as a small boat in distress, but succeeded in escaping, and what would have been another atrocity, the loss of the king's letter, was averted. It runs as follows:

Dear Sir: Your communication addressed to my most gloriously physogastric consort, was duly received. Her majesty, being extremely busy with oviposition — she has laid an egg every three minutes for

251

the past four years — and fearing that an interruption of even twenty minutes might seriously upset the exquisitely balanced routine of the termitarium, has requested me to acknowledge your expression of anxiety concerning the condition of the society in which you are living and to answer your query as to how we termites, to quote your own words, "managed to organize a society which, if we accept Professor Barrell's recent estimates of geological time, based on the decomposition of radium, has not only existed but flourished for a period of at least a hundred million years."

I answer your question the more gladly because the history of our society has long been with me a favorite topic of study. As you know, the conditions under which I live are most conducive to sustained research. I am carefully fed, have all the leisure in the world, and the royal chamber is not only kept absolutely dark and at a constant and agreeable temperature even during the hottest days of the Ethiopian summer, but free from all noises except the gentle rhythmic dropping of her majesty's eggs and the soft footfalls of the workers on the cement floor as they carry away the germs of future populations to the royal nurseries. And you will not wonder at my knowledge of some of the peculiarities of your society when I tell you that in my youth I belonged to a colony that devoured and digested a well selected library belonging to a learned missionary after he had himself succumbed to the appetite of one of the fiercest tribes of the Kamerun. If I extol the splendid solutions of sociological problems by my remote ancestors, I refrain from suggesting that your society would do well to imitate them too closely. This, indeed, would be impossible. I believe, nevertheless, that you may be interested in my remarks, for, though larger and more versatile, you and your fellow human beings are after all only animals like myself.

According to tradition our ancestors were descended in early Cretaceous times from certain kind-hearted old cockroaches that lived in logs and fed on rotten wood and mud. Their progeny, the aboriginal termites, although at first confined to this apparently unpromising diet, made two important discoveries. First, they chanced to pick up a miscellaneous assortment of Protozoa and Bacteria and adopted them as an intestinal fauna and flora, because they were able to render the rotten wood and mud more easily digestible. The second discovery, more important but quite as incidental, was nothing less than society. Our ancestors, like other solitary insects, originally set their offspring adrift to shift for themselves as soon as they hatched, but it was found that the fatty dermal secretions or exudates of the young were a delicious food, and that the parents could reciprocate with similar exudates as well as with regurgitated, predigested cellulose. Thenceforth parents and offspring no longer lived apart, for an elaborate exchange of exudates, veritable social hormones, was developed, which, continually circulating through the community, bound all its individuals together in one blissful, indissoluble, syntrophic whole, satisfied to make the comminution and digestion of wood and mud the serious occupation of existence, but the swapping of exudates the delight of every leisure moment. It may be said, therefore, that our society did not arise, like yours, from a combination of selfish predatism and parasitism but from a coöperative mutualism, or symbiosis. In other words, our ancestors did not start society because they thought they loved one another, but they loved one another because they were so sweet, and society supervened as a necessary and unforeseen by-product.

You will admit that no society could have embarked on its career through the ages with more brilliant prospects. The world was full of rotten wood and mud, and no laws interfered with distilling and imbibing the social hormones. But in the Midcretaceous our ancestors struck a snag. Not only had all the members of society begun to reproduce in the wildest and most unregulated manner, but their behavior toward one another had undergone a deterioration most shocking to behold. The priests, pedagogues, politicians, and journalists, having bored their way up to the highest strata of the society, undertook to influence or control all the activities of its members. The priests tried to convince the people that if they would only give up indulging in the social hormones and confine themselves to a diet of pure mud, they would in a future life eat nothing but rose-wood and mahogany, and the pedagogues insisted that every young termite must thoroughly saturate himself with the culture and languages of the Upper Carboniferous cockroaches. Some suspected that the main value of this form of education lay in intensifying and modulating the stridulatory powers, but for several thousand years most termites implicitly believed that ability to stridulate, both copiously and sonorously, was an infallible indication of brain-power. The politicians and the journalists — well, were it not that profanity has been considered to be very bad form in termite society since the Miocene, I might make a few comments on *their* activities. Suffice it to say that they consumed even more cellulose than the priests and pedagogues and secreted such a quantity of buncombe and flapdoodle that they well nigh asphyxiated the whole termitarium. Meanwhile, in the very foundations of the commonwealth, anarchists, syndicalists, I.W.W., and bolsheviki were busy boring holes and filling them with dynamite, while the remainder of society was largely composed of profiteers, grafters, shysters, drug-fiends, and criminals of all sizes, interspersed with beautifully graduated series of wowsers, morons, feeble-minded, idiots, and insane. (At this point the king has introduced a rather trivial note on the word "wowser." This word, he says, was first employed by the termites of Australia but later adopted by the human inhabitants of that continent to designate an individual who makes a business of taking the joy out of life, one who delights in pouring cold water into his own and especially into other peoples' soup. The term appears to be onomatopoeic to judge from a remark by one of our postcretaceous philologists who asserts that "whenever the wowser saw termites dancing, swearing, flirting, smoking, or over-indulging in the social hormones, he sat up on his hind legs, looked very solemn, swelled out his abdomen, and said 'WOW!' ")

To such depths, my dear sir, the letter continues, had termite society fallen in the Midcretaceous. The few sane termites still extant were on the point of giving up social life altogether and of returning to the solitary habits of the Palaeodictyoptera, but a king, Wuf-wuf IV, of the 529th dynasty, succeeded in initiating those reforms which led our ancestors to complete the most highly integrated social organization on the planet. He has aroused the enthusiastic admiration and emulation of every sovereign down to the present time. I can best describe him by saying that in his serious moments he displayed the statesmanship of a Hammurabi, Moses, Solomon, Solon, and Pericles rolled into one and that in his moments of relaxation he was a delightful blend of Aristophanes, Lucian, Rabelais, Anatole France, and Bernard Shaw. This king had the happy thought to refer the problems of social reform to the biologists. They were unfortunately few in number and difficult to find, because each was sitting in his hole in some remote corner of the termitarium, boring away in blissful ignorance of the depravity of the society to which he belonged. In obedience to the king's request, however, they were finally rounded up and persuaded to meet together annually just after the winter solstice for the purpose of stridulating about the relations of biology to society. After doing this for ten million years they adopted a program as elegant as it was drastic for the regeneration of termite society, and during the remaining fifteen million years of the Cretaceous they succeeded in putting their plan into operation. I can give you only the baldest outline of this extraordinary achievement.

Our ancient biological reformers started with the assumption that a termite society could not be a success unless it was constructed on the plan of a superorganism, and that such a superorganism must necessarily conform to the fundamental laws of the individual organism. As in the case of the individual, its success would have to depend on the adequate solution of the three basic problems of nutrition, reproduction, and protection. It was evident, moreover, that these problems could not be solved without a physiological division of labor among the individuals composing the society, and this, of course, implied the development of classes, or castes. Termite society was therefore divided into three distinct castes, according to the three fundamental organismal needs and functions, the workers being primarily nutritive, the soldiers defensive, and the royal couple reproductive. Very fortunately our earliest social ancestors had not imitated our deadly enemies, the ants, who went crazy in the early Cretaceous on the subject of parthenogenesis and developed a militant suffragette type of society, but insisted on an equal representation of both sexes in all the social activities. Our society is therefore ambisexual throughout, so that, unlike the ants, we have male as well as female workers and soldiers. It was early decided that these two castes should be forbidden to grow wings or reproduce and that the royal caste should be relieved from all the labor of securing food and defending the termitarium in order to devote all its energies to reproduction. The carrying out of this scheme yielded at least two great advantages: first, the size of the population could be automatically regulated to correspond with the food supply, and second, the production of perfect offspring was greatly facilitated.

During the late Cretaceous period of which I am writing our practical geneticists, in obedience to a general demand for a more varied diet, made two important contributions to our social life. The plant breeders found that what was left of the comminuted wood after its passage through the intestines of the worker termites could be built up in the form of elaborate sponge-like structures and utilized as gardens for the growth of mushrooms. Cultivation was later restricted to a few selected varieties of mushrooms which the biochemists had found to contain vitamins that accelerated the growth of the tissues in general and of the spermatocytes and oöcytes in particular. And for this reason only the royal caste and the young of the other castes were permitted to feed on this delicious vegetable food. The animal breeders of that age made a more spectacular though less useful contribution when they persuaded our ancestors to adopt a number of singular beetles and flies and to feed and care for them till they developed exudate organs. Owing to the stimulating quality of their

255

exudates these creatures, the termitophiles, added much variety to the previously somewhat monotonous social hormones. This quality, however, made it necessary to restrict the number of termitophiles in the termitarium for the same reason that your society would find it advisable to restrict the cattle industry if your animal breeders had succeeded in producing breeds of cows that yielded highballs and cocktails instead of milk.

It is, of course, one thing to have a policy and quite another to carry it out. The anarchistic elements in our late Cretaceous society were so numerous and so active that great difficulty was at first experienced in putting the theories of the biological reformers into practise, but eventually, just before the Eocene Tertiary, a very effective method of dealing with any termite that attempted to depart from the standards of the most perfect social behavior was discovered and rigorously applied. The culprit was haled before the committee of biochemists who carefully weighed and examined him and stamped on his abdomen the number of his colloidal molecules. This number was taken to signify that his conduct had reduced his social usefulness to the amount of fat and proteids in his constitution. He was then led forth into the general assembly, dismembered, and devoured by his fellows.

I describe these mores reluctantly and very briefly because I fear that they may shock your sensibilities, but some mention of them is essential to an appreciation of certain developments in our society within recent millennia. So perfectly socialized have we become that not infrequently a termite who has a slight indisposition, such as a sore throat or a headache, or has developed some antisocial habit of thought, or is merely growing old, will voluntarily resort to the committee of biochemists and beg them to stamp him. He then walks forth with a radiant countenance, stridulating a refrain which is strangely like George Eliot's "O, may I join the choir invisible!" and forthwith becomes the fat and proteid "Bausteine" of the crowd that assembles on hearing the first notes of his petition. If you regard this as an even more horrible exhibition of our mores, because it adds suicide to murder and cannibalism, I can only insist that you are viewing the matter from a purely human standpoint. To the perfectly socialized termite nothing can be more blissful or exalted than feeling the precious fats and proteids which he has amassed with so much labor, melting, without the slightest loss of their vital values, into the constitutions of his more vigorous and socially more efficient fellow beings.

Now I beg you to note how satisfactory was our solution of the many problems with which all animals that become social are confronted. I need hardly emphasize the matter of nutrition, for you would hardly contend that animals that can digest rotten wood and mud, grow perennial crops of mushrooms on their excrement, domesticate strange animals to serve as animated distilleries, and digest not only one anothers' bodies but even one anothers' secretions, have anything to learn in dietetics or food conservation. Our solution of the great problems of reproduction, notably those of eugenics, is if anything, even more admirable, for by confining reproduction to a special caste, by feeding it and the young of the other castes on a peculiarly vitaminous diet and by promptly and deftly eliminating all abnormalities, we have been able to secure a physically and mentally perfect race. You will appreciate the force of this statement when I tell you that in a recent census of the 236,498 individuals comprising the entire population of my termitarium, I found none that had hatched with more than the normal number of antennal joints or even with a misplaced macrochaeta. The only anomaly seen was one of no social significance, a slightly defective toenail in three workers. Rigid eugenics combined with rigid enforcement of the regulations requiring all antisocial, diseased, and superannuated individuals promptly to join the choir invisible, at the same time solved the problems of ethics and hygiene, for we were thus enabled, so to speak, to ram virtue and health back into the germ-plasm where they belong. And since we thus compelled not only our workers and soldiers but even our kings and queens to be born virtuous and to continue so throughout life, the Midcretaceous wowser caste, finding nothing to do, automatically disappeared. The problem of social protection was solved by the creation of a small standing army of cool-headed, courageous soldiers, to be employed not in waging war but solely for defensive purposes, and the development on the part of the soldiers and workers of ability to construct powerful fortifications. It may be said that the formation of the soldier caste as well as the invention of our cement subway architecture — an architecture unsurpassed in magnitude, strength, and beauty, considering the small stature of our laborers and the simple tools they employ — was due to the repeated failures, extending over many million years, of our politicians to form a league of nations with our deadly enemies, the ants. After a recent review of the army and an inspection of the fortifications of my termitarium, I agree with several of the kings of the present dynasty who believed that we ought really to be very grateful to our archenemies for their undying animosity.

Such was our society at the beginning of the Eocene, and such, with slight improvements in detail, it has remained for the past fifty million years, living and working with perfect smoothness, as if on carefully lubricated ball-bearings. Nor does it, like human society, live and work for itself alone, but with a view to the increase and maintenance of other types of life on the planet. On our activities depend the rapid decomposition of the dead vegetation and the rapid formation of the vegetable mould of the tropics. We are so numerous and our operations of such scope that we are a very important factor in accelerating the growth of all the vegetation, not only of the dry savannahs and pampas but even of huge rain-forests like those of the Congo, the Amazon, and the East Indies. And when you stop to consider that the animal and human life of the tropics absolutely depends on this vegetation you will not take too seriously the reports of our detractors who are forever calling attention to our destructive activities. One author, I am told, asserts that certain South American nations can never acquire any culture because the termites so quickly eat up all their libraries, and another gives an account of a gentleman in India who went to bed full of whisky and soda and awoke in the morning stark naked, because the termites had eaten up his pyjamas. How very unfair to dwell on the loss of a few books and a suit of pyjamas and not even to mention our beneficent and untiring participation in one of the most important biocoenoses!

You will pardon me if after this hasty sketch of our history I am emboldened to make a few remarks about your society, and in what I say you will, I hope, make due allowance both for the meagerness of my sources of information and the limitations of my understanding. I must confess that to me your society wears a strangely immature and at the same time senile aspect, the appearance, in fact, of a chimera, composed of the parts of an infant and those of a white-haired octogenarian. Although your species has been in existence little more than one hundredth of the time covered by our evolution, you are nevertheless such huge and gifted animals that it is surprising to find you in so imperfect a stage of socialization. And although every individual in your society seems to crave social integration with his fellows, it seems to be extremely difficult to persuade him to abate one tittle of all his natural desires and appetites, and every individual resists to the utmost any profound specialization of his structure and functions such as would seem to be demanded by the principle of the division of labor in any perfect society. Hence all the attempts which your society is continually making to form classes or castes are purely superficial and such as depend on the accumulation and transmission of property, and on vocation. And owing to the absence of eugenics and birth-control, and to your habit of fostering all weak and inefficient individuals, there is not even the dubious and slow working apparatus of natural selection to provide for the organic fixation of castes through heredity. So immature is your society in these respects that it might be described as a lot of cave-men and cave-women playing at having a perpetual pink tea or Kaffeeklatsch.

But the senile aspect of your society impresses me as even more extraordinary because our society — and the same is true of that of all other social insects — is perennially youthful and vigorous, owing to our speedy elimination of the old and infirm. And this brings me to a matter that interests me greatly and one on which I hope we shall have much further correspondence. To be explicit, it seems that though your society has no true caste system, it is, nevertheless, divided into what might be called three spurious castes, the young, the mature, and the aged. These, of course, resemble our castes only in number and in consisting of individuals of both sexes. They are peculiar in being rather poorly defined, temporary portions of the life-cycle, so that a single individual may belong to all of them in succession, and in the fact that only one of them, comprising the mature individuals, is of any great economic value to society and therefore actually functions as the host of the two others, which are, biologically speaking, parasitic. To avoid shocking your human sensibilities, I am willing to admit that both these castes may be worth all the care that is bestowed on them, the young on account of their promise and the old on account of past services. And I will even admit the considerable value of the young and the old as stimuli adapted to call forth the affection of the mature individuals. But, writing as one animal to another, I confess that I am unable to understand why you place the control of your society so completely in the hands of your aged caste. Your society is actually dominated by the superannuated, by old priests, old pedagogues, old politicians, and no end of old wowsers of both sexes who are forever suppressing or regulating everything from the observance of the Sabbath and the wearing of feathers on hats to the licking of postage stamps and the grievances and tribulations of stray tom-cats.

I notice that your educators, psychologists, and statisticians have much to say on human longevity, and you seem all to crave for nothing so much as an inordinate protraction of your egos. Psychologically, this is, of course, merely another manifestation of your fundamentally unsocial and individualistic appetites. Your writers make much of your long infancy, childhood, and adolescence as being very conducive to educability and socialization, and this is doubtless true, but the fact seems to be overlooked that the great lengthening of the initial phases of your life cycle is also attended by a grave danger, for it also increases the dependence of the young on the adult and the aged elements of society, especially on the parents, and this means intensifying what the Freudian psychologists call the father and mother complexes and therefore also an increased subservience to authority, a cult of the conservative, the stable and the senile. The deplorable effects of intensifying these complexes have long been only too evident in your various religious systems and are already beginning to show in the all too ready acceptance on the part of your society of the visionless policies and confused and hesitating methods of administration of your statesmen.

Unless I am much mistaken this matter of the domination of the old in your society deserves careful investigation. Unfortunately very little seems to be known about senility. In our society it cannot be investigated, because we do not permit it to exist, and in your society it is said to be very poorly understood, because no one is interested in it till he actually reaches it and then he no longer has the ability or the time to investigate it. When the social significance of this stage in the human life-cycle comes to be more thoroughly appreciated some of your young biologists and psychologists will make it a subject of exhaustive investigation and will discover the secret of its ominous and persistent domination. It will probably be found that many of your aged are of no economic importance whatever, and that the activities of many others may even be mildly helpful or beneficial, but you will find, as we found in the Midcretaceous, a small percentage, powerful and pernicious out of all proportion to their numbers, who are directly responsible for the deplorable inertia of your institutions, especially of your churches, universities, and political bodies. These old individuals combine with a surprising physical vigor a certain sadistic obstinacy which consecrates itself to obstructing, circumventing, suppressing, or destroying not only everything young or new, but everything any other old individual in their environment may suggest. The eminent physician who recommended chloroform probably had this type of old man in mind. Certain economic entomologists have advocated some more vigorous insecticide, such as hydrocyanic acid gas. This is, however, a matter concerning which it might be better to defer recommendation till the physiology, psychology, and ethology of the superannuated have been more thoroughly investigated.

It has sometimes occurred to me that your social problem may be quite insoluble — that when your troglodyte ancestors first expanded the family and clan into society they were already too long-lived, too "tough," and too specialized mentally and physically ever to develop the fine adjustments demanded by an ideal social organization. I feel certain, nevertheless, that you could form a much better society than the present if you could be convinced that your further progress depends upon solving the fundamental, preliminary problems of nutrition, reproduction, and social defence, which our ancestors so successfully solved in the late Cretaceous. These problems are, of course, extremely complicated in your society. Under nutrition you would have to include raw materials and fuel, that is, food for your factories and furnaces as well as food for your bodies. Your problems of reproduction comprise not only those of your own species but of all your domesticated animals and plants, and your social defence problems embrace not only protection from the enemies of your own species (military science) but from the innumerable other organic species which attack your domesticated animals and plants as well as your own bodies (hygiene, parasitology, animal and plant pathology, economic entomology). Like our ancestors you will certainly find that these problems can be solved only by the biologists — taking the word "biologists" in its very broadest sense, to include also the psychologists and anthropologists — and that till they have put their best efforts into the solution your theologians, philosophers, jurists, and politicians will continue to add to the existing confusion of your social organization. It is my opinion, therefore, that if you will only increase your biological investigators a hundred fold, put them in positions of trust and responsibility much more often and before they are too old, and pay them at least as well as you are paying your plumbers and bricklayers, you may look forward to making as much social progress in the next three centuries as you have made since the Pleistocene. That some such opinion may also be entertained by some of your statesmen sometime before the end of the present geologic age, is the sincere wish of

Yours truly,

Wee-Wee

43rd Neotenic King of the 8429th Dynasty of the Bellicose Termites.

On reperusing this letter before deciding, after many misgivings, to read it to so serious a body of naturalists, I notice a great number of inaccuracies and exaggerations, attributable, no doubt, to his majesty's misinterpretation of his own and very superficial acquaintance with our society. His remarks on old age strike me as particularly inept and offensive. He seems not to be aware of the fact that at least a few of our old men have almost attained to the idealism of a superannuated termite, a fact attested by such Freudian confessions as the following, taken from a letter recently received by one of my colleagues from a gentleman in New Hampshire:

"I do not understand how it is that an insect so small as to be invisible is able to worry my dog and also at times sharply to bite myself. A vet. friend of mine in Boston advised lard and kerosene for the dog. This seemed to check them for a time, but what I need is extermination, for I am in my eighty-fourth year."

G. MURRAY LEVICK

Penguins

From *Antarctic Penguins*. 1914

[DR. LEVICK was Staff-Surgeon on the *Terra Nova* or British Antarctic Expedition of 1910–13, on which Captain Scott lost his life. Levick's monograph on arctic penguins is written with such detail, accuracy, and charm that after being published in the official technical account of the expedition it was reprinted in amplified form in a most entertaining volume.]

THE PENGUINS of the Antarctic regions very rightly have been termed the true inhabitants of that country. The species is of great antiquity, fossil remains of their ancestors having been found, which showed that they flourished as far back as the eocene epoch. To a degree far in advance of any other bird, the penguin has adapted itself to the sea as a means of livelihood, so that it rivals the very fishes. The proficiency in the water has been gained at the expense of its power of flight, but this is a matter of small moment, as it happens.

In few other regions could such an animal as the penguin rear its young, for when on land its short legs offer small advantage as a means of getting about, and as it cannot fly, it would become an easy prey to any of the carnivora which abound in other parts of the globe. Here, however, there are none of the bears and foxes which inhabit the North Polar regions, and once ashore the penguin is safe.

The reason for this state of things is that there is no food of any description to be had inland. Ages back, a different state of things existed: tropical forests abounded, and at one time, the seals ran about on shore like dogs. As conditions changed, these latter had to take to the sea for food, with the result that their four legs, in course of time, gave place to wide paddles or "flippers," as the penguins' wings have done, so that at length they became true inhabitants of the sea. . . .

When seen for the first time, the Adelie penguin gives you the impression of a very smart little man in an evening dress suit, so absolutely immaculate is he, with his shimmering white front and black back and shoulders. He stands about two feet five inches in height, walking very upright on his little legs.

His carriage is confident as he approaches you over the snow, curiosity in his every movement. When within a yard or two of you, as you stand silently watching him, he halts, poking his head forward with little jerky movements, first to one side, then to the other, using his right and left eye alternately during his inspection. He seems to prefer using one eye at a time when viewing any near object, but when looking far ahead, or walking along, he looks straight ahead of him, using both eyes. He does this, too, when his anger is aroused, holding his head very high, and appearing to squint at you along his beak.

After a careful inspection, he may suddenly lose all interest in you, and ruffling up his feathers sink into a doze. Stand still for a minute till he has settled himself to sleep, then make sound enough to wake him without startling him, and he opens his eyes, stretching himself, yawns, then finally walks off, caring no more about you. . . .

The Adelie penguin is excessively curious, taking great pains to inspect any strange object he may see. When we were waiting for the ship to fetch us home, some of us lived in little tents which we pitched on the snow about fifty yards from the edge of the sea. Parties of penguins from Cape Royds rookery frequently landed here, and almost invariably the first thing they did on seeing our tents, was at once to walk up the slope and inspect these, walking all round them, and often staying to doze by them for hours. Some of them, indeed, seemed to enjoy our companionship. When you pass on the sea-ice anywhere near a party of penguins, these generally come up to look at you, and we had great trouble to keep them away from the sledge dogs when these were tethered in rows near the hut at Cape Evans. The dogs killed large numbers of them in consequence, in spite of all we could do to prevent this.

The Adelies, as will be seen in these pages, are extremely brave, and though panic occasionally overtakes them, I have seen a bird return time after time to attack a seaman who was brutally sending it flying by kicks from his sea-boot, before I arrived to interfere. . . .

The Adelie penguins spend their summer and bring forth their young in the far South. Nesting on the shores of the Antarctic continent, and on the islands of the Antarctic seas, they are always close to the water, being dependent on the sea for their food, as are all Antarctic fauna; the frozen regions inland, for all practical purposes, being barren of both animal and vegetable life.

Their requirements are few: they seek no shelter from the terrible Antarctic gales, their rookeries in most cases being open wind-swept spots. In fact, three of the four rookeries I visited were possibly in the three most windy regions of the Antarctic. The reason for this is that only wind-swept places are so kept bare of snow that solid ground and pebbles for making nests are to be found.

When the chicks are hatched and fully fledged, they are taught to swim, and when this is accomplished and they can catch food for themselves, both young and old leave the Southern limits of the sea, and make their way to the pack-ice out to the northward, thus escaping the rigors and darkness of the Antarctic winter, and keeping where they will find the open water which they need. For in the winter the seas where they nest are completely covered by a thick sheet of ice which does not break out until early in the following summer. Much of this ice is then borne northward by tide and wind, and accumulates to form the vast rafts of what is called "pack-ice," many hundreds of miles in extent, which lie upon the surface of the Antarctic seas.

It is to this mass of floating sea-ice that the Adelie penguins make their way in the autumn, but as their further movements here are at present something of a mystery, the question will be discussed at greater length presently.

When young and old leave the rookery at the end of the breeding season, the new ice has not yet been formed, and their long journey to the pack has to be made by water, but they are wonderful swimmers and seem to cover the hundreds of miles quite easily.

Arrived on the pack, the first year's birds remain there for two winters. It is not until after their first moult, the autumn following their departure from the rookery, that they grow the distinguishing mark of the adult, black feathers replacing the white plumage which has hitherto covered the throat.

The spring following this, and probably every spring for the rest of their lives, they return South to breed, performing their journey, very often, not only by water, but on foot across many miles of frozen sea.

For those birds who nest in the southernmost rookeries, such as Cape Crozier, this journey must mean for them a journey of at least four hundred miles by water, and an unknown but considerable distance on foot over ice. . . .

The first Adelie penguins arrived at the Ridley Beach rookery, Cape Adare, on October 13. A blizzard came on then, with thick drift which prevented any observations being made. The next day, when this subsided, there were no penguins to be seen. . . .

By the morning of October 19 there had been a good many more arrivals, but the rookery was not yet more than one-twentieth part full. All the birds were fasting absolutely. Nest building was now in full swing, and the whole place waking up to activity. Most of the pebbles for the new nests were being taken from old nests, but a great deal of robbery went on nevertheless. Depredators when caught were driven furiously away, and occasionally chased for some distance, and it was curious to see the difference in the appearance between the fleeing thief and his pursuer. As the former raced and ducked about among the nests, doubling on his tracks, and trying by every means to get lost in the crowd and so rid himself of his pursuer, his feathers lay close back on his skin, giving him a sleek look which made him appear half the size of the irate nest-holder who sought to catch him, with feathers ruffled in indignation. This at first led me to think that the hens were larger than the cocks, as it was generally the hen who was at home, and the cock who was after the stones, but later I found that sex makes absolutely no difference in the size of the birds, or indeed in their appearance at all, as seen by the human eye. After mating, their behaviour as well as various outward signs serve to distinguish male from female. Besides this certain differences in their habits, which I will describe in another place, are to be noted.

The consciousness of guilt, however, always makes a penguin smooth his feathers and look small, whilst indignation has the opposite effect. Often when observing a knoll crowded with nesting penguins, I have seen an apparently under-sized individual slipping quietly along among the nests, and always by his subsequent proceedings he has turned out to be a robber on the hunt for his neighbours' stones. The others, too, seemed to know it, and would have a peck at him as he passed them.

At last he would find a hen seated unwarily on her nest, slide up behind her, deftly and silently grab a stone, and run off triumphantly with it to his mate who was busily arranging her own home. Time after time he would return to the same spot, the poor depredated nest-holder being quite oblivious of the fact that the side of her nest which lay behind her was slowly but surely vanishing stone by stone.

Here could be seen how much individual character makes for success or failure in the efforts of the penguins to produce and rear their offspring. There are vigilant birds, always alert, who seem never to get robbed or molested in any way: these have big high nests, made with piles of stones. Others are unwary and get huffed as a result. There are a few even who, from weakness of character, actually allow stronger natured and more aggressive neighbours to rob them under their very eyes.

In speaking of the robbery which is such a feature of the rookery during nest building, special note must be made of the fact that violence is never under any circumstances resorted to by the thieves. When detected, these invariably beat a retreat, and offer not the least resistance to the drastic punishment they receive if they are caught by their indignant pursuers. The only disputes that ever take place over the question of property are on the rare occasions when a *bona-fide* misunderstanding arises over the possession of a nest. These must be very rare indeed, as only on one occasion have I seen such a quarrel take place. The original nesting sites being, as I will show, chosen by the hens, it is the lady, in every case, who is the cause of the battle, and when she is won her scoop goes with her to the victor.

As I grew to know these birds from continued observation, it was surprising and interesting to note how much they differed in character, though the weaker-minded who would actually allow themselves to be robbed, were few and far between, as might be expected. Few, if any, of these ever could succeed in hatching their young and winning them through to the feathered stage.

When starting to make her nest, the usual procedure is for the hen to squat on the ground for some time, probably to thaw it, then working with her claws to scratch away at the material beneath her, shooting out the rubble behind her. As she does this she shifts her position in a circular direction until she has scraped out a round hollow. Then the cock brings stones, performing journey after journey, returning each time with one pebble in his beak which he deposits in front of the hen who places it in position.

Sometimes the hollow is lined with a neat pavement of stones placed side by side, one layer deep, on which the hen squats, afterwards building up the sides about her. At other times the scoop would be filled up indiscriminately by a heap of pebbles on which the hen then sat, working herself down into a hollow in the middle.

Individuals differ, not only in their building methods, but also in the size of the stones they select. Side by side may be seen a nest composed wholly of very big stones, so large that it is a matter for wonder how the birds can carry them, and another nest of quite small stones.

Different couples seem to vary much in character or mood. Some can be seen quarrelling violently, whilst others appear most affectionate, and the tender politeness of some of these latter toward one another is very pretty to see.

I may here mention that the temperatures were rising considerably by October 19, ranging about zero F.

During October 20 the stream of arrivals was incessant. Some mingled at once with the crowd, others lay in batches on the sea-ice a few yards short of the rookery, content to have got so far, and evidently feeling the need for rest after their long journey from the pack. The greater part of this journey was doubtless performed by swimming, as they crossed open water, but I think that much of it must have been done on foot over many miles of sea-ice, to account for the fatigue of many of them.

Their swimming I will describe later. On the ice they have two modes of progression. The first is simple walking. Their legs being very short, their stride amounts at most to four inches. Their rate of stepping averages about one hundred and twenty steps per minute when on the march.

Their second mode of progression is "tobogganing." When wearied by walking or when the surface is particularly suitable, they fall forward on to their white breasts, smooth and shimmering with a beautiful metallic lustre in the sunlight, and push themselves along by alternate powerful little strokes of their legs behind them.

When quietly on the march, both walking and tobogganing produce the same rate of progression, so that the string of arriving birds, tailing out in a long line as far as the horizon, appears as a well-ordered procession. I walked out a mile or so along this line, standing for some time watching it tail past me and taking photographs. . . . Most of the little creatures seemed much out of breath, their wheezy respiration being distinctly heard.

First would pass a string of them walking, then a dozen or so tobogganing. Suddenly those that walked would flop on their breasts and start tobogganing, and conversely strings of tobogganers would as suddenly pop up on to their feet and start walking. In this way they relieved the monotony of their march, and gave periodical rest to different groups of muscles and nerve-centres.

The surface of the snow on the sea-ice varied continually, and over any very smooth patches the pedestrians almost invariably started to toboggan, whilst over "bad going" they all had perforce to walk. . . .

On October 21 many thousands of penguins arrived from the northerly direction, and poured on to the beach in a continuous stream, the snaky line of arrivals extending unbroken across the sea-ice as far as the eye could see. . . .

Although squabbles and encounters had been frequent since their arrival in any numbers, it now became manifest that there were two very different types of battle; first, the ordinary quarrelling consequent on disputes over nests and the robbery of stones from these, and secondly, the battles between cocks who fought for the hens. These last were more earnest and severe, and were carried to a finish, whereas the first named rarely proceeded to extremes.

In regard to the mating of the birds, the following most interesting customs seemed to be prevalent.

The hen would establish herself on an old nest, or in some cases scoop out a hollow in the ground and sit in or by this, waiting for a mate to propose himself. She would not attempt to build while she remained unmated. During the first week of the nesting season, when plenty of fresh arrivals were continually pouring into the rookery, she did not have long to wait as a rule. Later, when the rookery was getting filled up, and only a few birds remained unmated in that vast crowd of some three-quarters of a million, her chances were not so good.

For example, on November 16 on a knoll thickly populated by mated birds, many of which already had eggs, a hen was observed to have scooped a little hollow in the ground and to be sitting in this. Day after day she sat on looking thinner and sadder as time passed and making no attempt to build her nest. At last, on November 27, she had her reward, for I found that a cock had joined her, and she was busily building her nest in the little scoop she had made so long before, her husband steadily working away to provide her with the necessary pebbles. Her forlorn appearance of the past ten days had entirely given place to an air of occupation and happiness.

As time went on I became certain that invariably pairing took place after arrival at the rookery. On October 23 I went to the place where the stream of arrivals was coming up the beach, and presently followed a single bird, which I afterwards found to be a cock, to see what it was going to do. He threaded his way through nearly the whole length of the rookery by himself, avoiding the tenanted knolls where the nests were, by keeping to the emptier hollows. About every hundred yards or so he stopped, ruffled up his feathers, closed his eyes for a moment, then "smoothed himself out" and went on again, thus evidently struggling against desire for sleep after his journey. As he progressed he frequently poked his little head forward and from side to side, peering up at the knolls, evidently in search of something.

Arrived at length at the south end of the rookery, he appeared suddenly to make up his mind, and boldly ascending a knoll which was well tenanted and covered with nests, walked straight up to one of these on which a hen sat. There was a cock standing at her side, but my little friend either did not see him or wished to ignore him altogether. He stuck his beak into the frozen ground in front of the nest, lifted up his head and made as if to place an imaginary stone in front of the hen, a most obvious piece of dumb show. The hen took not the slightest notice nor did her mate.

My friend then turned and walked up to another nest, a yard or so off, where another cock and hen were. The cock flew at him immediately, and after a short fight, in which each used his flippers savagely, he was driven clean down the side of the knoll away from the nests, the victorious cock returning to his hen. The newcomer, with the persistence which characterizes his kind, came straight back to the same nest and stood close by it, soon ruffling his feathers and evidently settling himself for a doze, but, I suppose, because he made no further overtures the others took no notice of him at all, as, overcome by sheer weariness, he went to sleep and remained so until I was too cold to await further developments. On my way back to our hut I followed another cock for about thirty yards, when he walked up to another couple at a nest and gave battle to the cock. He, too, was driven off after a short and decisive fight. Soon there were many cocks on the war-path. Little knots of them were to be seen about the rookery, the lust of battle in them, watching and fighting each other with desperate jealousy, and the later the season advanced the more "bersac" they became. . . . The roar of battle and thuds of blows could be heard continuously, and of the hundreds of such fights, all plainly had their cause in rivalry for the hens.

When starting to fight, the cocks sometimes peck at each other with their beaks, but always they very soon start to use their flippers, standing up to one another and raining in the blows with such rapidity as to make a sound which, in the words of Dr. Wilson, resembles that of a boy running and dragging his hoop-stick along an iron paling. Soon they start "in-fighting," in which position one bird fights right-handed, the other left-handed; that is to say, one leans his left breast against his opponent, swinging in his blows with his right flipper, the other presenting his right breast and using his left flipper. My photographs of cocks fighting all show this plainly. It is interesting to note that these birds, though fighting with one flipper only, are ambidextrous. Whilst battering one another with might and main they use their weight at the same time, and as one outlasts the other, he drives his vanquished opponent before him over the ground, as a trained boxing man when "in-fighting" drives his exhausted opponent round the ring.

Desperate as these encounters are, I don't think one penguin ever kills another. In many cases blood is drawn. I saw one with an eye put out, and that side of its beak (the right side) clotted with blood, while the crimson print of a blood-stained flipper across a white breast was no uncommon sight.

Hard as they can hit with their flippers, however, they are also well protected by their feathers, and being marvellously tough and enduring the end of a hard fight merely finds the vanquished bird prostrate with exhaustion and with most of the breath beaten out of his little body. The victor is invariably satisfied with this, and does not seek to dispatch him with his beak.

It was very usual to see a little group of cocks gathered together in the middle of one of the knolls squabbling noisily. Sometimes half a dozen would be lifting their raucous voices at one particular bird, then they would separate into pairs, squaring up to one another and emphasizing their remarks from time to time by a few quick blows from their flippers. It seemed that each was indignant with the others for coming and spoiling his chances with a coveted hen, and trying to get them to depart before he went to her.

It was useless for either to attempt overtures whilst the others were there, for the instant he did so, he would be set upon and a desperate fight begin. Usually, as in the case I described above, one of the little crowd would suddenly "see red" and sail into an opponent with desperate energy, invariably driving him in the first rush down the side of the knoll to the open space surrounding it, where the fight would be fought out, the victor returning to the others, until by his prowess and force of character, he would rid himself of them all. Then came his overtures to the hen. He would, as a rule, pick up a stone and lay it in front of her if she were sitting in her "scoop," or if she were standing by it he might himself squat in it. She might take to him kindly, or, as often happened, peck him furiously. To this he would submit tamely, hunching up his feathers and shutting his eyes while she pecked him cruelly. Generally after a little of this she would become appeased. He would rise to his feet, and in the prettiest manner edge up to her, gracefully arch his neck, and with soft guttural sounds pacify her and make love to her.

Both perhaps would then assume the "ecstatic" attitude, rocking their necks from side to side as they faced one another, and after this a perfect understanding would seem to grow up between them, and the solemn compact was made.

It is difficult to convey in words the daintiness of this pretty little scene. I saw it enacted many dozens of times, and it was wonderful to watch one of these hardy little cocks pacifying a fractious hen by the perfect grace of his manners. . . .

In various places through the course of these pages, reference is made to the "ecstatic" attitude of the penguins. This antic is gone through by both sexes and at various times, though much more frequently during the actual breeding season. The bird rears its body upward and stretching up its neck in a perpendicular line, discharges a volley of guttural sounds straight at the unresponding heavens. At the same time the clonic movements of its syrinx or "sound box" distinctly can be seen going on in its throat. Why it does this I have never been able to make out, but it appears to be thrown into this ecstasy when it is pleased; in fact, the zoologist of the "Pourquoi Pas" expedition termed it the "Chant de Satisfaction." I suppose it may be likened to the crowing of a cock or the braying of an ass. When one bird of a pair starts to perform in this way, the other usually starts at once to pacify it. Very many times I saw this scene enacted when nesting was in progress. The two might be squatting by the nest when one would arise to assume the "ecstatic" attitude and made the guttural sounds in its syrinx. Immediately the other would get close up to it and make the following noise in a soft soothing tone:

A-ah

Always and immediately this caused the musician to subside and settle itself down again.

On November 3 several eggs were found, and on the 4th these were beginning to be plentiful in places, though many of the colonies had not yet started to lay.

Let me here call attention to the fact that up to now not a single bird out of all those thousands had left the rookery once it had entered it. Consequently not a single bird had taken food of any description during all the most strenuous part of the breeding season, and as they did not start to feed till November 8 thousands had to my knowledge fasted for no fewer than twenty-seven days. Now of all the days of the year these twenty-seven are certainly the most trying during the life of the Adelie.

With the exception, in some cases, of a few hours immediately after arrival (and I believe the later arrivals could not afford themselves even this short respite) constant vigilance had been maintained; battle after battle had been fought; some had been nearly killed in savage encounters, recovered, fought again and again with varying fortune. They had mated at last, built their nests, procreated their species, and, in short, met the severest trials that Nature can inflict upon mind and body, and at the end of it, though in many cases blood-stained and in all caked and bedraggled with mire, they were as active and as brave as ever. . . .

By November 7, though many nests were still without eggs, a large number now contained two, and their owners started, turn and turn about, to go to the open water leads about a third of a mile distant to feed, and as a result of this a change began gradually to come over the face of the rookery. Hitherto the whole ground in the neighbourhood of the nests had been stained a bright green. This was due to the fasting birds continually dropping their watery, bile-stained excreta upon it. (The gall of penguins is bright green). These excreta practically contained no solid matter excepting epithelial cells and salts.

The nests themselves are never fouled, the excreta being squirted clear of them for a distance of a foot or more, so that each nest has the appearance of a flower with bright green petals radiating from its centre. Even when the chicks have come and are being sat upon by the parents, this still holds good, because they lie with their heads under the old bird's belly and their hindquarters just presenting themselves, so that they may add their little decorative offerings, petal by petal! Now that the birds were going to feed, the watery-green stains upon the ground gave place to the characteristic bright brick-red guano, resulting from their feeding on the shrimp-like euphausia in the sea; and the colour of the whole rookery was changed in a few days, though this was first noticeable, of course, in the region of those knolls which had been occupied first, and which were now settled down to the peaceable and regular family life which was to last until the chicks had grown. . . .

During the fasting season, as none of the penguins had entered the water, they all became very dirty and disreputable in appearance, as well may be imagined considering the life they led, but now that they went regularly to swim, they immediately got back their sleek and spotless state.

From the ice-foot to the open water, the half mile or so of sea-ice presented a lively scene as the thousands of birds passed to and fro over it, outward bound parties of dirty birds from the rookery passing the spruce bathers, homeward bound after their banquet and frolic in the sea. So interesting and instructive was it to watch the bathing parties, that we spent whole days in this way.

As I have said before, the couples took turn and turn about on the nest, one remaining to guard and incubate while the other went off to the water.

On leaving their nests, the birds made their way down the ice-foot on to the sea-ice. Here they would generally wait about and join up with others until enough had gathered together to make up a decent little party, which would then set off gaily for the water. They were now in the greatest possible spirits, chattering loudly and frolicking with one another, and playfully chasing each other about, occasionally indulging in a little friendly sparring with their flippers.

Arrived at length at the water's edge, almost always the same procedure was gone through. The object of every bird in the party seemed to be to get one of the others to enter the water first. They would crowd up to the very edge of the ice, dodging about and trying to push one another in. Sometimes those behind nearly would succeed in pushing the front rank in, who then would just recover themselves in time, and rushing round to the rear, endeavour to turn the tables on the others. Occasionally one actually would get pushed in, only to turn quickly under water and bound out again on to the ice like a cork shot out of a bottle. Then for some time they would chase one another about, seemingly bent on having a good game, each bird intent on finding any excuse from being the first in. Sometimes this would last a few minutes, sometimes for the better part of an hour, until suddenly the whole band would change its tactics, and one of the number start to run at full tilt along the edge of the ice, the rest following closely on his heels, until at last he would take a clean header into the water. One after another the rest of the party followed him, all taking off exactly from the spot where he had entered, and following one another so quickly as to have the appearance of a lot of shot poured out of a bottle into the water.

A dead silence would ensue till a few seconds later, when they would all come to the surface some twenty or thirty yards out, and start rolling about and splashing in the water, cleaning themselves and making sounds exactly like a lot of boys calling out and chaffing one another.

So extraordinary was this whole scene, that on first witnessing it we were overcome with astonishment, and it seemed to us almost impossible that the little creatures, whose antics we were watching, were actually birds and not human beings. Seemingly reluctant as they had been to enter the water, when once there they evinced every sign of enjoyment, and would stay in for hours at a time.

As may be imagined, the penguins spent a great deal of time on their way to and from the water, especially during the earlier period before the sea-ice had broken away from the ice-foot, as they had so far to walk before arriving at the open leads.

As a band of spotless bathers returning to the rookery, their white breasts and black backs glistening with a fine metallic lustre in the sunlight, met a dirty and bedraggled party on its way out from the nesting ground, frequently both would stop, and the clean and dirty mingle together and chatter with one another for some minutes. If they were not speaking words in some language of their own, their whole appearance belied them, and as they stood, some in pairs, some in groups of three or more, chattering amicably together, it became evident that they were sociable animals, glad to meet one another, and, like many men, pleased with the excuse to forget for a while their duties at home, where their mates were waiting to be relieved for their own spell off the nests.

After a variable period of this intercourse, the two parties would separate and continue on their respective ways, a clean stream issuing from the crowd in the direction of the rookery, a dirty one heading off towards the open water, but here it was seen that a few who had bathed and fed, and were already perhaps half-way home, had been persuaded to turn and accompany the others, and so back they would go again over the way they had come, to spend a few more hours in skylarking and splashing about in the sea.

In speaking of these games of the penguins, I wish to lay emphasis on the fact that these hours of relaxation play a large part in their lives during the advanced part of the breeding period. They would spend hours in playing at a sort of "touch last" on the sea-ice near the water's edge. They never played on the ground of the rookery itself, but only on the sea-ice and the ice-foot and in the water, and I may here mention another favourite pastime of theirs. I have said that the tide flowed past the rookery at the rate of some five or six knots. Small ice-floes are continually drifting past in the water, and as one of these arrived at the top of the ice-foot, it would be boarded by a crowd of penguins, sometimes until it could hold no more. This "excursion boat," as we used to call it, would float its many occupants down the whole length of the ice-foot, and if it passed close to the edge, those that rode on the floes would shout at the knots of penguins gathered along the ice-foot who would shout at them in reply, so that a gay bantering seemed to accompany their passage past the rookery.

Arrived at the farther end, some half a mile lower down, those on the "excursion boat" had perforce to leave it, all plunging into the tide and swimming against this until they came to the top again, then boarded a fresh floe for another ride down. All day these floes, often crowded to their utmost capacity, would float past the rookery. Often a knot of hesitating penguins on the ice-foot, on being hailed by a babel of voices from a floe, would suddenly make the plunge, and all swim off to join their friends for the rest of the journey, and I have seen a floe so crowded that as a fresh party boarded it on one side, many were pushed off the other side into the water by the crush.

Once, as we stood watching the penguins bathing, one of them popped out of the water on to the ice with a large pebble in its mouth, which it had evidently fetched from the bottom. This surprised me, as the depth of the sea here was some ten fathoms at least. The bird simply dropped the stone on the ice and then dived in again, so that evidently he had gone to all the trouble of diving for the stone simply for the pleasure of doing it. Mr. J. H. Gurney in his book on the gannet, says they (gannets) are said to have got themselves entangled in fishing-nets at a depth of 180 ft. and that their descent to a depth of 90 ft. is quite authentic, so that perhaps the depth of this penguin's dive was not an unusual one.

The tide at the open water leads where they bathed ran a good six knots, but the Adelies swam quite easily against this without leaving the surface.

In the water, as on the land, they have two means of progression. The first is by swimming as a duck swims, excepting that they lie much lower in the water than a duck does, the top of the back being submerged, so that the neck sticks up out of the water. As their feet are very slightly webbed, they have not the advantages that a duck or gull has when swimming in this way, but supplement their foot-work by short quick strokes of their flippers. This they are easily able to do, owing to the depth to which the breast sinks in the water.

The second method is by "porpoising."

This consists in swimming under water, using the wings or "flippers" for propulsion, the action of these limbs being practically the same as they would be in flying. As their wings are beautifully shaped for swimming, and their pectoral muscles extraordinarily powerful, they attain great speed, besides which they are as nimble as fish, being able completely to double in their tracks in the flash of a moment.

In porpoising, after travelling thirty feet or so under water, they rise from it, shooting clean out with an impetus that carries them a couple of yards in the air, then with an arch of the back they are head first into the water again, swimming a few more strokes, then out again, and so on.

Perhaps the most surprising feat of which the Adelie is capable is seen when it leaps from the water on to the ice. We saw this best later in the year when the sea-ice had broken away from the ice-foot, so that open water washed against the ice cliff bounding the land. This little cliff rose sheer from the water at first, but later, by the action of the waves, was under-cut for some six feet or more in places, so that the ledge of ice at the top hung forwards over the water. The height of most of this upper ledge varied from three to six feet.

Whilst in the water, the penguins usually hunted and played in parties, just as they had entered it, though a fair number of solitary individuals were also to be seen. When a party had satisfied their appetites and their desire for play, they would swim to a distance of some thirty to forty yards from the ice-foot, when they might be seen all to stretch their necks up and take a good look at the proposed landing-place. Having done this, every bird would suddenly disappear beneath the surface, not a ripple showing which direction they had taken, till suddenly, sometimes in a bunch, sometimes in a stream, one after the other they would all shoot out of the water, clean up on to the top of the ice-foot. Several times I measured the distance from the surface of the water to the ledge on which they landed, and the highest leap I recorded was exactly five feet. The "take off" was about four feet out from the edge, the whole of the necessary impetus being gained as the bird approached beneath the water.

The most important thing to note about this jumping from the water was the accuracy with which they invariably rose at precisely the right moment, the exact distance being judged during their momentary survey of a spot from a distance, before they dived beneath the water, and carried in their minds as they approached the ice. I am sure that this impression was all they had to guide them, as with a ripple on the water, and at the pace they were going, they could not possibly have seen their landing place at all clearly as they approached it, besides which, in many cases, the ledge of ice on which they landed projected many feet forwards from the surface; yet I never saw them misjudge their distance so as to come up under the over-hanging ledge.

During their approach they swam at an even distance of about three or four feet beneath the surface, projecting themselves upwards by a sudden upward bend of the body, at the same time using their tail as a helm, in the manner well shown in one of my photographs, in which one of the birds is seen in the air at the moment it left the water, the tail being bent sharply up towards the back.

Their quickness of perception is shown very well as they land on the ice. If the surface is composed of snow, and so affords them a good foothold, they throw their legs well forward and land on their feet, but should they find themselves landing on a slippery ice-surface, they throw themselves forward, landing on their breasts in the tobogganing position.

The Adelies dive very beautifully. We did not see this at first, before the sea-ice had gone out, because to enter the water they had only to drop a few inches, but later, when entering from the ice terraces, we constantly saw them making the most graceful dives.

At the place where they most often went in, a long terrace of ice about six feet in height ran for hundreds of yards along the edge of the water, and here, just as on the sea-ice, crowds would stand near the brink. When they had succeeded in pushing one of their number over, all would crane their necks over the edge, and when they saw the pioneer safe in the water, the rest followed.

When diving into shallow water they fall flat, but into deep water, and from any considerable height, they assume the most perfect positions and make very little splash. Occasionally we saw them stand hesitating to dive at a height of some twenty feet, but generally they descended to some lower spot, and did not often dive from such a height, but twelve feet was no uncommon dive for them.

The reluctance shown by each individual of a party of intending bathers to be the first to enter the water may partly have been explained when, later on, we discovered that a large number of sea-leopards were gathered in the sea in the neighbourhood of the rookery to prey on the penguins. These formidable animals used to lurk beneath the overhanging ledges of the ice-foot, out of sight of the birds on the ice overhead. They lay quite still in the water, only their heads protruding, until a party of Adelies would descend into the water almost on top of them, when with a sudden dash and snap of their great formidable jaws, they would secure one of the birds.

It seemed to me then, that all the chivvying and preliminaries which they went through before entering the water, arose mainly from a desire on the part of each penguin to get one of its neighbours to go in first in order to prove whether the coast was clear or not, though all this manoeuvring was certainly taken very lightly, and quite in the nature of a game. This indeed was not surprising, for of all the animals of which I have had any experience, I think the Adelie penguin is the very bravest. The more we saw of them the fonder we became of them and the more we admired their indomitable courage. The appearance of a sea-leopard in their midst was the one thing that caused them any panic. With dozens of these enemies about they would gambol in the sea in the most light-hearted manner, but the appearance of one among them was the signal for a stampede, but even this was invariably gone through in an orderly manner with some show of reason, for, porpoising off in a clump, they at once spread themselves out, scattering in a fan-shaped formation as they sped away, instead of all following the same direction.

As far as I could judge, however, the sea-leopards are a trifle faster in the water than the Adelies, as one of them occasionally would catch up with one of the fugitives, who then, realizing that speed alone would not avail him, started dodging from side to side, and sometimes swam rapidly round and round in a circle of about twelve feet diameter for a full minute or more, doubtless knowing that he was quicker in turning than his great heavy pursuer, but exhaustion would overtake him in the end, and we could see the head and jaws of the great sea-leopard rise to the surface as he grabbed his victim. The sight of a panic-stricken little Adelie tearing round and round in this manner was a sadly common sight late in the season. . . .

When the sea-ice had gone out, leaving open water right up to the ice-foot, a ledge of ice was left along the western side of the rookery, forming a sort of terrace or "front," with its sides composed of blue ice, rising sheer out of the water to a height of some six feet or more in places. From this point of vantage it was possible to stand and watch the penguins as they swam in the clear water below, and some idea was formed of their wonderful agility when swimming beneath the surface. As they propelled themselves along with powerful strokes of their wings, they swerved from side to side to secure the little prawn-like euphausia which literally swarm everywhere in the Antarctic seas, affording them ample food at all times. Their gluttonous habits here became very evident. They would gobble euphausia until they could hold no more, only to vomit the whole meal into the water as they swam, and so enlightened start to feast again. As they winged their way along, several feet beneath the surface, a milky cloud would suddenly issue from their mouths and drift slowly away down stream, as, without the slightest pause in their career, they dashed eagerly along in the hunt for more.

When a penguin returned to his mate on the nest, after his jaunt in the sea, much formality had to be gone through before he was allowed to take charge of the eggs. This ceremony of "relieving the guard" almost invariably was observed.

Going up to his mate, with much graceful arching of his neck, he appeared to assure her in guttural tones of his readiness to take charge. At this she would become very agitated, replying with raucous staccato notes, and refusing to budge from her position on the eggs. Then both would become angry for a while, arguing in a very heated manner, until at last she would rise, and, standing by the side of the nest, allow him to walk on to it, which he immediately did, and after carefully placing the eggs in position, sink down upon them, afterwards thrusting his bill beneath his breast to push them gently into a comfortable position. After staying by him for a little while, the other at length would go off to bathe and feed. . . .

When the chicks began to appear all over the rookery, a marked change was noticed in the appearance of the parents as they made their way on foot from the water's edge to the nests. Hitherto they had been merely remarkable for their spotless and glistening plumage, but now they were bringing with them food for the young, and so distended were their stomachs with this, that they had to lean backward as they walked, to counterbalance their bulging bellies, and in consequence frequently tripped over the inequalities of the ground which were thus hidden from their gaze.

What with the exertion of tramping with their burden across the rookery, and perhaps on rare occasions one or two little disputes with other penguins by the way, frequently they were in some distress before they reached their destination, and quite commonly they would be sick and bring up the whole offering before they got there. Consequently, little red heaps of mashed up and half digested euphausia were to be seen about the rookery. Once I saw a penguin, after he had actually reached the nest, quite unable to wait for the chick to help itself in the usual manner, deposit the lot upon the ground in front of his mate. When this happens the food is wasted, as neither chick nor adult will touch it however hungry they may be, the former only feeding by the natural method of pushing his head down the throat of a parent, and so helping himself direct from the gullet.

When the chicks are small they are kept completely covered by the parent who sits on the nest. They grow, however, at an enormous rate, gobbling vast quantities of food as it is brought to them, their elastic bellies seeming to have no limit to their capacity; indeed,

when standing, they rest on a sort of tripod, formed by the protuberant belly in front and the two feet behind. . . .

To see an Adelie chick of a fortnight's growth trying to get itself
covered by its mother is a most ludicrous sight. The most it can hope
for is to get its head under cover, the rest of its body being exposed
to the air; but the downy coat of the chick is close and warm, and
suffices in all weathers to protect it from the cold. . . .

Some way back I made allusion to the way in which many of the
penguins were choosing sites up the precipitous sides of the Cape
at the back of the rookery. Later I came to the conclusion that this
was purely the result of their love of climbing. There was one colony
at the very summit of the Cape, whose inhabitants could only reach
their nests by a long and trying climb to the top and then by a walk
of some hundred yards across a steep snow slope hanging over the
very brink of a sheer drop of seven hundred feet on to the sea-ice.

During the whole of the time when they were rearing their young,
these mountaineers had to make several journeys during each twenty-
four hours to carry their enormous bellyfuls of euphausia all the
way from the sea to their young on the nests — a weary climb for
their little legs and bulky bodies. The greater number who had
undertaken this did so at a time when there were ample spaces un-
occupied in the most eligible parts of the rookery.

I have mentioned that large masses of ice were stranded by the
sea along the shores of the rookery. These fragments of bergs, some
of them fifteen to twenty feet in height, formed a miniature mountain range along the shore. All day parties of penguins were to be
seen assiduously climbing the steep sides of this little range. Time
after time, when half way up, they would descend to try another
route, and often when with much pains one had scaled a slippery
incline, he would come sliding to the bottom, only to pick himself up
and have another try.

Generally, this climbing was done by small parties who had
clubbed together, as they generally do, from social inclination. It
was not unusual for a little band of climbers to take as much as an
hour or more over climbing to the summit. Arrived at the top they
would spend a variable period there, sometimes descending at once,
sometimes spending a considerable time there, gazing contentedly
about them, or peering over the edge to chatter with other parties
below.

Again, some half a mile from the beach, a large berg some one
hundred feet in height was grounded in fairly deep water, accessible
at first over the sea-ice, but later, when this had gone, surrounded by

open water. Its sides were sheer except on one side, which sloped steeply from the water's edge to the top.

From the time when they first went to the sea to feed until the end of the season, there was a continual stream of penguins ascending and descending that berg. As I watched them through glasses I saw that they had worn deep paths in the snow from base to summit. They had absolutely nothing to gain by going to all this trouble but the pleasure they seemed to derive from the climb, and when at the top, merely had a good look round and came down again.

JOHN BURROUGHS (1837–1921)

Old Friends in New Places

From *Under the Apple Trees*. 1916

[I DO NOT KNOW what future generations will think of this quiet New England essayist, so gentle, so friendly; and my own judgment is too warped by a boy's adulation. He added little to the sum of natural history, and his unconscious avoidance of humor is a regrettable lack. Nevertheless one's heart warms at the very sight of the little, olive-green volumes. Even if we read him less and less, by his omission from this Anthology there would be left a minor but unfillable gap between Thoreau and some of the modern naturalists, writers about nat-ure, more voluble, much more perspicacious and recondite, but far less restful.]

LAST WINTER and early spring in central Georgia I had great pleasure in the little glimpses of wild life, mostly bird-life, that I got from the windows of the cabin study which my friend built for me in one corner of an old unused building situated in a secluded place near a bushy spring run and a grove of pine- and oak-trees. Many of our more northern birds — such as song sparrows, bluebirds, juncoes, and white-throats — winter in Georgia and impart a sort of spring air to the more secluded places at all times. The mockingbird, the brown thrasher, the cardinal, the meadowlark, the crested titmouse, the Carolina wren, the blue jay, the downy woodpecker, and a few others are there the year round.

February in Georgia is like April in New York or New England, and March has many of the features of early May. In late February or early March the red maples are humming with honey-bees and the elms are beginning to unpack their floral budgets.

The sparrows — white-throats and song sparrows — were at home in the weedy and bushy ground around my little hermitage, and I soon encouraged them to come under my window by a plentiful sprinkling of finely cracked corn and bird-seed. They were always very shy, but they soon learned to associate me with the free lunch, so that, very soon after my appearance, — about nine o'clock in the morning, — they would begin to gather from the near-by coverts, one to two dozen white-throats, with four or five song sparrows, and now and then a female chewink. The chewinks remain there the year round, but the song sparrows and the white-throats, like myself, were only there for a season.

By easy stages from one covert to another, traveling mostly at night, the birds were soon to begin the return journey northward. I think the same birds lingered with me day after day, though one cannot be sure in such a matter. The individual units in a stream of slowly passing birds of the same species do not differ from one another in appearance any more than do the separate ripples in a stream of flowing water. Outside of man's influence, the individuals of a species of wild creatures or wild flowers do not seem to differ from one another by as much as one hair or one feather or one petal. They are like coin stamped with the same die, and the wonder of it is that each and all, among the birds, at least, seem like new coin — not one blurred or imperfect impression. This fact always strikes one in gazing upon a flock of wild birds of any kind in the fall or in the spring. The wear and tear of life seems to leave no mark upon them. Take a hundred snow buntings in winter, or robins or bluebirds in the spring, and each individual seems up to the standard of its kind. Indeed, Nature has standardized them all.

Among the song sparrows and white-throats that gathered for their daily lunch under my window, I noted differences between male and female and between old and young, yet each individual seemed at the top of its condition. How free from spot or blemish they were, not one disheveled or unkempt, not one vagabond or unfortunate among them. How neatly groomed they were, every feather perfect and every feather in its place. How bright and distinct the pencilings of the song sparrows' backs! The surplices of the white-throats had just come from the laundry. Among all the wild creatures it is the same. Nature deals evenly and impartially with them. They differ markedly in this respect from birds and mammals under domestication. A brood of newly hatched chickens are fresh and clean enough, but they very soon deteriorate in appearance; but a brood of young grouse or quail keep as clean and bright as shells upon the beach. Then consider the chipmunks and red squirrels — how rarely is one of them below the standard of its kind! how rarely one shows any indication of hard luck, or a loss of standing among his fellows! None are poor; all are equally prosperous. Success is written on every one of them. Rarely is a single hair out of place.

How wise the white-throats are about cracked corn, taking nothing above a certain size! They pick up the larger pieces and test them with their beaks and drop them, then pick them up and feel them again to be quite sure they have made no mistake. Their little gizzards cannot grind the flinty corn except when taken in very small bits. The fruit- and insect-eating birds that sometimes came about your door in winter or spring with the white-throats and juncoes came daily to the dooryard of a friend of mine near New York City. She sprinkled the ground with rolled oats and hominy grits and her visitors made the most of her bounty. One morning there was a newcomer — a thrush evidently hard put for food. He hopped about amid the feeding sparrows with drooping wings, picking up the seeds and grains and dropping them again, apparently wondering what the others found that was so appetizing. The bird was in desperate straits; he ate the snow, but I fancy it only aggravated his hunger.

The newcomer turned out to be a hermit thrush. I told my friend to take any dried fruit she happened to have — raisins, dried currants, dried cherries, or dried berries, and cut them up and sprinkle them among the seeds. She did so, and it was not long before the thrush began to examine them and taste them doubtingly, but very soon he was eating them. That afternoon his drooping wings were getting back to their normal place, and in a day or two he was a changed bird, brisk and bold, domineering over the other birds, — in a very courteous way, however, — and very much set up in life.

A bird never appears emaciated; it will starve and retain its plump appearance. Robins will famish amid a world of seeds and grains. They must have fruit or worms. Three years ago, while spending the winter in Georgia, I had evidence that a vast number of robins starved to death in March. People picked them up in their yards and in the fields and along the edge of the woods. They seem to have started north from Florida and the Gulf States too soon. A sudden cold snap kept the worms and insects below the surface of the ground, and there was no fruit but the white, dry china-berries, and these appear to poison or to paralyze the robins when they eat them. In my walk one morning I picked up a cock robin that was unable to fly. As it did not appear to have been injured in any way, and was of very light weight, I concluded it was starving. I took it into the house and let it perch on the back of a chair in the study. It showed little signs of fear and made no effort to escape. I dug a handful of earthworms, and dangled one of them before its beak. After eyeing it a moment it opened its beak and I dropped the worm into its mouth. Others soon followed, and still others. The bird began to wake up and come to itself. In a little while it was taking the food eagerly and without any signs of fear. I could stroke it with one hand while I

285

fed it with the other. It would sit on my knee or arm and take the food that was offered it. I was kept pretty busy supplying its wants till in the afternoon it began to fly and to run about the room and utter its call-note. Before night it had become so active and so clamorous for its freedom that we opened the window. With a dash and a cry it was out of the house and on the wing to a near-by tree. I trust, with the boost I had given it, it was soon safely on its northward journey.

The incident shows how extreme hunger in a wild creature banishes fear. One March day, when I was a boy, I found a raccoon wandering about the meadow so famished that he allowed me to pick him up by the tail and carry him to the house. He ate ravenously the food I offered him.

The struggle for life among the birds and other wild creatures is so severe that the feeble and malformed, or the handicapped in any way, quickly drop out. Probably none of them ever die from old age. They are cut off in their prime. A weeding-out process goes on from the time they leave the nest. A full measure of life, the perfection of every quill and feather, and unerring instinct carry them along. They are always in the enemy's country; they are always on the firing-line; eternal vigilance and ceaseless activity are the price of life with them. The natural length of life of our smaller birds is probably eight or ten years, but I doubt if one in a thousand reaches that age. Not half a dozen times in my life have I found the body of a dead bird that did not show some marks of violence.

Next to the trim, prosperous, well-dressed appearance of a flock of wild birds, one is struck with their caution and watchfulness, not to say nervousness, at all times, especially when feeding in the open. My band of sparrows were apprehensive of danger every moment. Here are some notes made on the spot:

"Now there are over two dozen sparrows, among them a solitary female chewink, feeding on the ground in front of my window. An ever-present fear possesses every one of them. They pick up the seeds hurriedly, looking up every few seconds. Suddenly they all stop, and, crouching, look toward the near-by weeds and bushes. Some vague alarm has seized them. Then two of them dart away; then the whole flock rushes to cover. I see no cause for the panic; there is none; the strain has become too great to be longer borne. Though no danger is near, yet their instinct, developed and sharpened by the experiences of untold generations, tells them danger might be near — a hawk, a cat, or other enemy — and that safety demands a frequent rush to cover. After a few minutes they return, one by one, flying from weed-stalk to weed-stalk, and dropping upon the ground where the seed is scattered, with many a suspicious flip of wing and flirt of tail. A dozen or more are soon hurriedly feeding again, now and then running spitefully at one another, as if the aggressors felt a prior claim, but not actually coming to blows.

"When the dry grass and weeds cover the seed a song sparrow may be seen now and then executing a quick movement upon it with both feet, a short double jump forwards and backwards. This is the way the sparrow scratches — a crude and awkward way, certainly. She has not yet learned to stand alternately upon one foot and scratch with the other, as do the hen and all other true scratchers, and she probably never will. The sparrows, and many other birds, move the two feet together. They are hoppers, and not walkers or runners. Such birds make a poor show of scratching. The chewink scratches in the same way, but being a much larger bird, she rakes or kicks obtruding weeds about quite successfully.

"In less than two minutes the birds again take the alarm and dart away to their weedy refuge."

This is the habit of all birds that feed in numbers in this way in open places. Snow buntings, juncoes, sparrows, reed-birds, blackbirds — all are haunted by a vague sense of impending danger when they are feeding, and are given to sudden flights to cover, or to circling in the air.

I remember that the flocks of passenger pigeons that I used to see in my youth would burst up from the ground when they were feeding, at short intervals, in the same sudden, alarmed way. It is easy to see how the fear of all ground-feeders has become so developed and fixed. Hawks are doubtless the main cause of it. The hawk comes suddenly and strikes quickly, and is doubtless as old an enemy as the birds have. For ages he had been wont to swoop down from the air or from the cover of a tree, or has skimmed over the hill and in a twinkling snatched a feeding bird. I have seen the sharp-shinned hawk in winter sweep over a garden fence and snatch an English sparrow from a flock feeding in the street. I have seen one of the smaller hawks pick up a high-hole feeding in the fields in the same way. Birds feeding singly are less easily alarmed than when feeding in flocks, just as you and I would be. Fear is contagious, and a bird feeding alone has no alarms or suspicions but its own to disturb it.

Since these birds left Canada and northern New England last October they have probably traveled over two thousand miles, beset by their natural enemies at all times and places — in fields and marshes and woods; in danger of hawks and shrikes and cats by day, and of owls and other prowlers by night; compelled to hustle for food at all times, and to expose themselves to a thousand dangers. Is it any wonder that they are nervous and watchful?

In returning they will be exposed to the same dangers. Their traveling is mostly done by night and it is probably by easy stages. But just how long any single flight is we have no accurate means of knowing. It would be interesting to know if the song sparrows and juncoes traveled in company with the white-throats, as they are usually found together by day. If they do, the song sparrows would begin to drop out of the procession by the time they reached the Potomac, and continue dropping out more and more all through New York and New England, but some of them keeping on well into Canada. The juncoes would begin to drop out in the Catskills, where they breed, and a few white-throats may do so likewise, as I have found them in midsummer in some of the higher regions of these mountains.

Fear and suspicion are almost constant companions of most of the wild creatures. Even the crow, who has no natural enemies that I know of, is the very embodiment of caution and cunning. That peculiar wing-gesture when he alights or walks about the fields — how expressive it is! It is a little flash or twinkle of black plumes that tells you how alert and on his guard he is. It is a difficult problem to settle why the crow is so suspicious and cunning, since he has few or no natural enemies. No creature seems to want his flesh, tough and unsavory as it evidently is, and we can hardly attribute it to his contact with man, as we can the wildness of the hawk, because, on the whole, mankind is rather friendly to the crow. His suspicion seems ingrained, and probably involves some factor or factors in his biological history that we are ignorant of.

On the whole, it is only the birds and animals which are preyed upon that show excessive caution and fear. One can well understand how the constant danger of being eaten does not contribute to the ease and composure of any creature, and why these which are so beset are in a state of what we call nervousness most of the time. Behold the small rodents — rats, mice, squirrels, rabbits, woodchucks, and the like; they act as if they felt the eyes of the mink or the weasel or the cat or the hawk upon them all the time.

Among the birds some are much more nervous and "panicky" than others. The woodpeckers are less so than the thrushes and finches; the jays less than the starlings and the game-birds. The seed-eaters and fruit-eaters are probably preyed upon much more than the purely insectivorous birds, because doubtless their flesh is sweeter.

Birds of prey have few enemies apart from man. Among the land animals we ourselves prefer the flesh of the vegetable-eaters, and the carnivora do the same. We all want to get as near to the vegetable as we can, even in our meat-eating.

The birds, even the prettiest of them, are little savages. In watching from my window the feeding white-throats and song sparrows, I cannot help noticing how ungenerously they behave toward one another — apparently not one of them willing to share the feast with another. Each seems to think the food his or her special discovery and that the others are trespassers. They charge spitefully upon one another, but rarely come to blows. Just what makes one give way so readily before another, without any test of strength, is a puzzle. Is the authority in the eye, in the bearing, or is it just a matter of audacity and self-assertion? There may be timid and retiring souls among the birds as well as among other folk. I am inclined to think that usually it is the males bullying the females. Occasionally two males, known by their more conspicuous markings, confront each other and rise in the air a yard or two beak to beak, and then separate.

During the mating season there is mutual aid and cooperation between the sexes, the male bird often feeding the female. But at other times there is little friendliness, certainly no gallantry. The downy woodpecker in winter will drive the female spitefully away from the bone or the suet on the tree in front of my window till he is first served. I have never seen crows quarrel or strive with one another over their food. On the contrary, if the crow discovers food in winter, he seems glad to be joined by a companion or several of them. The crow is a generous bird; he has the true social instinct. He will watch while his fellow feeds; he cheerfully shares his last morsel with a comrade. How different from any of the hawk tribe! A farm-boy living near me brought up four young sparrow hawks in a cage. They were as jealous of one another over their food as cats are, and when they were nearly full-grown, and the food was insufficient, they proceeded to devour one another I kept two of the survivors a few days, but they were so utterly cruel and savage that I was glad to let them escape.

Most of our rodents are as free from guile as our birds; they have none of the subtlety and cunning of their enemies the fox and the wolf; they are simply wild and shy. The rabbit has little wit, yet she manages to run the gantlet of her numerous enemies. Some of her arts of concealment are as old as mankind — the art of hiding where no one would think of looking — concealment where there is little to conceal her. One March day I started a rabbit from her form in a broad, open cultivated field. She had excavated a little place in the soft ground just deep enough to admit the hind part of her body and there she crouched in the open sunlight with only a little dry grass partly screening her. When I was within two paces of her she bounded away like the wind and directed her course toward a bushy ravine several hundred yards away. The advantage of her position was that she commanded all approaches; nothing could steal a march upon her, and she could flee in any direction. In a tangle of weeds or bushes she would have been where every one of her natural enemies prowl or beat about, and where concealment would have been more or less confinement. A few yards farther along I came upon another vacant form — the perfection of art without any art. When the rabbit builds her nest and has her young she does not seek out a dense cover, but comes right out into the clear open spaces where you would never think of looking. She excavates a little cradle in the ground, gathers some dry grass, weaves a little blanket of dry grass and fur from her own body, just large enough to cover it, and her secret is well kept — most hidden when hidden the least. Quail and grouse know something of the same art, and never make their nests in a thick tangle. I have seen a quail's nest with twenty eggs in it on the edge of a public highway. The brooding bird allowed me almost to touch her with my hand before she flew away.

If every bushy and weedy spring run in Georgia embracing not more than an acre or two of ground has two dozen sparrows, to say nothing of a pair or two of cardinals, Carolina wrens, and mocking-birds, one can get some idea of what a vast number of birds such a large State — over three hundred miles long and two hundred miles wide — holds. With two pairs of birds to the acre, a fair estimate, it would count up to over seventy millions. The farm of about one hundred and thirty acres upon which I passed February and March probably held several dozen sparrows and as many juncoes, a score or two blue jays, and two or three dozen meadow-larks, a pair each of cardinals, Carolina wrens, and brown thrashers, besides other birds. In one ploughed field I saw, day after day, ten or fifteen killdee plovers. Their wild cries, their silver sides glancing in the sun, and their long powerful wings were always a welcome sight and sound.

Probably more kinds of birds feed on insects than upon seeds and fruits, though the seed- and fruit-eaters are the more numerous, and abide with us more months in the year. It is true also that the seed-eaters nearly all eat insects at times, and start their young in life upon insect food. One can easily see, then, what an inevitable part the birds play in keeping down the insect pests that might otherwise overwhelm us.

JOHN FARRER (1880–1920)

Chagola

From *On the Eaves of the World*. 1917

[IN THE FEW ACCOUNTS I can find of the life of Farrer it is emphasized that he is an artist, not a scientist; a horticulturist, not a botanist; but he seems to me preëminently a superb writer about wild places of the earth. His descriptions are not travelogues or narratives, but words that make vivid the beauty and charm of such places as western China and the eastern Himalayas. They combine sound natural history with a vocabulary which lifts his style to a very high plane.

Hunters of plants might, at first thought, be considered to have a soft job in the activities of exploration, but as a matter of fact, the opposite is true. A collector of insects or birds or animals usually establishes a semi-permanent camp from which he makes daily excursions in various directions and returns each night. An efficient plant-hunter soon exhausts the desirable flora of any one spot and must keep on the move, seeking out new and lesser-known growths in jungle or on mountain-side. This actually invites discomforts and dangers and hence I have been glad to include in Dr. Farrer's account an adventure or two, having little direct connection with his object, but illustrating the difficulties he had to overcome to find and collect the roots and stems, the flowers and seeds of rare plants.

In fact, he lost his life in the heart of Burma at a comparatively early age, otherwise he might easily have become the best author in his field in the English language.]

OUR FIRST Tibetan inn at Second Look (Di-er-kan) [was] a dim-yarded, storied place, not, in fact, purely Tibetan, but wholly different from the Chinese style. The population was inquisitive, but not friendly; they refused to sell us anything, and their picturesque bedizened women were coy to all attempts at conversation. In the big lower room, smoky and dark, there was a noble open hearth, at which, for the first time since leaving England, I was able to enjoy the luxury of warming my feet at a fire, while in and out of the shadows ran grunting and routing a tiny pet black pig, who

291

was ultimately captured and put to bed in his basket like any puppy or kitten. Round the hearth gathered the staff, Rembrandtesque in the effects of the firelight, busily engaged in chat about Chago and other perils of the way. It seems hard to realise that here we are now actually at the very foot of this elusive mountain that we have so long been chivying through the untracked wilds of Western China. To-morrow we dare the perils of the pass and the Abbey. The population talks of nothing else. Even here, just over the other side of the mountain wall, but in a country more tinctured with China, the place has an almost fabulously evil reputation; you would fancy that no Chinese could manage to pass through it without being burned alive in a bonfire of brushwood. It is quite certain, anyhow, that nobody does pass through it, except on the most urgent necessity, and then in bands of as many as can be got together. The tale was all of Chago and local broils; at Wen Hsien we had said farewell for weeks to the outer world behind us, and now for a long time all the happenings in China and elsewhere were to be a complete blank.

The momentous day came up glorious and without a cloud. A little farther we followed the Eastway River, now clear as a green jewel; but in the rocky angle of a gorge it turned away and left us, winding south-west into the Marches of Szechwan. We crossed, and mounted a steep spur, to a little stone porch at the top, from which there was a memorable vista backwards over the river, the valley, and the village, smoking blue amid its verdure in the slanting early sunshine, with big hills and forests and high snows overtopping the dry downs of the glen. And then we turned away, down over the shoulder and along a hillside wooded in holm-oak, till we reached the depths of a ravine and struck away straight upwards to the right, beneath lovely hanging woods now blushing with scarlet and green and gold and crimson in a haze of spring rejoicing. Up and up, and up and up, the path almost as steep as a stairway, through more and more woodland glens, with glades of open grass and rosy Peonies sprouting amid the rocks. At last we came into a beck-bottom that was really alpine, enveloped in coppice, where the water purled gurgling through the brakes, and a crimson cuckoo-pint shone brilliant among the sere stalks and the pale withered wreckage of last year's ferns. And then, far overhead, so sheer and high that it seemed impossible we should ever attain it, appeared the wall of Chagola itself, crested all along with snow. Upwards coil over coil we wound, from pinnacle to ascending spur of the climb, and down below us sank deeper and deeper the cold forested folds of the hill where snow lay dappled chill on the pale moss-beds down among the lavender haze of the red birch, here a mere winter cobweb of deadness with no apparent hope of resurrection. Only the stately old spruces, crowded and sombre, that punctuated the skeleton pallors of the birch, gave a look of ominous and ancient life to the slopes.

Now the woodland was left below, and we embarked on the main climb, which ascends the wall of the pass in very short abrupt zig-zags, much more ferocious than the ample and leisurely hairpins of the Alps. It was all grass, coarse at first and yellow with winter, but ere long growing more alpine with every yard one climbed. Life had hardly begun to awake on that Sixth of May. A little rue-buttercup twinked glistering many-rayed stars of bright snow at intervals in the dank black slopes of silt at the path-side, and we had our first sight of that tedious little ugly Anemone, like a small white buttercup itself, which is such a universal weed all the way up these borders at least as far as the alps of the Koko-nor. Meanwhile, looking back, we had real mountain-ranges at last to feast our eyes. Gradually over the intervening hills unfolded high and jagged splendours of rock and ice and snow; one behind the other arose the huge ranges, surging like the frozen foam of an arising Aphrodite from the surf of lesser masses at their feet, with the curving course of the Eastway notable far down into the south-western distance, and a level line of great mountains closing it in all along the horizon. We were now well up into the alpine zone; the turf was a fine brown sward, still soaked and sodden and dark from the winter; but the keen and glorious daylight seemed to be calling life out of it every moment in a magic exhalation of the awakening hillside. The breath of the mountain had a sharp and mystic sweetness, heady and aromatic, almost like that heady aromatic fragrance of the mountain clover which sweeps across the brown highlands of the Cottians and Graians in the spring, like a breath of etherealised fairy wine. All over the slopes a big Narcissus Anemone was sprouting in myriads of fluffy tuffets, but the only other flower that promised yet was another rue-buttercup, whose large daisy-like blooms lay helpless on the ground, sodden with late storms, and apparently chilled to a cold and chilblainy blue. For now we were actually in the limit of the late storms, and all the hill was a solid sheet of melting snow, in the laps of which it was just possible to discover the dank wide wads of a perennial Meconopsis, from the heart of whose rosettes were pushing the hairy globules of green that were to be the shoots of the year. Otherwise, the crest was a white wilderness, in which stood up the black skeleton seed-heads of various louseworts, but nowhere of any Primula; all the forest lay far away below us now, and at these heights there was only a scanty coppice of big Rhododendron in the dells, gaunt and scraggy, with a heathery scrub of wee Potentilla and wee Rhododendrons giving such a heathered effect to the open hillside that at any moment one expected grouse to rise and scatter with their harsh metallic whirr. And so at last, in an ecstasy with the day and the air and the alps, we achieved the long ascent, and stood on the very neck of Chagola, looking down upon its northern wall and out across the whole plan of the Kansu-Tibetan March.

I do not know of any view in our own mountains that can compare with this sudden revelation. It is like nothing I remember, at once much more episodic and much more regular than any effect in the European Alps. For these Tibetan ranges descend to China in so well-arranged a series of parallel sweeps that here, on the neck of one chain, you are looking far out across the intervening folds of hill and down and forest to another exactly opposite your own. And also, away to the left, one behind another, you see the successive links of the chain on which you yourself are standing, rising up in isolated masses, seen end-on, with the effect of a series of gigantic icebergs floating on a faintly rippled ocean that is the forested hill-country round their feet; while in front of you, perfectly straight and regular, rises the long stark wall which is the last westerly effort of the Min S'an, culminating in the stupendous mass which we shall learn to know as Thundercrown ("Lei-gor S'an"), with other masses behind it and behind it away to the left, till the eye loses itself in that archipelago of titanic ice-islands which is Eastern Tibet. You are looking out, too, into a country wholly different from what we have left. Here are no bony barren valleys, no loess desolations of ochre, but the sides of the mountains fall away into steep over steep of dense and ancient forest, down in the deep heart of which roars the ice-grey little torrent of the Satanee Hor; while all the intervening hills between range and range are mollified by the influence of the big ranges into clothing themselves with vegetation and woodland. From the crest of Chagola you can luxuriate in the glories of a typically alpine country.

Not the faintest notion can you form that away out there, beneath the very feet of Thundercrown, you will meet once more the Blackwater, storming down a blazing valley as dry and sere as that in which you left it. All the finer mountains, in fact, are away and away, like fair Inez, into the west; you might have guessed that from the comparative proximity of Chagola to Chinese districts and influence. There is, of course, no border between China and Tibet, no neatly defined line with outposts and douanes; but China gently fades as

the mountains become more and more unnegotiable and unprofitable, and here you are at the first point where the Central Asiatic
ranges begin to be really tiresome and distasteful to the Chinese.
Both the Satanee Alps, in fact, and the Min S'an, are dying ranges
at this point, where China begins to take possession; almost abruptly
they sink eastward into the downlands of Western Kansu. Immediately above you now, on the right, towers the naked leonine head of
limestone which is the last outbreak of the Satanee range. (You may
call it Chagoling if you want a name, Chagola being the right Tibetan
word for a pass, while the Chinese "ling," for pass or mountain, seems
to be of less precise application.) In the same way, Thundercrown is
the last spasm of the Min S'an, after which, in a higher and more
magnificent wall than this of Chagoling, the range continues hardly
any farther eastward than this, severed between its final vertebrae by
the deep cutting of the Nan Hor, which the enfeebled spine is no
longer able to resist.

The northern face of Chagola is very different from the southern.
As soon as you are sated with the glory of that prospect and the dazzling air, you sink immediately into a black midnight of tortured
matted old spruces and glossy tree Rhododendrons, gnarled and co-
agulated with the storms and snows of this grim elevation, and even
now heavy with snow that in their depths is still an unbroken deep
bed. The track descends with vertiginous abruptness; it is like nothing so much as a sheer cataract of ice and snow in the desolate darkness of that jungle. Down and down the ice-shoot we slither and
slide and flounder. It was best not to think of the mules and ponies
descending behind us; but on they faithfully came, all difficulties notwithstanding, sitting down to a glissade where the ice was more than
usually sheer and impracticable. Gradually the firs gained ground
over the Rhododendron, and increased in stature; the snow continued very far down in the fastnesses of that solemn primeval forest,
where no breath stirred in the dense stillness, ultimately and gradually turning to a slough of unutterable slush. There are no flowers
in that dark night; there is only a little pale dead thing like a ghost
that has been sodden for years in water. Until, turning a corner, I
suddenly gave tongue upon the first of our flowering Primulas.[1]

In the mossy bank it gleamed and glittered here and there, and in
the dappled darkness of the forest its amethystine pinkness seemed
to shine like so many stars of soft flame. I did not revel in it the less,
either, because it so exactly recalled to me a big primrose or oxlip,
its sturdy stems, each unfolding a loose head of big flowers daintily

[1] *P. hylophila,* sp. *nova.*

lilac-rose in colour, with an elaborate ten-rayed eye of white diffusing from the greenish throat. It was like a primrose, too, in its tastes, growing copiously in just the same places, in rich woodland soil, and especially on trunks of long-fallen and quite rotten timber, all down the track-side, and in all the more open woodland glades and at the edges of the coppice in this region, between some five to nine thousand feet. We were lucky to see its beauty, for only here, at its topmost extension, was it still lingering in bloom, and when we reached the region of its main abundance lower down in the woods there were none of the pink stars still shining, but stout little green scapes standing up in seed, with the stem of each separate flower in the head so swollen and stiff that they made the effect of so many outstanding trumpets of palest green swelling to the five-toothed mouth of the calyx. It had no scent when I saw it, but that, perhaps, may have been the result of the recent storm; nothing could have appreciably added to the charm of its beauty, and I was in a continual course of raptures as I came floundering down the interminable mud-slides and sloughs of that coiling path, perpetually, as we descended more and more into the region of deciduous trees, delighting in a richer and richer abundance, in all the opener places, of that cheery little lovely primrose, springing everywhere from its crinkly primrose-like starfishes of foliage, crisp as lettuce, and clothed in an almost microscopic and invisible coat of dense emerald-green fur. None of its race, in fact, has more fascinating foliage, even apart from the bloom; even in late autumn it is always with fresh joy that you see its crumpled star unfurling from the mossy face of some long-rotten tree-trunk across your path.

So we had our fill of the Wood-nymph, sparkling at us from all the banks as we came. We were not the only people on the road, though it was a marvel to meet, even on that last appalling piece at the top of the pass, whole ascending trains of wretched carriers, staggering painfully upwards, bowed down beneath huge burdens of meal and flour that they were conveying over to the starving dry valleys of the Eastway. It shows the desperation of that luckless land, that provisions should find it worth while to undertake so arduous and wild a journey. The descent of that first wall from Chagola brings you down through all the alpine zones of woodland, from the chaos of storm-driven Rhododendrons at the top, through that in which enormous firs and pines stand motionless above the gloom of the jungle, down to the dappled glades and spring-coloured undulations of the deciduous woodland; finally, the track debouches into a most beautiful alpine plain of grass, from which several converging glens, now so many rolling seas of lavender and tenderest green and pink and gold with the approach of spring, lead up into the fastnesses of the naked rocks that look so impregnably sheer and far overhead that it seems impossible you can so immediately have come down from such heights.

Here, according to their unvarying habit, the "boys," in a minute or two, had a cheerful bonfire blazing on the sward, and, as its blue smoke went coiling up against the background of hazy forest, we ourselves rejoiced in the prospect of a camping centre so ideal in every respect. Certainly a good part of our summer should be spent in this green little vale of peace, that afforded such good access to so many portions of the range; no doubt we could get on to terms with the people of Chago for our supplies, and all would be well. So murmured hope; meanwhile, we sauntered and browsed about amid the peaceful beauties of the spot, enraptured by our first sight of the universal sky-blue Fumitory [2] of these ranges, which was here sprouting amid the shingles of the beck, and astounded us with its intense blazing azure. The floor of the woodland, too, was a surf of snowy wood anemones, that seemed to dance in the flickering lights and shades.

Now it was time to proceed. The track led onward through an immemorial stretch of flat woodland. The ground was an unbroken snow-starred carpet of wood anemones, beneath the soft green light that filtered through the vast entangled tree-trunks overhead, whose branches made an interarching canopy looped up with ferns. It seemed a very ancient watchful place, haunted with silence; the long furry boughs reached down everywhere towards us like monstrous arms of apes. The windless green calm, the profound stillness of the forest, were indescribably solemn and soothing. Common among the Anemones was Pachysandra, in its ordinary wild leathern green form, as common as dog's mercury and quite as ugly; and here and there the brown and bronzy mounds of the giant lily broke the level, here rather poor, but becoming more common as we descended. The Wood-nymph Primula, now gone quite out of flower, accompanied us in abundance, and the brilliant purple cuckoo-pint, blazing here and there in the dark distances of the coppice was always getting itself mistaken for its betters, and especially for some new Primula of notable brilliancy and port. The gentle descent gradually grew wilder, though; and we began dropping more steeply through a narrowing gorge of high precipices, choked with vast old boulders clothed in moss, where yet another woodland Primula was just begin-

[2] *Corydalis curvioflora.*

ning to uncoil its leaves from their winter sleep. In a tangled riot the aged Rhododendrons wove arbours above our heads, and in the moss amid the boulders shone the greenish-golden orbs of an Adonis. Wilder and wilder grew the gorge, steeper and rockier the track; the way seemed endless, till in a damp moss cushion close at hand we saw a budding Primula whose mealy leaves showed clearly a different species. Evidently it was a strayed outlier from some main colony, which I now accordingly set myself to discover. But it was not till I had climbed a hundred feet or so, up a ruin of fallen boulders covered in brown moss, that on the damp precipice above I saw the glinting rosy-purple of the Rock-nymph Primula. Here, though, she was poor and frail in growth, with scanty inferior flowers, and we shall nowhere see her full beauty except on the dark rockwalls of the Bastion Gorges opposite Satanee.[3]

Gradually, now, the woodland thinned out and the descent grew calm; flowers of lower levels appeared, more and different wood anemones, abundance of the giant lily, a coppiced tangle of Spiraeas, and numbers of a most beautiful thing which is called the dusky Disporon, because its long hanging bells of blossom, in loose clusters on six-inch stems, are of the most ravishing waxy white, recalling little Lapagerias or gigantic Solomon's seals. At last even the lower coppice tailed away, and we came out into open fields, rounding and rounding the long bays of the descent. Stalwart Tibetans, clothed in what looks like ancient sacking, were urging their yaks in front of the plough, and evidently we were now down in the culture zone, and out of hope of any more Primulas in this fat land of tillage and hedgerows. Hardly had I said the word than I turned my head, and on a bank above a little rill beheld the very Primula of which I had just so reasonably despaired. I was so far right, though, that the Bankside Primula [4] does not suggest being a wild plant, nor at all an alpine; it is a small poor cousin of the vicious *P. obconica,* and not a thing of merit, of value, or any eminent beauty — although, be it never so humble, a Primula is always a Primula, and as such to be venerated, even if its flowers be small and mean and magenta like the squinny little stars of *P. riparia.*

The Mafu and the Mee had long since ridden on ahead to secure us lodgings at Chago; we ourselves began to wonder if we should ever arrive at this singularly evasive place. The glorious day was drawing on to a calm and radiant evening, and still at each brow of the hill we only saw another below us like the last. Very far up behind

---

[3] *P. scopulorum,* sp. *nova.*
[4] *P. riparia,* sp. *nova.*

us now rose the forested wall of Chagola, with the bare precipice of the mountain looming overhead. The stream of the gorge now deserted us, and went plunging down in deep wild cañons of woodland far below, while we ourselves continued breasting the fell sides, and leisurely descending between hedges all ablow with pear. Finally, for the last straw of the day, we had a long and unexpected rise out of the glen by a coppiced bank to a rocky headland, turned the corner beneath a stone porch, and came suddenly into full view of Chago, not a mile away, lying around the last bay of the hill, with its abbey squatting out beyond on the headland, from which the mountain side evidently fell finally now to the far-off invisible depths of the Satanee Hor.

Two tracks led on to Chago. One was a high, steep, and stony climb that ascended right round the whole cirque of the hillside; the other was smooth and short and pleasant, cutting straight across it through the sprouting fields of corn. Never noting that Purdom and the mules were gratuitously choosing the high and difficult course, the Go-go and I embarked immediately on the low and easy one, making short work of a bunch of brushwood that lay across the entrance to the track. We had hardly advanced half a dozen yards, though, ere I was aware of a buzzing in the village ahead of me, and howls and hoots. I was paying no attention to these, and certainly putting none of them to my account, when I found myself being urgently reclaimed by Purdom, and bidden to follow in that upper way. Not only in the sphere of morals is the high and stony way more recommendable than the short and smooth. It seems that in the season of the sprouting corn the Tibetans of the March, conscious that their living margin is very small, and that the least accident of weather may beggar them for a year, in half an hour of hail, hedge round the hopes of their budding crops with an intricate network of taboos. This must not be done, that not be said, such and such a path on no account be trodden, during the period of peril, on pain of wakening the anger of the powers that be against the corn and the cattle. Inadvertently, then, I had been straying on to one of these forbidden paths, open to pedestrians, indeed, but not to ponies, and had thus innocently played exactly the part I could least have wished, in provoking the enmity of Chago against us from the very start.

Not understanding this at present, it was in a sore mood, after so long a day, that I returned into the high and stony path; not even a most charming little dainty honeysuckle [5] bush, with showers of dropping rosy trumpets all along its flattened sprays of tiny leaves,

[5] *Lonicera Farreri, sp. nova.*

could quite banish my sense of grievance, as still we toiled upwards round the encircling hills, and so down at last upon Chago, a shingled huddle of grey chalets along the headland, as if it had been some village of the Italian Alps.

A tired mule had just fainted in the track ahead of me, and the consequent block and delay put the copingstone on my annoyance. It was long after Purdom's own arrival that I and the weary caravan came winding down into Chago, and up a lane knee-deep in the black mud of ages, to where dense crowds of assembled people indicated that some sort of lodging awaited us. It was my first sight of a true Tibetan crowd, and with innocent interest I scanned their stately forms and big-boned faces. These were people of a very massive type, crude giants compared with the Chinese to whom one had grown accustomed. Sacking was the common wear, and many of the burly forbidding-looking women wore upon their breasts immense charms and circular amulet-boxes of silver or brass containing texts or charms, or relics of saint or buddha. Here and there among them, loomed the dim purple of a monk and his austere round head. And they all stared hard in a dark and non-committal manner; one young monk alone gave me a friendly smile.

We shall probably never know what exactly did happen before our arrival, and whether any rashness was perpetrated by the Mafu or the Mee to crystallise the unsettled evil mood of the Chagolese towards strangers in general into a particular and definite hostility towards our unfortunate selves. Long, long afterwards there floated a tale, handed down through a lengthy descent of repetitions, that one of them had made some trouble, and insisted on a welcome being prepared for us. Neither of the officious Yamun-trained Mee nor of the rough-and-ready Mafu would that be wholly improbable. All we can say is that we gave them strictest orders to be as conciliatory as possible to the villagers, and that the Mafu, at least, knew perfectly well the vital importance of being so in such circumstances; also that, at the time, no complaint of any sort was made, nor was there the smallest sign of any trouble having arisen. Be this as it may, our reception at Chagola was blackly and sullenly unfriendly from the very first, and it was only with the utmost difficulty that accommodation had been arranged for us for the night. Buying was impossible, the people would not even answer a question, and if occasionally some bluff giant seemed less disinclined for pleasantness than the rest, his grim sulky wife would pluck him by the sackcloth sleeve and immediately scold him out of all notion of such civilities.

Not quite aware of the situation and its full difficulties, I passed into the dark central hall and sat down at the hearth. The people of the house gathered round, and while the man was not unreconcilable, his two large dingy women eyed us with unconcealed dislike, and withered every attempt at conversation with an impenetrable sulkiness. In the middle of these overtures, too, it was discovered that I was drying my boots on the throne of the Hearth Spirit, which did not tend to improve matters. But now flowed in upon the scene such a flood of monks as filled the hall to overflowing. At their head came the Prior, primed with questions and thoroughly suspicious and unfriendly. We had to undergo a most exhaustive catechism on our actions, intentions, and motives; none of our answers carried any conviction. How could they? How is a Tibetan monk to understand anybody's voluntarily coming so far after useless weeds? It is perfectly plain to him that this pretence merely masks the search for gold, which is the present monopoly of the Church, and protected accordingly with all the legends and taboos by which the monks can guard the solitude of the alps, and not only prevent the peasant from prowling up there himself, but also reinforce all his rage against strangers, who, by doing so, will surely bring the wrath of Heaven on the village that has allowed them thus to affront the powers of the air. And if foreigners come here after gold, they will certainly soon be followed by other foreigners with fire and sword, escorting alien creeds and discords, attacking the inviolable authority of the Church alike in this world and the next. Perish the thought! The Church and its flock see eye to eye on the point, for not even a catholic Church is so much at one with its people as the Lamaist, where at least one son taken from every family for the monastic life welds the whole Tibetan population, secular and lay, into an indissoluble community of sympathies and interests. A peasant boy of Central Europe *may* become a monk or a priest; from every household of Tibet one boy, to say no more, *must* become a priest and a monk. Hence the enormous solidarity of Lamaism, and its profound intimate hold on the people with whom it is thus intertwined.

The Prior accordingly proceeded grimly with his catechism, to an accompaniment of growls from the dark recesses of the room behind. One felt that every answer only made a blacker impression than its predecessor. They denied possession of an Abbot, a bad sign in itself, and not even our cards duly sent up to the Presiding Elder produced much relaxation. Gradually, however, as the dialogue continued, a certain thawing seemed to appear. The news of my own creed made a certain effect, and I was proclaimed as a sort of Western Lama engaged in procuring Tibetan flowers to beautify the shrines of Europe.

This thin tale was listened to with more attention than the rest, and was followed by an invitation to me to come up to service to-morrow. I accepted this with bows, and presented the Prior with candles and a book of pictures. Distinctly mollified in manner he turned to go, and all his train followed after, leaving us with the pleasing notion that we had now smoothed over our difficulties most successfully and made ourselves really popular. Hugging this happy idea we sat and made plans, and praised the Prior for a prudent and friendly person. And meanwhile this estimable ecclesiastic was busily issuing a proclamation that we were all to be murdered in the night with as little fuss and unpleasantness as might reasonably be. Unconscious of these tokens of friendliness in preparation, we ourselves retired to sleep. I had a tiny little room off the main hall, while Purdom and the boys arranged their rolls of bedding on various of the big lockers that ran all round the three sides of the hall, each with a smaller lower one in front of it convenient for rolling off onto.

We had ample leisure, through that long night, for wondering what curious scrabblings those might be that went on through the small hours round the outer walls. In point of fact, they were the work of the pious peasantry endeavouring to get into the house and carry out the Prior's prescription without drawing too much attention to the fact. There was evidently a peace party in the place, — of which, indeed, we found subsequent and satisfactory evidence; if the job could be put through quietly and silently and anonymously, well and good, but the Prior and his following had no desire to affront the weaker vessels of his flock with open bloodshed. We, however, ignorant of these intricacies, consumed the darkness in wonderings, whenever the onslaughts of the bugs left an interval. I, indeed, aloft on my camp-bed, suffered comparatively little, but Purdom and the boys were so ceaselessly assaulted that finally they gave up all pretence of slumber, cast aside their bedding, and finished the night astride on the mule-packs in the middle of the floor.

On our grey mood the dawn came grey and weeping: Chagoling and Chagola had retired into an impenetrable veil of cloud and rain. No further sign of hostility was shown, but we were left quite alone, and it was a bad omen that none of the inhabitants either cared or dared to satisfy their natural curiosity by coming to see us, if it were only to ask for some pill of medicine. The two bun-faced women of the house, engrained in the filth of many years, sat and made a whispered conversation together by the hearth, and refused either notice or answer to all our efforts, shutting us off into a more rigid isolation than ever. If more urgently spoken to they feigned sickness, and sighed and grumped and faintly groaned, declining speech, immediately afterwards returning to the eager sibilations of their own tête-à-tête.

Our morning passed in the dismal task of trying to rid ourselves of unwelcome visitors. The bug is an unlovely beast. There is a certain humanity about the flea, who always reminds me of a brisk little curate hopping round after subscriptions. But the bug, obese and squalid, is more like some horrible old blood-sucking Mrs. Warren in decline, waddling with a fat slow sureness towards her unholy purposes. Even her death lacks the odour of sanctity. And in Tibetan houses the bug abounds. Beware, all along the Border, of villages where the room-fittings are of wood. The loess lands give you complete immunity, but where there is woodwork, with crevices and cracks, there you will do well to hedge yourself about betimes with oilcloth, and take refuge on the safe elevation of a camp-bed. For with the development of darkness, forth from all the chinks and cavities march phalanxes of bugs, the progeny of many generations established, like county families, in the immemorial grime and darkness of those uninvestigated cavities. Their name is legion, so, probably, are their species, and I make little doubt that an entomologist could do much profitable searching in his shirt after one night of a Tibetan house. Even the flea is not our old familiar friend. There is a legend that once a Rothschild offered ten thousand pounds for a specimen of the white Tibetan flea. Alas and alas! on the morning after my arrival at Chago I might have become a millionaire at that rate; for assuredly "ten thousing times ten thousing" would have been the value of the game I bagged in my own coverts — strange little pearly diaphanous things, like minutest white grains of corn, each with a black bloblet in the middle which is their belly, darkened with its burden of one's own blood.

In these scientific researches the hours were consumed, and we concluded not to face a second night downstairs, but to retire to the big hayloft up above, where the gaping shingles and the wilderness of accumulated hay could not weigh against its blessed security from vermin. After lunch I set to work diagnosing yesterday's haul of Primulas, while Purdom, more energetic, went out for a walk in the mournful wet up towards the hill at the back, where a mass of purple lilac flared from afar amid the copse. When my task was ended I sank into a doze, from which I was stirred by occasional violent and prolonged howls in the street outside, which I took to proceed from pedlars or some such passer-by. But at last these so increased in number and volume that I was completely roused, and went out to the stairway-head of the loft to see what it might all be.

About a quarter of a mile away, on the hillside, I saw the meeting-place of the village was crowded with burly figures in sackcloth, leaping and hopping and bounding in a sort of demoniac dance, to an accompaniment of wild yells irregularly but constantly emitted. I watched this spectacle with an innocent ethnological interest. Clearly these good people were celebrating the spring or the crops in this quaint primeval rite; those disorderly jumpings and that choir of dissonant howls aroused in me quite a pang of emotion, as I sat and realised that just from some such ritual must the drama originally have been born. Pursuing the train of these elevating thoughts, I was planning an article on the primitive practices of Tibetan nature-worship, when I was suddenly awakened to the realities of life by pale and shaking voices that summoned me in whispers to descend from my exposed post of observation. Thunderclapped out of my meditations, I swiftly and silently crept downstairs.

I found the big room full of stir and turmoil, and soon learned the truth of that interesting demonstration on the hill. For Purdom, it seems, accompanied by the Go-go, had been discerned quite peacefully proceeding along the pathway up the hill behind the village. It was an open path, without any taboo, but the Prior evidently thought this a good opportunity to carry out his purpose of yesterday night, now that the force of the foreigners was divided. Accordingly, as Purdom strolled, he was hailed from behind with an increasing pandemonium of howls, and, turning, saw all the male population of the village gathered in a disorderly mob, with arquebuses at rest and fuses lit, in the obvious intention of shooting him then and there. To pause would have been fatal, for the Tibetan, though useless at a moving mark, is a dead shot at a still one; flight up the hill would have been useless, besides abandoning his base. It only remained to turn round and proceed back to the village, passing the gauntlet of the mob. It so happened that that day, for the first and only time in our travels, he had gone out without so much as a revolver.

Armed only, then, with his fascinating smile, and with hands thrown wide to show his harmlessness, he braced himself to face those smoking guns and that howling horde of demons. Down the path he steadily came, with the Go-go giggling at his heels, and thinking it all a fine joke; in an avenue on either hand, banked up along the hedgerow, the villagers gnashed and bellowed upon him as he came, with faces and gestures so devilish that it was long before they deserted his dreams. But it was probably the very absence of the revolver that sufficed to turn the hair-balance of the situation, and avert yet another of those Tibetan tragedies that come and pass in the twinkling of an eye, arising in an instant, and in an instant determining for life or death. For the peace party was not unrepresented in that crowd, and Purdom's unarmed condition gave them a chance of pressing home their point, and insisting on the insanity of destroying anyone so obviously inoffensive, whose destruction would assuredly in time attract the notice of China to Chago in a most undesirable manner. Accordingly, they managed to hold the situation poised; the least untoward incident would have precipitated matters, and our expedition would have ended, at its very beginning, in a sudden flare like those which accounted for Brooke and Margary and Dutreuil de Rins. Fortunately, the nice balance held; a young monk outpoured smooth words on the howling sea, an old patriarch held out hands of peace to Purdom, and a venerable grandam urged him most unconvincingly to have no fear. So, supported from point to point, he passed at last through the peril and left it behind him. Resisting all temptations to dwell on those guns now pointing at him from the rear, any one of which might fire at any moment, he preserved a proper dignity of pace, and with unhastening step and deliberate pantomime of pipe-lighting, proceeded indifferently homeward, and arrived unscathed.

Even now, of course, it was obvious that the situation was by no means saved. In fact, the most critical moments might now be at hand. So we put the house in the best posture for defence, posted the boys and primed our weapons, and were prepared to do or die as the case might demand. Furious at having let their prey escape, we thought it possible that now the villagers might try to rush the place *en masse*. But when, after an uneasy ten minutes, we were at last invaded, it was not by murderers, but by monks, who swept into the room in a tumultuous flood. Among them came the young monk, claiming merit for having saved Purdom's life — a claim I am not wholly inclined to disallow; and this time the monastery did produce an Abbot, who, however, added little to the conversation, being a small-pocked young shock-headed Peter, frowzy and stockish, stupid in face, squat in figure, and of a general toad-like conformation. A deafening pandemonium ensued. Evidently the murder party, having failed of its object, had now retired completely out of existence, for the talk was all of peace and love and friendliness; never was there a warmer esteem and affection than that in which, as it now appeared, we were held by all that holy brotherhood. In a loud unanimous gabble, however, they deplored their inability to impress the same views on their lawless and obstinate parishioners. Convinced though they themselves were of our exhaustive virtues, it was impossible to

305

get the agreement of the village on this point. In fact, in plainer words, these protesting people no longer dared to desire our deaths, but were quite determined on our departure. For the sum of all they said was that we should be best advised to clear off as soon and as quietly as possible, so as no more to affront the susceptibilities of the village; and henceforth rigidly avoid these mountains and their peoples.

It was a depressing but unanswerable manoeuvre. It was not possible for anyone to establish themselves in that district against the declared ill-will of the population and the professed inability of the Church to control it. Sooner or later there is bound to be trouble — trouble with the people, trouble with the monks, and trouble, ultimately, for China and the Legations. Even if you have the inclination and the force of arms, you cannot impose yourself on the Tibetans permanently with guns and rifles; they are a tenacious and irreconcilable race, as excellent in craft and patience as in open attack. The largest and best-armed expeditions do well to treat them with extreme care, while a small one, helpless and unofficial, has no choice whatever but to follow the wind of Tibetan public opinion. Accordingly, with gloom in our hearts, we prepared to take the advice of the monks, ironical though it was, for all parties concerned knew perfectly well that they, and they alone, were the real masters and origins of the situation, however glibly they might deplore their powerlessness to restrain the popular animosity that they themselves had kindled and assuredly meant to maintain. There was no help for it; in preparation for an early and untroubled start we retired betimes to the hay-loft, having carefully enjoined the muleteers, on danger of provoking perhaps a fatal riot such as is so apt to spring up at a moment's notice in the tortuosities of a Tibetan street, to bring round their animals in the first dawn as quietly as might be, without the usual jangling caparison of bells. To which, impressed with the urgency of the occasion, they consented, and we then made ourselves couches in the hay and curled up, our bedding being packed ready for the start.

No one who knows Chinese muleteers will need to be told, accordingly, that the day was well up, and we ourselves had been waiting impatiently for some hours, before the mules did at last arrive, and even then it was with a full-throated jangle of all their bells that must have roused every Tibetan within a mile. No, whatever danger pressed in the visible world, nothing was to deprive those mules of their protection against those of the spiritual one, nor drag their owners a moment sooner from the warm comfort of their bed. However, Chago was like a silent place of the dead as we threaded its muddy streets in the breathless radiance of the morning. We had forecasted stonings and opprobrium, and perhaps even a rear attack, as we descended the slope below the village. And a few loud whoops that preceded our departure from some inner alley put us, indeed, on the alert; but the alarm was followed by no overt act, and even here and there we met a bland smile as we defiled through the lanes, until at length we were out of Chago, embarked on the long descent of the mountain-side to the depths of the Satanee Hor.

VILHJALMUR STEFANSSON (1879–   )

Foxes, Owls, and Polar Bears

From *The Friendly Arctic*. 1921

[STEFANSSON has made the arctic regions more familiar than any other explorer, and has proved that white men can exist and even live healthily on the wild game of ice, air, and water. He has added materially to our knowledge of the geography, meteorology, archæology, and ethnology of the far north, and these preoccupations have reduced his natural-history communications to relatively short but significant passages.]

THE SECOND DAY after the gale we were able to commence traveling. The ice was under no pressure now, for the storm had blown it offshore and had drifted our island against the edge of the pack where it had stuck fast. The temperature, to our great distress, continued warm — never below zero, Fahrenheit. Still, as there was no pressure, the mush solidified enough in two nights to permit crossing in several places, although we were able to make only three miles the first traveling day. In some cases where the cracks between floes were no more than three to five yards wide, we used to bridge them by chopping ice for an hour or two with our pickaxes and throwing the fragments into the water until their combined buoyancy was enough to support the sled during the crossing. And the farther from shore we got, the fewer cracks we had to cross.

A lead of open water appeared in front of us on April 4th. We could have crossed it by using the sled boat, but because in half a dozen such crossings the mush ice would have chafed holes in the canvas we did not do so. Furthermore, the pack was in motion and we expected the lead to close at any time, giving an easy crossing. So we did no traveling that day.

To encourage the men, and to demonstrate to them how easy it was to make a living at sea, I shot a number of seals and so did Storkerson and some of the others. A few animals sank but we recovered six. When there seemed no use in killing more, I oiled the barrel of my rifle, as I always do when the temperature is not low, put it in its case and strapped the case on a sled. Meantime the men had made a bonfire of blubber and cooked some fresh seal meat. While we were feasting there was a sudden commotion among the dogs, which were still hitched to the sleds, for we expected to cross the lead at any moment. The sled with my rifle strapped on it was about six feet from the water, the other sleds only a little farther away, while the fire over which we were cooking was about twenty yards.

The cause of the barking was a polar bear, the first one that some of the men had seen. By the time he arrived the lead had closed to a width of not more than five yards and on the very brink of it was the bear, pacing up and down, trying to make up his mind to plunge in, like a bather reluctant to take a dive into cold water. I don't know what it really was made him hesitate. It can hardly have been the chill of the water, though he gave distinctly that impression. But even while I theorized about his motives and behavior, there came to mind the need for instant action, for some of the excited dogs might jump into the water to get at him, dragging a sled after them. Were the bear to cross the lead to our side the dogs, all tangled in their harness, would doubtless attack him. He would probably run away, but there was no certainty of it. Clearly he bore no hostility towards them nor had he any fear of their barking, or of the shouting of the six men who ran back and forth telling each other what to do.

According to his own account McConnell must have been one of the coolest of us, for he said afterwards that he immediately ran for his camera, asking us to wait until he got a picture. To get at my rifle I had to run around to the side of the sled nearest the lead, and while I was unstrapping the case my back was towards the bear about five yards from me. Storkerson's rifle was on the sled next to mine, and while he was getting it I noticed that I was in direct line between him and the bear. He had his rifle first, for it had not been lashed to the sled, and seeing that he was likely to fire I requested him to be careful to get the bear and not me. There was doubtless no likelihood of the mistake, but I thought a word of caution wouldn't hurt. When it came the explosion was so close to my ear as to leave me partly deaf for some time. The bullet struck the bear, of course, and probably surprised him as much as it hurt. He was leaning over the water just getting ready to dive and was startled into falling on his back in the lead, splashing water over me as he fell. The water was perfectly clear and looking around I saw him going down like a sounding lead, with his feet at first uppermost, though he soon straightened out, rose to the surface and scrambled up on the far side. As he was struggling out, Storkerson gave him a second shot and a moment later as he was running away a third; but the rifle was only a .30-30 and, although he was bleeding profusely, the bear was making off with considerable speed. For the further encouragement of the party, to prove that no bear could come as close to us as this and get away, I thought I had better try the Mannlicher. This shot rolled him over and I took the story to be ended. After I had turned away to put the rifle back in the case he got unsteadily to his feet and disappeared behind an ice cake.

The lead had been gradually closing, and Crawford with a rifle, and McConnell with a camera, were able to follow and find him about two hundred yards away, trying to cross a second lead. They fired several times, but when I got over he had crawled out on the ice, so that one more shot was necessary. It is always so when a group becomes excited — there is a hullabaloo and a fusillade of wasteful shooting. One bullet near the heart does a great deal more damage than a dozen badly placed, as many of these were, for some were in the paws, some in the neck and some in other fleshy parts. An exciting bear hunt may be interesting to read about but it is a poor hunt. One properly located Mannlicher bullet is all that should be necessary.

On shore polar bears are ordinarily timid animals, afraid of men, and afraid of dogs and wolves. But the behavior of this visitor was typical of bears far from shore. There they have no enemy to fear. Besides their own kind they are familiar on the ice-pack with only three living things — the seals, on which they live, the white foxes which they unintentionally provide with food but which never come near enough to be caught themselves, and the gulls which cry loudly and flutter about them at their meals. Zoologists know, but it is not commonly realized by the laity, that the white fox is almost as much of a sea animal as the polar bear, for probably 90 per cent. of white foxes spend their winters on the ice. They are not able at sea to provide their own living, so several will be found following a bear wherever he goes. When the bear kills a seal he eats all he wants, usually from a quarter to half of the carcass. In many cases he touches none of the meat, but eats merely a portion of the blubber and the skin that goes with it. After this satiating meal he probably feels as if he will never care to eat again and goes away to sleep under a neighboring hummock, leaving for the foxes what is left. It is not likely that he will come back, but if he did, the foxes would hop and the gulls flutter away. From long experience he gets the impression that these creatures are not the least bit dangerous, but too elusive to be caught.

Without doubt the bear is able to tell the difference between a living seal and the meat of a dead one when he sniffs them in the air. There is always seal meat in our baggage and the smell is always about our camp. When a bear passes to leeward he must perceive the many camp odors, but the only one which interests him is that of the seal meat. Knowing no fear, he comes straight into camp, walking leisurely because he does not expect the dead seals which he smells to escape him; neither has he in mind any hostility or disposition to attack, for, through long experience with foxes and gulls, he expects any living thing he meets to make way for him. But if on coming within a hundred or two hundred yards of camp he happens to see a sleeping dog, and especially if the dog were to move slightly, as is common enough, the bear apparently thinks, "Well, that is a live seal, after all!" He then instantly makes himself unbelievably flat on the ice, and with neck and snout touching the snow advances almost toboggan-fashion toward the dogs, stopping dead if one of them moves, and advancing again when they become quiet. If there is any unevenness in the ice, as there nearly always is in the vicinity of our camps — we choose such camping places — he will take cover behind a hummock and advance in its shelter.

Our dogs are always tied, for in the dead of night a good dog may be killed or incapacitated in their fights with one another in less time than it takes a sleepy man to wake up and interfere. But we know the danger from approaching polar bears and endeavor to scatter the dogs in such a way that while a bear is approaching one dog in an exposed situation, another will get the animal's wind. Usually, too, we tie the dogs to windward of the camp, so that the bear shall have to pass us before he comes to them. When one dog sees or smells the bear he commences barking, and in a second every other dog is barking. At once the bear loses interest. He apparently thinks, "After all, this is not a seal, but a fox or a gull." His mind reverts to the seal meat he has been smelling, he gets up from his flat position and resumes his leisurely walk toward the camp. By that time, even though we may have been asleep, one of us will be out with a rifle, and a properly placed bullet ends the story.

When the bear comes as this one did in broad daylight, with the dogs awake and the men moving about, he apparently takes the dogs and us for a variety of gull, noisier perhaps than any he has heard, but no more dangerous. In a party used to bears the men stand with guns ready, while the one who is to do the killing sits quietly and waits until in his natural zigzag approach the bear exposes one side or the other so as to give a chance for the shot near the heart. . . .

On June 29th we came upon the only caribou seen on this trip along the Melville Island coast. It was a yearling and therefore thin, so we made no serious attempt to get it. That day also we saw the first owl since the preceding 20th February when we noticed one just north of Cape Kellett. We had noted in the fall of 1914 that the owls which were very numerous in the summer became gradually fewer towards Christmas and seeing one in February really surprised us. So far as we know, their main food is the lemming and these must be hard to get in winter time. Still, we occasionally see lemming tracks in any month of winter and it is doubtless these stragglers the owls live on.

Watching the owls in their lemming hunts I have marveled at their intelligence but equally at their stupidity. An instance is a short autumn day when I sat for several hours on a hill in southwest Banks Island and studied through my field glasses the white foxes and owls all about. Within a circle of a few miles were several foxes, now hidden by hills or in ravines, now visible in the open, hunting lemmings. On knolls here and there sat owls watching the foxes.

There had been a four- or six-inch fall of snow which lay as yet untouched by wind, level and fluffy. Under this snow, tunneling it and fondly believing themselves unobserved, the lemmings were everywhere. The foxes moved about at a leisurely, elastic trot. Every few minutes I could see one of them stop, cock his head on one side, and listen. Possibly the senses of sight and smell were also active, but certainly they gave primarily the impression of listening. After a moment or two of alert attention the fox gave a high leap in the air like a diver from a springboard and came down in the snow with nose and forepaws together. In half the cases the lemming was caught at that instant, in half the remainder he was caught a moment later, but in a few instances he escaped — probably into a hole in the frozen ground. If left undisturbed, the fox would kill the lemming with a sharp nip or two, drop it on the snow, look at it contemplatively for a moment, pick it up again and bury it in the soft snow to trot off and — I have no doubt — forget all about it. For days and days the lemming catch would be far in excess of appetite, and before the fox became hungry a hundred miles might intervene. If these buried lemmings are ever found and eaten it is probably by a wolf or some other fox.

But the fox rarely buried the lemming undisturbed. From a nearby knoll an owl was watching with eyes and interest as keen as mine. When the fox paused, alert for a sound beneath the snow, an owl on a nearby hill half-turned and part-crouched for flight; while yet the fox was on its springboard leap and dive through the air the owl's broad wings were spreading; and before the fox had buried its kill the owl was upon him. This must have been the thousand and first experience of the sort for the fox but it acted as if completely surprised. No doubt its attention had been so focussed on the business of securing the lemming that owls were temporarily forgotten. At the wing swish and approaching shadow the fox cringed as if in abject fear, but nevertheless evidently half realized that the object of the owl was robbery rather than murder, for with the very cringing and slinking motion of fear and flight the fox picked up the lemming (if it had been dropped). Then came a dash away, fast for a fox but slow as compared with the easy glide of the owl, at the end of a short second of which the owl was directly over the fox, reaching for it with its talons but never touching, for evidently discretion was part of its campaign. After two or three sharp doublings and vain attempts to get away from the owl the fox would turn on his pursuer and make a great leap in the air towards her. Apparently the owl's object was to make the fox snap at her, thus in excitement dropping the lemming from its mouth. In this I never saw the owl successful, for in every case watched by me the owl gave up worrying the fox after half an hour or so, but I was told by Eskimos that they had seen foxes drop their lemmings in snapping at the owls, whereupon the owl snatched the lemming from the snow and was up and away. Such outcomes now and then must account for the cheerful optimism with which the owls keep up their watching and worrying of the foxes.

But this ingenuity of the owl is more than matched by her stupidity. Why doesn't she wait till the fox buries the lemming under four or five inches of fluffy snow and trots off? With a scratch or two of her claws in the snow the owl could now have the lemming. Just that much increase of intelligence would certainly make the owl's struggle for existence during the northern winter far simpler. As it is, it must be a severe struggle, which accounts for most of them going south during midwinter, if not before. It is only rare owls, like rare ravens, that spend the whole winter far north of the treeline.

CARL AKELEY (1864–1926)

Elephant Friends and Foes

(From *In Brightest Africa.* 1923)

[AKELEY refuses to be catalogued, so confusing is his versatility as taxidermist, inventor, sculptor, artist, photographer, and explorer. But fortunately he was also a writer, and in his writing he also excelled, and did much to dispell the fog of ignorance which for long had enveloped the continent of Africa.

If I have been generous to him in the matter of quotation it is because he was an exception to the large aggregation of explorers and big-game hunters who were more at home with the rifle than the pen, and whose ultimate satisfaction rested less on additions to the realm of natural history, than to a glassy-eyed trophy head attached to the wall of a "den."]

I HAVE sat in the top of a tree in the middle of a herd a quarter of a mile from a native village in Uganda in a last desperate effort to inspect the two hundred and fifty elephants which had been chevying me about so fast that I had not had a chance to see whether there were any desirable specimens among them or not. I have spent a day and a night in the Budongo Forest in the middle of a herd of seven hundred elephants. I have stood on an ant-hill awaiting the rush of eleven elephants which had got my wind and were determined to get me. I have spent a day following and fighting an old bull which took twenty-five shots of our elephant rifles before he succumbed. And once also I had such close contact with an old bull up on the slopes of Mt. Kenia that I had to save myself from being gored by grabbing his tusks with my hands and swinging in between them.

I have spent many months studying elephants in Africa — on the plains, in the forests, in the bamboo, up on the mountains. I have watched them in herds and singly, studied their paths, their feeding grounds, everything about them I could, and I have come to the conclusion that of all the wild animals on this earth now, the African elephant is the most fascinating, and that man, for all the thousands of years he has known of elephants, knows mighty little about him.

314

I am speaking only of the African elephant. He has not been domesticated as his Indian cousin has. The two are different in size and different in shape and different in habits. The low point of an African elephant's back line is the highest point of that of the Indian elephant. The African elephant's ears and tusks are larger, and his tusks usually spread wider at the points instead of coming together. Unless one studies him in his native haunts, one cannot get to know him. His disposition is held to be wilder than that of the Indian elephant, but the infrequency of his appearance in circuses and in zoological parks may be attributed to the ease with which tamed elephants may be obtained from India rather than to a difference of temper in the two beasts. An African elephant at Washington and one in the Bronx zoological park are the only ones I know of in this country, and no animal in captivity can give one more than a slight idea of his natural habits in his jungle home.

Very few people have studied African elephants in the field. Ninety-five per cent of those who have followed them have been purely hunters and their desire has been, not to study, but to shoot — to see the elephant the shortest possible time. Time to judge the ivories and get a bead on the brain was all that they wanted. Of other elephant knowledge all that they needed was the simple facts of how to follow and find them. The comparatively few men who have tried to study the elephant have not gained as much knowledge as one would imagine, because without trying it one cannot realize how extremely difficult it is to study the live African elephant.

For example, as I said before, I spent a day with seven hundred elephants in the Budongo Forest, but although I heard them all the time and was very acutely conscious that they were near me, I do not believe that I actually had my eyes on an elephant more than half an hour, all told, during the day. It happened this way.

One night about dark, after a week or two of hunting, we heard the squeal of an elephant while we were sitting at dinner. A little later there were more squeals and occasional trumpeting — more and more, clearer and clearer — and by the time we had finished dinner the noise was only a mile or so away. It was a continuous row which suggested a tremendous herd. We went to bed early with elephants getting closer to camp all of the time. There is little danger of elephants attacking a camp, and, as there is no way to study them at night, about the only thing left to do was to go to bed and get in good shape for the next day. Along about midnight Mrs. Akeley came over to my tent and said that she had loaded my guns and that they were all ready. She could not sleep; so she went out to sit by the fire.

315

The elephants were then within a hundred yards of our tents and there was a continuous roar made up of trumpetings, squealing, and the crashing of bushes and trees.

I got up in the morning and had breakfast before daybreak. The elephants had moved on down the edge of the forest. What had been a jungle of high grass and bush the day before was trampled flat. There were at least seven hundred elephants in the herd — government officials had counted them on the previous day as they came down. I followed the trails to the edge of the forest but saw none. I started back to cross a little *nullah* (a dry water course), but felt suspicious and decided to look the situation over a little more closely. I ran up on a sloping rock and, almost under me on the other side, I saw the back of a large elephant. Over to one side there was another one, beyond that another, and then I realized that the little *nullah* through which I had planned to pass was very well sprinkled with them. I backed off and went up to a higher rock to one side. Elephants were drifting into the forest from all directions. The sun was just coming up over the hills and was shining upon the forest, which sparkled in the sunlight — morning greetings to the forest people. The monkeys greeted one another with barks and coughs. Everything was waking up — it was a busy day. There was not a breath of air. I had gone back a million years; the birds were calling back and forth, the monkeys were calling to one another, a troop of chimpanzees in the open screamed, and their shouts were answered from another group inside the forest. All the forest life was awake and moving about as that huge herd of elephants, singly and in groups, flowed into the forest from the plain. There was one continuous roar of noise, all the wild life joining, but above it all were the crashing of trees and the squealing of the elephants as they moved into the forest on a front at least a mile wide. It was the biggest show I ever saw in Africa.

Then an old cow just at the edge of the forest suddenly got my wind, and wheeling about, she let out a scream. Instantly every sound ceased, everything was quiet. The monkeys, the birds — all the wild life — stopped their racket; the elephants stood still, listening and waiting. For a moment I was dazed. The thought came through my mind — "What does it all mean? Have I been dreaming?" But soon I heard the rustling of the trees as though a great storm were coming. There was no movement of the air, but there was the sound of a wind storm going through a forest. It gradually died away, and I realized that the elephants had made it as they moved off. It was the rustling of the dry leaves on the ground under their feet and the rubbing of their bodies through the dried foliage of the forest. I never heard a noise like that made by elephants — before or since. The conditions were unique, for everything was thoroughly parched, and there had not even been a dew. Ordinarily, if there is any moisture, elephants when warned can travel through a forest without the slightest noise. In spite of their great bulk they are as silent and sometimes as hard to see in their country as a jack rabbit is in his. I remember on one occasion being so close to an old cow in the jungle that I could hear the rumbling of her stomach, and yet when she realized my presence the rumbling ceased, as it always does when they are suspicious, and she left the clump of growth she was in without my hearing a sound.

But going back to the big herd. From the time I had seen the first elephant until the last of them disappeared in the forest it had been perhaps fifteen minutes — fifteen minutes in which to see the sight of a lifetime, a thing to go to Africa a dozen times to get one glimpse of. But what did I learn about the habits of the elephant in that fifteen minutes? A little perhaps but not much. It takes a long time and much patience to get at all intimate with old Tembo, as the Swahilis call him, on his native soil.

After the herd disappeared in the forest I watched for ten or fifteen minutes and heard the squeal of the elephants and the noise of the monkeys again. Their suspicions were over. I followed into the forest where the trails showed me that they had broken up into small bands. I followed along on the trail of one of these bands until I got a glimpse of an elephant about fifty yards ahead of me in the trail. You don't see a whole elephant in the forest. What you do see is just a glimpse of hide or tusk or trunk through the trees. And if you want to get this glimpse without disturbing him you must do your glimpsing from down the wind.

There was a little open space ahead of the group I was following. I worked around until I got.to a place where I could see them as they passed through this open space. They were moving along slowly, feeding. Two or three came out into the opening, then they became suspicious and wheeled into the forest again. I followed cautiously. I had gone only a short distance when I saw a very young calf about twenty yards ahead of me. As I halted, the mother came trotting back down the trail looking for the baby. I froze to the side of a tree with my gun ready. She came to the baby and turning, boosted it along with her trunk after the rest of the herd. I followed along after them into an opening where I found them rounded up in a patch of burned-over ground. They were milling around in a

317

rather compact mass seemingly preparing for defence. I could not see very plainly, for a cloud of dust rose from the burned ground as they shuffled about. I stood watching them a little time and suddenly caught sight of a fine tusk — an old bull and just what I wanted for the group I was working on for the Museum of Natural History. I ran up behind a bush at the edge of the clearing and peeked through it. There, not more than twenty yards from me, was my bull, partially exposed and partially covered by the other animals. I could not get a shot at his brain as he was standing, but the foreleg on my side was forward exposing his side so that I had a good shot at his heart — a shot I had never made before. The heart is eighteen or twenty inches long and perhaps a foot up and down — a good mark in size if one's guess at its location is accurate. If you can hit an elephant's vertebrae and break his back you can kill him. You can kill him by hitting his heart, or by hitting his brain. If you hit him anywhere else you are not likely to hurt him much and the brain and heart shots are the only safe bets. I fired at his heart with both barrels and then grabbed my other gun from the gun boy, ready for their rush, but the whole herd, including the old bull, made off in the other direction, raising a cloud of dust. I ran around and climbed an ant-hill four or five feet high to keep them in sight. When I caught sight of them they had gone about fifty yards and stopped. And then I *did* learn something about elephants. My old bull was down on the ground on his side. Around him were ten or twelve other elephants trying desperately with their trunks and tusks to get him on his feet again. They were doing their best to rescue their wounded comrade. They moved his great bulk fifteen or twenty feet in their efforts, but were unable to get him up. I don't know of any other big animals that will do this. I had heard stories that elephants had the chivalry to stick by their wounded and help them, but I was never sure myself until I had actually seen this instance. Some time later Major Harrison, a very experienced elephant hunter and a keen observer, told me of an even more remarkable instance that he had seen. He was shooting in the Congo and came upon four big bulls. One he killed and another he wounded. The wounded one went down but the two survivors helped him regain his feet, and with one on each side helping him the three moved off. Although Major Harrison followed the rest of the day he was not able to catch up with them.

I did not see the end of their efforts to raise the bull I had shot, for those that were not helping him began to circle about with their ear out to hear anything of their enemy and with their trunks up feeling for my wind. They were moving in ever-increasing circles which threatened to envelop my ant-hill, and I beat a hasty retreat. Not long after they evidently were convinced that the bull was dead and all together they moved away. I then went to the body. He was dead, but as we approached there was a reflex action which twitched his trunk from time to time. This frightened the gun boys so that I went up and slapped the elephant's eye, the customary test, and. as there was no reaction the boys were convinced. When I looked the carcass over I was disappointed to find that only one of his tusks was big and well developed. The other was smaller, and out of shape from an injury; consequently I decided not to take him for the museum group. He was, however, a good deal of a temptation, for he was one of the largest elephants I had ever seen, measuring eleven feet four inches to the top of his shoulders, and the circumference of his front foot was sixty-seven and a half inches. To the best of my knowledge this is a record size by about four inches. I did not even skin him but contented myself with taking his tusks, which I sold for nearly $500 without even going down to Nairobi.

The phenomenon of elephants helping each other when wounded is not general by any means. Only a few days after shooting the big bull I had an instance of elephants abandoning one of their number that was wounded and not very badly wounded, either.

I had gone into the forest again, and had come upon another bunch in very thick country. I could only get little glimpses of a patch of hide or ivory once in a while. After working along with them for a while in the hope of getting into more open ground I tried the experiment of beating on the tree trunks with sticks. This was new to them as it was to me. I felt sure it would make them run but I wasn't sure whether they would go toward it or away from it. Happily they bolted from the forest into the high grass, grumbling all the while. I followed as closely as I dared until finally, in hope of getting a view over the top of the high grass, I started to climb a tree. Just then they rushed back into the forest, fortunately to one side of me. I thought it was time to quit, so we started back to camp. At that moment I heard another group of elephants. They were coming out of the forest into the grass. I climbed up an ant-hill where I could see them as they passed over a ridge. There were eleven of them and not a specimen that I wanted among them. I stood watching to see what would happen next. They were about three hundred yards away when they got my wind. Back they came, rumbling, trumpeting, and squealing. I knew that I had trouble on my hands. The only thing for me to do was to stick, for if I got down in the tall grass I couldn't see anything at all. They came up over a hill, but they were not coming straight toward me and it looked as if they would pass me at forty or fifty yards; but, unfortunately, the cow in front saw me standing in full view on my ant-hill pedestal. They turned straight at me. When the leading cow was as close as I wanted her to get — about twenty-five yards — I fired. She hesitated but again surged on with the others. A second shot knocked her down. The rest surged past her, turned, smelled of her, and ran off into the forest. After a few minutes she got upon her feet and rather groggily went off after them.

Elephants have the reputation of having very bad eyesight. I personally am of the opinion that their sight is pretty good, but on this subject, as on most others about elephants, information is neither complete nor accurate. But my experience makes me think that they can see pretty well. In this case the cow that saw me was only about fifty yards away, but at another time on the Uasin Gishu Plateau an elephant herd charged me from 250 yards with the wind from them to me. The behaviour of this particular herd gave me a clue to their reputation for bad eyesight. The elephant is not afraid of any animal except man, and consequently he is not on the alert for moving objects as are animals that are hunted. Neither does he eat other animals, so he is not interested in their movements as a hunter. In fact, he isn't normally particularly interested in moving objects at all. He pays no attention. When we first came up with this herd on the Uasin Gishu Plateau we could move around within fifty yards of them without attracting their attention. However, after they got our wind and recognized us as enemies, they were able to see us at a distance of 250 yards, and charge us.

But however good the elephant's sight, it is nothing in comparison with his smelling ability. An elephant's trunk is probably the best smelling apparatus in the world, and he depends on his sense of smell more than on any other sense. When he is at all suspicious he moves his trunk around in every direction so that he catches the slightest taint in the air, from whichever way it comes. I have often seen elephants, when disturbed, with their trunks high in air reaching all around for my wind. I likewise, on one occasion, had an intimate view of a very quiet smelling operation by which an old cow escaped me. I was on an elephant path one day on Mt. Kenia looking for an elephant I had heard, when my gun-bearer gripped my shoulder and pointed into the forest. I looked and looked but could see nothing but the trees. Finally I noticed that one of the trees diminished in size toward the ground and I recognized an elephant's trunk.

My eyes followed it down. At the very tip it was curled back, and this curled-back part, with the nostrils distended was moving slowly from side to side quietly fishing for my wind. She was waiting concealed beside the trail to pick me up as I came along. She was no more than forty feet away, but when she decided to give up and moved away, I could not hear her going although it was a dense forest and she was accompanied by two youngsters. Very often in the forest where there is very little air stirring it is hard to tell the direction of the wind. I used to light wax taper matches as tests, for they could be struck without any noise and the flame would show the direction of the slightest breath of air.

In many other ways besides its smelling ability the elephant's trunk is the most extraordinary part of this most extraordinary animal. A man's arm has a more or less universal joint at the shoulder. The elephant's trunk is absolutely flexible at every point. It can turn in any direction and in whatever position it is, and has tremendous strength. There is no bone in it, of course, but it is constructed of interwoven muscle and sinew so tough that one can hardly cut it with a knife. An elephant can shoot a stream of water out of it that would put out a fire; lift a tree trunk weighing a ton and throw it easily; or it is delicate enough to pull a blade of grass with. He drinks with it, feeds himself with it, smells with it, works with it, and at times fights with it. Incidentally, a mouse that endeavoured to frighten an elephant by the traditional nursery rhyme method of running up his trunk would be blown into the next county. There is nothing else like an elephant's trunk on earth.

And for that matter, there is nothing else like the elephant. He has come down to us through the ages, surviving the conditions which killed off his earlier contemporaries, and he now adapts himself perfectly to more different conditions than any other animal in Africa.

He can eat anything that is green or ever has been green, just so long as there is enough of it. He can get his water from the aloe plants on the arid plains, or dig a well in the sand of a dry river bed with his trunk and fore feet, and drink there, or he is equally at home living half in the swamps of better-watered regions. He is at home on the low, hot plains of the seacoast at the equator or on the cool slopes of Kenia and Elgon. So far as I know, he suffers from no contagious diseases and has no enemies except man. There are elephants on Kenia that have never lain down for a hundred years. Some of the plains elephants do rest lying down, but no one ever saw a Kenia elephant lying down or any evidence that he does lie down to rest.

The elephant is a good traveller. On good ground a good horse can outrun him, but on bad ground the horse would have no chance, and there are few animals that can cover more ground in a day than an elephant. And in spite of his appearance, he can turn with surprising agility and move through the forest as quietly as a rabbit.

An elephant's foot is almost as remarkable as his trunk. In the first place, his foot is encased in a baglike skin with a heavy padded bottom, with some of the characteristics of an anti-skid tire. An elephant walks on his toes. His toes form the front part of his foot and the bones of his foot run not only back but up. Underneath these bones at the back of his foot is a gelatine-like substance, which is a much more effective shock absorber than rubber heels or any other device. One of the curious things about this kind of a foot is that it swells out when the weight is on it and contracts when the weight is removed. As a consequence an elephant may sink four feet into a swamp but the minute he begins to lift his legs, his feet will contract and come out of the hole they have made without suction. The elephant's leg, being practically a perpendicular shaft, requires less muscular effort for him to stand than it does for ordinary animals. This is one of the reasons why he can go for a century without lying down.

A country that elephants have long inhabited takes on some of the particular interest of the animals themselves. I believe that before the white man came to East Africa the elephant was nearly as much a plains animal as a forest animal, but he now tends to stay in the forests where the risk is not so great. On the plains, there are no elephant paths now, if there ever were, for in open country elephants do not go in single file. But in the forests there are elephant paths everywhere. In fact, if it were not for the elephant paths travel in the forest would be almost impossible, and above the forests in the bamboo country this is equally true. One travels practically all the time on their trails and they go everywhere except in the tree ferns. Tree fern patches are not very extensive, but I have never seen an elephant track or an elephant in them. The elephants are constantly changing the paths for various reasons; among others, because the natives are in the habit of digging elephant pits in the trails. But there are some trails that have evidently been used for centuries. One time we had followed a band of elephants on the Aberdare Plateau and had devilled them until they began to travel away. We followed until the trail led through a pass in the mountains and we realized that they were going into a different region altogether. That trail in the pass was a little wider than an elephant's foot and worn six inches deep in the solid rock. It must have taken hundreds of years for the shuffling of elephants to wear that rock away.

At another place on Kenia I found an elephant passage of a stream where the trail was twenty feet wide. Single paths came in from many directions on one side of the stream and joined in this great boulevard, which crossed the stream and broke up again on the other side into the single paths radiating again in every direction. In many places where the topography of the ground is such that there is only one place for a trail there will be unmistakable evidence that the trails have stayed in the same place many years — such as trees rubbed half in two by the constant passing of the animals or damp rocks polished by the caress of their trunks. And along all the trails, old and new, are elephant signs, footprints, dung, and gobs of chewed wood and bark from which they have extracted the juices before spitting them out.

But finding the elephants is not so frequent or easy as the multiplicity of the signs would indicate. One reason is that the signs of elephants — tracks, rubbed trees, and so forth — are more or less enduring, many of them being very plain in places where the elephants have not been for months or even years. If, however, you come on fresh elephant tracks, not more than a day old, you can usually catch up with the elephants, for as they feed along through the country they do not go fast. Only if they are making a *trek* from one region to another it may take much longer to catch them.

Once up with an elephant, if you are shooting, you are pretty sure that, even if he is charging you, a bullet from an elephant gun, hitting him in the head, will stop him even if it does not hit him in a vital spot. Moreover, if you stop the leader of a bunch that is charging you, the bunch will stop. I never heard of a case in which the leader of an elephant charge was stopped and the others came on, and I doubt if we ever will hear of such a thing, for if it does happen there won't be any one to tell about it. It is unusual for an elephant to keep on after being hit even if the hit does not knock him down. The old cow that charged me at the head of ten others was rather the exception to this rule, for after my first shot stopped her she came on again until my second shot knocked her down. But I had one experience that was entirely at variance with this rule. One old bull took thirteen shots from my rifle and about as many from Mrs. Akeley's before he was content either to die or run away.

In Uganda, after six months in the up-country after elephants, we decided to go down to the Uasin Gishu Plateau for lion spearing, for the rainy season was beginning and the vegetation was growing so thick that elephant hunting was getting very difficult. On the way down we came one morning upon the fresh trail of a herd of elephants. We followed for about two hours in a high bush country over which were scattered clumps of trees. Finally we came upon the elephants at the time of their mid-day siesta. The middle of the day is the quietest time of the twenty-four hours with elephants. If they are in a herd, they will bunch together in the shade. They do not stand absolutely still, but mill about very slowly, changing positions in the bunch but not leaving. They are neither feeding nor travelling but, as nearly as they ever do, resting. I even saw a young bull once rest his tusks in the crotch of a tree during this resting period. We got up to within twenty-five yards of them behind some bushes down the wind. We finally decided upon one of the bulls as the target. Mrs. Akeley studied carefully and shot. The bull went down, apparently dead. Ordinarily, we should rush in for a finishing shot, but in this case the rest of the herd did not make off promptly, so we stood still. When they did go off we started toward the apparently dead animal. As we did so, he got upon his feet and, in spite of a volley from us, kept on after the herd. We followed, and after half an hour's travel we caught sight of him again. We kept along behind him, looking for a place where we could swing out to one side and get abreast to fire a finishing shot at him. He was moving slowly and groggily. It was hard to move anywhere except in his trail without making a noise, and I suddenly discovered that the trail was turning so that the wind was from us to him.

Immediately we swung off to one side, but it was too late. I didn't see him when he got our wind but I knew perfectly he had it for there was the sudden crash of his wheel in the bushes and a scream. An elephant's scream is loud and shrill and piercing. And it is terrifying, too — at least to any one who knows elephants — for it means an angry animal and usually a charge. Then came a series of grunts and rumblings. A second or two later he came into sight, his ears spread out twelve feet from tip to tip, his trunk up and jerking fiercely from side to side. There is no way of describing how big an elephant looks under these conditions, or the speed at which he comes. At about thirty yards I shot, but he took it. He stopped, seemingly puzzled but unhurt. I shot the second barrel and looked for my other gun which was thirty feet behind me. The boy ran up with it and I emptied both barrels into the elephant's head, and still he took it like a sand hill. In the meanwhile, Mrs. Akeley had been firing, too. And then he turned and went off again. I went back to Mrs. Akeley. Everything that I knew about elephant shooting had failed to apply in this case. I had stopped him with one shot. That was normal enough. But then I had put three carefully aimed shots into his head at short range, any one of which should have killed him. And he had taken them with only a slight flinch and then had gone off. I felt completely helpless. Turning to Mrs. Akeley, I said:

"This elephant is pretty well shot up, and perhaps we had better wait for developments."

"No, we started it; so let's finish it," she said.

I agreed as we reloaded, and we were about to start following when his screaming, grunting, roaring attack began again. Exactly the same thing happened as the first time except that this time Mrs. Akeley, the boy, and I were all together. We fired as we had before. He stopped with the first shot and took all the others standing, finally turning and retreating again. Apparently our shots had no, effect except to make him stop and think. I was sick of it, for maybe next time he wouldn't stop and evidently we couldn't knock him down. We had about finished reloading when we heard him once more. There was nothing to do but stand the charge, for to run was fatal. So we waited. There was an appreciable time when I could hear his onrush but couldn't see him. Then I caught sight of him. He wasn't coming straight for us, but was charging at a point thirty yards to one side of us and thrashing back and forth a great branch of tree in his trunk. Why his charge was so misdirected I didn't know, but I was profoundly grateful. As he ran I had a good brain shot from the side. I fired, and he fell stone dead. With the greatest sense of relief in the world I went over to him. As I stood by the carcass I felt very small indeed. Mrs. Akeley sat down and drew a long breath before she spoke.

"I want to go home," she said at last, "and keep house for the rest of my life."

Then I heard a commotion in the bush in front of the dead elephant and as I looked up a black boy carrying a cringing monkey appeared. Only the boy wasn't black. He was scared to an ashen colour and he was still trembling, and the monkey was as frightened as the boy. It was J. T. Jr., Mrs. Akeley's pet monkey, and Alli, the monkey's nurse. They had followed to see the sport without our knowledge, and they had drawn the elephant's last charge.

This experience with an animal that continued to make charge after charge was new to me. It has never happened again and I hope never will, but it shows that with elephants it isn't safe to depend on any fixed rule, for elephants vary as much as people do. This one was the heaviest-skulled elephant I ever saw, and the shots that I had fired would have killed any ordinary animal. But in his case all but the last shot had been stopped by bone.

I couldn't measure his height, but I measured his ear as one indication of his size. It was the biggest I ever heard of. And his tusks were good sized — 80 pounds. He was a very big animal, but his foot measurement was not so large as the big bull of the Budongo Forest. Later I made a dining table of his ear, supporting it on three tusks for legs. With the wooden border it was eight feet long and seated eight people very comfortably.

Most wild animals, if they smell man and have an opportunity to get away, make the most of it. Even a mother with young will usually try to escape trouble rather than bring it on, although, of course, they are quickest to fight. But elephants are not always in this category. In the open it has been my experience that they would rather leave than provoke a fight; if you hunt elephants in the forest, you are quite likely to find that two can play the hunting game, and find yourself pretty actively hunted by the elephants. If the elephants after you are making a noise, it gives you a good chance. When they silently wait for you, the game is much more dangerous.

The old bull, who is in the centre of the elephant group in the Museum of Natural History now, tried to get me by this silent method. I was out on a trail and I saw that a big bunch of animals were near. I wasn't following any particular trail for they had moved about so that signs were everywhere and much confused. Finally I came to a gully. It wasn't very broad or very deep, but the trail I was on turned up it to where a crossing could be made on the level. The forest here was high and very thick, and consequently it was quite dark. As I looked up the trail I saw a group of big shapes through the branches. I thought they were elephants and peered carefully at them, but they turned out to be boulders. A minute later I saw across the gully another similar group of boulders, but as I peered at them I saw through a little opening in the leaves, plain and unmistakable, an elephant's tusk. I watched it carefully. It moved a little, and behind it I caught a glimpse of the other tusk. They were big, and I decided that he would do for my group. I couldn't get a glimpse of his eye or anything to sight by, so I carefully calculated where his brain ought to be from the place where his tusk entered his head, and fired. Then there was the riot of an elephant herd suddenly starting. A few seconds later there was a crash. "He's down," I thought, and Bill, the gun boy, and I ran over to the place where the animals had been. We followed their tracks a little way and found where one of the elephants had been down, but he had recovered and gone on. However, he had evidently gone off by himself when he got up, for while the others had gone down an old trail he had gone straight through the jungle, breaking a new way as he went. With Bill in the lead, we pushed along behind him. It was a curious trail, for it went straight ahead without deviation as if it had been laid by compass. One hour went by and then another. We had settled down for a long *trek*. The going wasn't very good and the forest was so thick that we could not see in any direction. We were pushing along in this fashion when, with a crash and a squeal, an elephant burst across our path within fifteen feet of us. It was absolutely without warning, and had the charge been straight on us we could hardly have escaped. As it was, I fired two hurried shots as he disappeared in the growth on the opposite side of the trail. The old devil had grown tired of being hunted and had doubled back on his own trail to wait for us. He had been absolutely silent. We hadn't heard a thing, and his plan failed, I think, only because the growth was so thick that he charged us on scent or sound without being able to see us. I heard him go through the forest a way and then stop. I followed until I found a place a little more open than the rest, and with this between me and the trees he was in I waited. I could hear him grumbling in there from time to time. I didn't expect him to last much longer so I got my lunch and ate it while I listened and watched. I had just finished and had a puff or two on my pipe when he let out another squeal and charged. He evidently had moved around until he had wind of me. I didn't see him but I heard him, and grabbing the gun I stood ready. But he didn't come. Instead I heard the breaking of the bushes as he collapsed. His last effort had been too much for him.

The efforts of the next elephant who tried the quiet waiting game on me were almost too much for me.

We had just come down from the ice fields seventeen thousand feet up on the summit of Mt. Kenia, overlord of the game regions of British East Africa, and had come out of the forest directly south of the pinnacle and within two or three miles of an old camping ground in the temperate climate, five or six thousand feet above sea level, where we had camped five years before and again one year before. Instead of going on around toward the west to the base camp we decided to stop here and have the base camp brought up to us. Mrs. Akeley was tired, so she said she would stay at the camp and rest; and I decided to take advantage of the time it would take to bring up the base camp to go back into the bamboo and get some forest photographs.

There was perfectly good elephant country around our camp but I wanted to go back up where the forests stop and the bamboo flourishes, because it was a bamboo setting that I had selected for the group of elephants I was then working on for the African Hall in the American Museum of Natural History. I started out with four days' rations, gun boys, porters, camera men, and so forth — fifteen men in all. The second day out brought me to about nine thousand feet above sea level where the bamboo began. Following a well-worn elephant trail in search of this photographic material, I ran on to a trail of three old bulls. The tracks were old — probably as much as four days — but the size was so unusual that I decided to postpone the photography and follow them. I did not expect to have to catch up their four days' travel, for I hoped that they would be feeding in the neighbourhood and that the trail I was on would cross a fresher trail made in their wanderings around for food. I had run upon their tracks first about noon. I followed until dark without finding any fresher signs. The next morning we started out at daybreak and finally entered an opening such as elephants use as a feeding ground. It is their custom to mill around in these openings, eating the vegetation and trampling it down until it offers little more, and then move on. In six months or so it will be grown up again eight or ten feet high and they are very apt to revisit it and go through the same process again. Soon after we entered this opening I came suddenly upon fresh tracks of the elephants I had been following. Not only were the tracks fresh but the droppings were still steaming and I knew that the animals were not far away; certainly they had been there not more than an hour before. I followed the trail amongst the low bush in the opening but it merely wandered about repeatedly bringing me back to the place where I had first seen the fresh tracks, and I realized that I might do this indefinitely without getting closer to the elephants. I decided to go outside the opening and circle around it to see if I could find the trail of my bulls as they entered the forest. This opening was at the point on the mountain where the forest proper and the bamboos merged. I followed an elephant path out of the opening on the bamboo side and had gone but a little way when I discovered fresh signs of my three bulls, who had evidently left the opening by the same path that I was following, and at about the same time I heard the crackling of bamboo ahead, probably about two hundred yards away. This was the signal for preparation for the final stalk.

I stood for a moment watching one of the trackers going up the trail to a point where it turned at right angles in the direction of the sounds I had heard. There he stopped at rest, having indicated to me by signs that they had gone in that direction. I·turned my back to the trail, watching the porters select a place to lay down their loads amidst a clump of large trees that would afford some protection in case of a stampede in their direction. The gun boys came forward presenting the guns for inspection. I took the gun from the second boy, sending him back with the porters. After examining this gun I gave it to the first boy and took his. When I had examined this I leaned it against my body while I chafed my hands which were numb from the cold mists of the morning, knowing that I might soon need a supple trigger finger. During this time the first gun boy was taking the cartridges, one by one, from his bandoleer and holding them up for my inspection — the ordinary precaution to insure that all the ammunition was the right kind, and an important insurance, because only a full-steel-jacketed bullet will penetrate an elephant's head. While still warming up my hands, inspecting the cartridges, and standing with the gun leaning against my stomach, I was suddenly conscious that an elephant was almost on top of me. I have no knowledge of how the warning came. I have no mental record of hearing him, seeing him, or of any warning from the gun boy who faced me and who must have seen the elephant as he came down on me from behind. There must have been some definite signal, but it was not recorded in my mind. I only know that as I picked up my gun and wheeled about I tried to shove the safety catch forward. It refused to budge, and I remember the thought that perhaps I had left the catch forward when I inspected the gun and that if not I must pull the triggers hard enough to fire the gun anyway. This is an impossibility, but I remember distinctly the determination to do it, for the all-powerful impulse in my mind was that I must shoot instantly. Then something happened that dazed me. I don't know whether I shot or not. My next mental record is of a tusk right at my chest. I grabbed it with my left hand, the other one with my right hand, and swinging in between them went to the ground on my back. This swinging in between the tusks was purely automatic. It was the result of many a time on the trails imagining myself caught by an elephant's rush and planning what I would do, and a very profitable planning, too; for I am convinced that if a man imagines such a crisis and plans what he would do, he will, when the occasion occurs, automatically do what he planned. Anyway, I firmly believe that my imaginings along the trail saved my life.

He drove his tusks into the ground on either side of me, his curled-up trunk against my chest. I had a realization that I was being

329

crushed, and as I looked into one wicked little eye above me I knew I could expect no mercy from it. This thought was perfectly clear and definite in my mind. I heard a wheezy grunt as he plunged down and then — oblivion.

The thing that dazed me was a blow from the elephant's trunk as he swung it down to curl it back out of harm's way. It broke my nose and tore my cheek open to the teeth. Had it been an intentional blow it would have killed me instantly. The part of the trunk that scraped off most of my face was the heavy bristles on the knuckle-like corrugations of the skin of the under side.

When he surged down on me, his big tusks evidently struck something in the ground that stopped them. Of course my body offered practically no resistance to his weight, and I should have been crushed as thin as a wafer if his tusks hadn't met that resistance — stone, root, or something — underground. He seems to have thought me dead for he left me — by some good fortune not stepping on me — and charged off after the boys. I never got much information out of the boys as to what did happen, for they were not proud of their part in the adventure. However, there were plenty of signs that the elephant had run out into the open space again and charged all over it; so it is reasonable to assume that they had scattered through it like a covey of quail and that he had trampled it down trying to find the men whose tracks and wind filled the neighbourhood.

Usually, when an elephant kills a man, it will return to its victim and gore him again, or trample him, or pull his legs or arms off with his trunk. I knew of one case where a man's porters brought in his arm which the elephant that had killed him had pulled off his body and left lying on the ground. In my case, happily, the elephant for some reason did not come back. I lay unconscious for four or five hours. In the meanwhile, when they found the coast was clear, the porters and gun boys returned and made camp, intending, no doubt, to keep guard over my body until Mrs. Akeley, to whom they had sent word, could reach me. They did not, however, touch me, for they believed that I was dead, and neither the Swahili Mohammedans nor the Kikuyus will touch a dead man. So they built a fire and huddled around it and I lay unconscious in the cold mountain rain at a little distance, with my body crushed and my face torn open. About five o'clock I came to in a dazed way and was vaguely conscious of seeing a fire. I shouted, and a little later I felt myself being carried by the shoulders and legs. Later again I had a lucid spell and realized that I was lying in one of the porter's tents, and I got clarity of mind enough to ask where my wife was. The boys answered that she was back in camp. That brought the events back to me how I had left her at camp, found the trail of the three old bulls, followed them and, finally, how I was knocked out. I was entirely helpless. I could move neither my arms nor legs and I reached the conclusion that my back was broken. I could not move, but I felt no pain whatever. However, my coldness and numbness brought to my mind a bottle of cocktails, and I ordered one of the boys to bring it to me. My powers of resistance must have been very low, for he poured all there was in the bottle down my throat. In the intervals of consciousness, also, I got them to give me hot Bovril — a British beef tea — and quinine. The result of all this was that the cold and numbness left me. I moved my arms. The movement brought pain, but I evidently wasn't entirely paralyzed. I moved my toes, then my feet, then my legs. "Why," I thought in some surprise, "my back isn't broken at all!" So before I dropped off again for the night I knew that I had some chance of recovery. The first time I regained consciousness in the morning, I felt that Mrs. Akeley was around. I asked the boys if she had come. They said no, and I told them to fire my gun every fifteen minutes. Then I dropped off into unconsciousness again and awoke to see her sitting by me on the ground.

When the elephant got me, the boys had sent two runners to tell Mrs. Akeley. They arrived about six in the evening. It was our custom when separated to send notes to each other, or at least messages. When these boys came on to say that an elephant had got me, and when she found that there was no word from me, it looked bad. Mrs. Akeley sent word to the nearest government post for a doctor and started her preparations to come to me that night. She had to go after her guides, even into the huts of a native village, for they did not want to start at night. Finally, about midnight, she got under way. She pushed along with all speed until about daybreak, when the guides confessed that they were lost. At this juncture she was sitting on a log, trying to think what to do next. And then she heard my gun. She answered, but it was more than an hour before the sounds of her smaller rifle reached our camp. And about an hour after the boys heard her gun she arrived.

She asked me how I was, and I said that I was all right. I noticed a peculiar expression on her face. If I had had a looking glass, I should probably have understood it better. One eye was closed and the forehead over it skinned. My nose was broken and my cheek cut so that it hung down, exposing my teeth. I was dirty all over, and from time to time spit blood from the hemorrhages inside. Altogether, I was an unlovely subject and looked hardly worth saving. But I did

irely over it all, although it took me three months in bed. The
that was serious was that the elephant had crushed several
of my ribs into my lungs, and these internal injuries took a long
time to heal. As a matter of fact, I don't suppose I would have pulled
through even with Mrs. Akeley's care if it hadn't been for a Scotch
medical missionary who nearly ran himself to death coming to my
rescue. He had been in the country only a little while and perhaps
this explains his coming so fast when news reached him of a man
who had been mauled by an elephant. The chief medical officer at
Fort Hall, knowing better what elephant mauling usually meant,
came, but he didn't hurry. I saw him later and he apologized, but
I felt no grievance. I understood the situation. Usually when an elephant gets a man a doctor can't do anything for him.

But this isn't always so. Some months later I sat down in the hotel
at Nairobi with three other men, who like myself had been caught
by elephants and had lived to tell the tale. An elephant caught Black
in his trunk, and threw him into a bush that broke his fall. The elephant followed him and stepped on him, the bush this time forming
a cushion that saved him, and although the elephant returned two
or three times to give him a final punch, he was not killed. However,
he was badly broken up.

Outram and a companion approached an elephant that was shot
and down, when the animal suddenly rose, grabbed Outram in his
trunk and threw him. The elephant followed him, but Outram
scrambled into the grass while the elephant trampled his pith helmet into the ground, whereupon Outram got right under the elephant's tail and stuck to this position while the elephant turned
circles trying to find him, until, becoming faint from his injuries,
Outram dived into the grass at one side. Outram's companion by this
time got back into the game and killed the elephant.

Hutchinson's story I have forgotten a little now, but I remember
that he said the elephant caught him, brushed the ground with him,
and then threw him. The elephant followed him and Hutchinson put
off fate a few seconds by somehow getting amongst the elephant's
legs. The respite was enough, for the gun boy, by this time, began
firing and drove the elephant off.

In all of these cases, unlike mine, the elephants had used their
trunks to pick up their victims and to throw them, and they had
intended finishing them by trampling on them. This use of the
trunk seems more common than the charge with the tusks that had
so nearly finished me. Up in Somaliland Dudo Muhammud, my gun
boy, showed me the spot where he had seen an elephant kill an

Italian prince. The elephant picked the prince up in his trunk and beat him against his tusks, the prince, meanwhile, futilely beating the elephant's head with his fists. Then the elephant threw him upon the ground, walked on him, and then squatted on him, rubbing back and forth until he had rubbed his body into the ground.

But elephants do use their tusks and use them with terrible effect. About the time we were in the Budongo Forest, Mr. and Mrs. Longdon were across Lake Albert in the Belgian Congo. One day Longdon shot a bull elephant and stood watching the herd disappear, when a cow came down from behind, unheard and unseen, ran her tusk clear through him, and, with a toss of her head, threw him into the bush and went on. Longdon lived four days.

But although the elephant is a terrible fighter in his own defense when attacked by man, that is not his chief characteristic. The things that stick in my mind are his sagacity, his versatility, and a certain comradeship which I have never noticed to the same degree in other animals. I like to think of the picture of the two old bulls helping along their comrade wounded by Major Harrison's gun; to think of several instances I have seen of a phenomenon, which I am sure is not accidental, when the young and husky elephants formed the outer ring of a group protecting the older ones from the scented danger. I like to think back to the day I saw the group of baby elephants playing with a great ball of baked dirt two and a half feet in diameter which, in their playing, they rolled for more than half a mile, and the playfulness with which this same group teased the babies of a herd of buffalo until the cow buffaloes chased them off. I think, too, of the extraordinary fact that I have never heard or seen African elephants fighting each other. They have no enemy but man and are at peace amongst themselves.

The Book of Naturalists
ed. William Beebe
Princeton Univ Press 1944
paperback   1988   p. 315

HENRY FAIRFIELD OSBORN (1857–1935)

The Migration of Elephants

First published in "The Elephants and Mastodonts Arrive in America,"
Natural History, 1925; reprinted in *Proboscidea: A Monograph . . .
of the Elephants of the World*, 1936

[IN THE CASE OF Henry Fairfield Osborn I experience the great difficulty of lack of perspective. From my point of view of him as a boy's
adviser, a master teacher, as associate in scientific undertakings, and
as intimate friend, any cold critical judgment is thwarted. In his writings he was always scientist and paleontologist first and naturalist
second. Like other eminent scientists, his emotional appreciation of
evolution, of the infinite stretches of past time, of the drama of the
origin and the majestic development of mammalian life, so evident in
conversation, became necessarily diluted as his thoughts crystallized
into written words. He marshaled his sentences to serve first and
last the need for scientific condensation and philosophical clarity and
brevity.

In the paragraph quoted, Professor Osborn let himself go and it
recalls vividly the spirit of his spoken conversation.]

A N INSATIABLE *Wanderlust* has always possessed the souls of elephants as it has those of the tribes and races of man. Not only
to overcome the changes and chances of this mortal life, but
also to gratify their intelligent curiosity ever to explore afresh forests, pastures, fields, rivers, and streams, they have gone to the very
ends of the earth and have far surpassed man in adapting their clothing and teeth to all possible conditions of life. Thus the romances
of elephant migration and conquest are second only to the romances
of human migration and conquest. Variety is the spice of elephant
life, as it is of human life, and the very longing for a change of scene
and of diet has been the indirect cause of what in scientific parlance
we term *adaptive radiation* — the reaching out in every direction for
every kind of food, every kind of habitat, in itself the *cause* of radiating or divergent evolution and adaptation. It is to this predisposi-
tion to local, continental or insular, and world-wide wanderings that
we attribute the many branches and sub-branches which have been
developed in this remarkable family.

334

BASSETT DIGBY (1888–   )

Mammoths in the Flesh

From *The Mammoth, and Mammoth-hunting in Northeast Siberia.* 1926

[As reporter and author Digby traveled widely and in Siberia he actually studied mammoths in the flesh as they thawed from glaciers.]

D O NOT IMAGINE that a flesh-and-blood mammoth, freshly yielded up by a landship on a thawing tundra, is anywhere a common object of the wayside, a mere incident in the hodge-podge of queer sights that any tourist can see by poking his head out of a *wagon-lit* on the Trans-Siberian Railway.

No mammoth in the flesh, and very few bones, has been found within a couple of thousand miles of the railway zone. The finds are made chiefly in the New Siberian Isles, and within a hundred miles or so of the Arctic seashore.

The Samoyedes, Ostiaks, Tunguses and Yakuts, who roam the tundra with their sledge-dogs and reindeer, are far from quiet in their minds about these uncanny great beasts. Most of them will summon up enough courage to lop the tusks from a specimen they encounter, knowing that the white men down south will give them vodka and fishhooks, shiny black tiles of brick-tea and yards of cloth for them; but there the matter ends. The nomad natives are not prone to rushing off to a Russian settlement, hundreds of miles away up one of the rivers, and blurting out the news of their find. "Do the dirty work, at the grave risk of being haunted by bogeys and evil spirits, and then let well enough alone!" is their motto. It is not probable that the few scattered Russian officials and traders of Northern Siberia hear of every find. Usually, when the news does leak through, so long a period has elapsed that the chances are that the carcass has been rotted away by the summer sunshine, or torn to bits by wolves and foxes.

However, several white men of education and a few intelligent natives have examined mammoths in the flesh, and one has even been brought down, in frozen packages, on a caravan of sledges, to the Trans-Siberian Railway, by which it was conveyed to Petrograd.

A vivid account of such a discovery was contained in a letter, said to have been written to a friend in Germany, by Benkendorf, a young Russian surveyor, who was employed by the Government to map out tracts of the Arctic coast around the estuaries of the Lena and Indigirka rivers.

"In 1846," he wrote, "there was unusually warm weather in the north of Siberia. Already in May abnormal rains poured over the swamps and bogs, storms shook the earth, and the streams not only carried broken ice to the sea, but also swept away large masses of soil thawed by the warm off-run of the southern rains. . . . We steamed up the Indigirka on the first favourable day, but we saw no signs of land. The landscape was flooded as far as the eye could see; we saw around us only a sea of dirty brown water, and knew that we were in the river only by the strength of the current. A lot of debris was coming downstream, uprooted trees, swamp litter, and large masses of buoyant tufts of grass, so that navigation was not easy. At the end of the second day we were only about forty versts upstream. Someone had to stand continually with the sounding rod in hand, and our little steamboat received many a shock that made her tremble to the keel. A wooden vessel would have been smashed. We saw nothing all around but floods. For eight days we met with hard going of this kind, until at last we reached the spot where our Yakuts were to have met us. Farther up was a place called Ujandina, whence people were to have come to us. No one had turned up, however. Evidently the floods had stopped them.

"As we had been here in former years, we knew the place. But how it had changed! The Indigirka, here about three versts wide, had torn up the land and made itself a fresh channel. When the floods subsided we saw, to our astonishment, that the old river-bed had become merely that of an insignificant stream. This enabled us to shove through the soft mud; and we went reconnoitring up the new stream, which had cut its way westward. Later we landed on the new bank, and surveyed the undermining and destructive work of the wild waters that were carrying away, with extraordinary rapidity, masses of peat and loam.

"It was then that we made a wonderful discovery.

"The land on which we were walking was turfy bog, covered thickly with young plants. Many lovely flowers rejoiced the eye in the warm radiance of a sun that shone for twenty-two out of the twenty-four hours. The stream was tearing away the soft sodden bank like chaff, so that it was dangerous to go near the brink. In a lull in the conversation we heard, under our feet, a sudden gurgling and movement in the water under the bank. One of our men gave a shout, and pointed to a singular shapeless mass which was rising and falling in the swirling stream. I had noticed it, but had not paid it any attention, thinking it only driftwood. Now we all hastened to the bank. We had the boat brought up close, and waited until the mysterious thing should again show itself.

"Our patience was tried. At last, however, a huge black horrible mass bobbed up out of the water. We beheld a colossal elephant's head, armed with mighty tusks, its long trunk waving uncannily in the water, as though seeking something it had lost. Breathless with astonishment, I beheld the monster hardly twelve feet away, with the white of his half-open eyes showing.

yike

" 'A mammoth! A mammoth!' someone shouted.

"I called out: 'Chains and ropes — quick!'

"I will tell you of our preparations for securing the monster that the river was trying to tear from us. As it again sank under the surface we had to wait for a chance to throw a rope over its head. This was accomplished only after many efforts. After a close examination I satisfied myself that the hind-legs of the mammoth were still embedded in the frozen mud, and that the waters that swept over the caved-in fall of soil from the bank would loosen them for us. We accordingly made a noose fast round its neck, threw a chain round the tusks, which were 8 ft. long, drove a stake into the ground about twenty feet from the bank, and made chain and rope fast to it.

"The day passed quicker than I expected; but still the time seemed long enough before the animal was free at last, which happened about twenty-four hours after the cave-in.

"The position of the beast interested me; it was standing in the earth, thus indicating the manner of its destruction, not lying on its side or back, as a dead animal naturally would. The soft peat or bog on to which it stepped, thousands of years ago, gave way under the weight of the giant, and he sank as he stood, on all four feet, unable to save himself. A severe frost came, turning into ice both him and the bog which overwhelmed him. The surface of the bog gradually became covered with driftwood, and sand and uprooted plants swept over it by each successive spring freshet. God only knows what agencies had worked for its preservation. Now, however, the stream had brought it once more to the light of day. And I, an ephemera of life compared with this primaeval giant, was sent here by Providence just at the right time to welcome him. You can imagine how I jumped for joy!

"During our evening meal our outposts announced the approach

337

of strangers. A group of Yakuts came up on their fast shaggy ponies. They were the people we had arranged to meet, and they seemed very glad to see us. They increased our party to about fifty persons.

"On our pointing out to them our wonderful capture, they hastened to the bank. It was amusing to hear them jabber at the sight. They were intensely excited.

"For a day I left them in possession, but when, on the following day, the ropes and chains gave a great jerk, a sign that the mammoth was quite free from the clutch of the frozen soil, I ordered them to exert all their strength to drag the beast ashore.

"At length, after much hard work, in which the horses came in extremely useful, the animal was brought ashore, and we were able to roll the carcass about twelve feet from the water. The rapidly decomposing effect of the warm air filled us all with astonishment.

"Picture to yourself an elephant with a body covered with thick fur, about 13 ft. in height and 15 ft. in length, with tusks 8 ft. long, thick and curving outward at their ends. A stout trunk 6 ft. long, colossal legs $1\frac{1}{2}$ ft. thick, and a tail bare up to the tip, which was covered with thick tufty hair.

"The beast was fat and well grown. Death had overwhelmed him in the fullness of his powers. His large, parchment-like, naked ears lay turned up over the head. About the shoulders and back he had stiff hair about a foot long, like a mane. The long outer hair was deep brown and coarsely rooted. The top of the head looked so wild and so steeped in mud that it resembled the ragged bark of an old oak. On the sides it was cleaner, and under the outer hair there appeared everywhere a wool, very soft, warm and thick, of a fallow brown tint. The giant was well protected against the cold.

"The whole appearance of the great beast was fearfully strange and wild. It had not the shape of our present elephants. As compared with the Indian elephant, its head was rough, the brain-case low and narrow, the trunk and mouth much larger. The modern elephant is an awkward animal, but compared with this mammoth he is an Arabian steed to a coarse, cumbersome dray-horse.

"I could not divest myself of a feeling of fear as I approached the head. The open eyes gave the beast a lifelike aspect, as though at any moment it might stir, struggle to its feet, and bear down upon us with a stentorian roar. . . .

"The bad smell of the carcass warned us that it was time to save of it what we could; the encroaching river, too, bade us hasten.

"First we hacked off the tusks and sent them aboard our boat. Then the natives tried to hew off the head, but notwithstanding their efforts this was slow work. . . . I had the stomach cut out and dragged aside. It was well filled. The contents were instructive and well preserved. The chief contents were young shoots of fir and pine. A quantity of young fir-cones, also in a chewed state, were mixed with the mass.

"As we were eviscerating the animal, I was as careless and forgetful as my Yakuts, who did not notice that the ground was sinking under their feet until a cry of alarm warned me of their predicament, as I was still groping in the beast's stomach. Startled, I sprang up and beheld how the undermined bank was caving in, to the imminent danger of our Yakuts and our laboriously rescued find. Fortunately our boat was close at hand, so our natives were saved in the nick of time. But the carcass of the mammoth was swept away by the swift current, and sank, never to appear to us again."

The first flesh-and-blood mammoth find made by a man of science was, I believe, the Adams discovery on the tundra, near the mouth of the Lena, though nomad natives had told the Russian travellers, Sarietchiev and Merck, in 1787, that about sixty-seven miles below the village of Alazeysk, on the little River Alazeja, a gigantic beast covered with hide and hair had been washed out of the sandy shore, in an upright position.

It was in 1799 that Schumarov, a wandering Tungus, happened upon a hairy monster, protruding from the side of a crumbling bank as if literally in the act of springing out of the earth in answer to the Trump of Doom. Schumarov had never seen anything like that before, and he was not anxious to see anything like it again. He invoked the gods to spare him such uncanny visions, and immediately headed for the horizon. Two years later, however, he regained his nerve, or his curiosity got the better of his trepidation, and he went back for another peep. What he saw served him right. The monster's gigantic curly tusks and one of its huge hairy flanks were exposed. To say that the poor chap wanted to have nothing to do with it understates the case. He hurried back to his tent home, on the shore of Lake Onkul, and when he told his family there was a painful scene. A few years ago, his wife reminded him, another hunter had made a similar find, and all his children had died shortly afterward.

But, finding themselves all alive and healthy two years later, the Schumarov family decided that perhaps the fate of the other family of mammoth finders was a case of *post hoc* rather than *propter hoc*. Schumarov *père* laid in a good stock of Dutch courage — a vodka bottle in each pocket — and, returning, actually had grit enough to hack off tusks, which, a few weeks later, he sold to a Russian trader for

339

fifty roubles. The great beast was so fat, he told the trader, that its belly hung below its knees. It was a male, with a long shaggy mane, and its legs had been badly torn about by bears, wolves and foxes.

Tidings of the find reached Petrograd, and in 1805, seven years after Schumarov's first discovery, Professor Adams, of the Imperial Academy of Arts and Sciences, turned up on the spot and collected what was left. About three-quarters of the skin was saved. It was a dark grey hide, an inch thick, covered in places with reddish wool and long black hairs. It measured 16 ft. 4 in. from the front of the skull to the end of the mutilated tail, and 9 ft. 4 in. it stood at the shoulder. The tusks, which were traced to Yakutsk, and purchased on Adams's return journey, ran 9 ft. 6 in. along the outer curve. They weighed 360 lb. the pair, and skull and tusks together turned the scale at 414 lb.

About 40 lb. of hair was collected, though much more had been trodden into the sand by wild animals. The mounted skeleton is preserved, I found, in the museum of the Imperial Academy, out on Vassili Ostrov, one of the Petrograd islands.

ERNEST THOMPSON SETON (1860–   )

The Sea Otter

From *Lives of Game Animals.* 1926

[THE WORK of this well-known naturalist may be divided into several categories: first, his tales of animals, authentic in every detail of habit but related as combined in the life of a single individual. An example of these is *Lives of the Hunted.* Another group includes his travel volumes, such as *Arctic Prairies;* and finally we have the monumental work *Lives of Game Animals,* part thorough compilation from reliable sources, part Seton's own contributions, while the monograph is rounded out with an abundance of excellent drawings by the gifted author. His method of presentation varies with each kind of work, but is always vivid and realistic.

The sea otter is one of the strangest, rarest, and most interesting of all mammals. In ages past its ancestors left the sea and became four-footed and terrestrial. It is now well on its way back to an aquatic life, with habits so unusual and numbers so reduced that every fact is important. Seton does full justice to this animal.

Seton's influence in arousing interest in wild animals, in conservation, and in woodcraft among the young people of our country has been wide in scope and of definite and permanent value.

I owe to Seton a suggestion that has been a real help to me many times. I was on the point of starting out on a two-year, round-the-world study of pheasants, and this was his advice:

"There will come a time, say in the heart of Borneo, when you think you have exhausted your study and observations of some particular area or subject, and will look around at a loss to add any more notes. Instantly close your eyes tight, and keep them closed for several minutes, while you tell yourself in a most convincing manner that you are now back in your laboratory in New York, never to return to Borneo again. Ask yourself if there is anything you wish you had noticed, or watched, or collected, or photographed. Five minutes of this self-questioning, this egocentric third degree, will always materialize a crop of the most common, apparent things or phenomena which you had overlooked, to which you should have paid special attention and didn't."

341

I have tried this scores of times and invariably my eyes pop open and I begin renewed, enthusiastic investigation. My advice is never to try it at night, because I once attempted it at Myitkyina, in northern Burma, and the stimulus was so immediate that I got up, lit the dak bungalow's candle, and wrote for three hours! Far from being a silly psychic game, it is a real spur to jaded eyes and mind.]

OF ALL THE CREATURES living in the cold North seas, the Sea-otter is alone in that he usually swims on his back. Sailing, paddling, shooting or diving he goes, with his back to the deep, and his shining breast to the sky. But his neck is doublebent, so his big soft eyes sweep the blue world above and around. Propelled almost wholly by his big finlike hind feet, he moves with easy sinuous sweeps through the swell with its huge broad fronds of kelp — with back first, he ever goes forward, until the moment comes to dive — supple as an eel he turns — back up like Seal or Beaver, and down he goes — down, down — long strings of silver bubbles mark the course, and by a strange atmospheric change, the colour of the black merman now is yellow-brown as a seaman's slicker, or golden as a bunch of kelp. Down 30 to 100 feet or more he goes, and gropes around in the gloom, until he finds some big fat squid or sea-urchin. He does not hurry, for he can stay under 4 or 5 minutes. Then up he comes with the prey in his jaws, back to the top, to the borderland, that eternal line between the two kingdoms of air and water, on which he lives.

Here again on his back he lies, as, using his broad chest for dinner table, he tears open the sea beastie, feeds on its meat, and flings its shell aside, if it have any. Then he repeats the dive, and the feasting, until his sleek round belly is well filled. Now, among the heaving, friendly kelp he lazes on his back, plays ball perhaps with a lump of the leathery weed, tossing it from paw to paw, taking keen delight in his cleverness at keeping it aloft, as a juggler does his balls; and sniffing in disappointment if he should foozle the ball and miss the catch.

Other Sea-otters are about him, for Amikuk, like most fishermen, is of a neighbourly spirit and loves good company. His mate may be there with her water baby in her big motherly lap. She tumbles it off into the deep for a swimming lesson; and round and under she swims to exercise "the kid," and make it learn. This is a very ancient game, this water-tag; the earliest monad that ever wriggled tail in the hot first seas, no doubt invented it. It is deep in everything that swims, or moves, and loves good company; so father Otter pitches in, and plays it, too. For half an hour they may keep it up. Father is still strong and frolicsome; so is mother; but the water baby is tired. Its big round eyes are blinking in weariness, and it is ready for sleep. Trust mother to look out for the little one. She curls up and takes it, not pick-a-back, but in the snug bed she makes by curling belly up, as she floats among the weed. Her four feet are the bedposts, and in some degree the coverlet, too, for she holds it to her breast, crooning softly to it, till its whimpering ceases and it sleeps. Its fur may be wet and cold, but its skin is dry and warm; and drifting like a log of drift among that helpful wrack, they float, and love the lives they live.

But father is full of energy. He is one of thirty or forty that herd along this bed of kelp, that marks a deep-down feeding ridge, where their shiny seafood swarms. And away they go, in a race that recalls the tremendous speed and energy of the Porpoise in the sea. Undulating like water serpents, or breaking from the side of a billow, to leap in a long curve, splash into the high wet bank of the next; one after another they go, racing round, in air or far below, diving, jumping, plunging, somersaulting, back up, belly up, or sides up — it matters not, so they speed, for the joy of rapid flight, for the wonderful pleasure of using their pent-up energy in mastery of the elements about them.

Strange as it may seem, these merry games have mostly place in rough and heavy weather, almost a storm. Swanson says that, on calmer days, they are never seen playing; that is, the bands of older ones are not. They need the stimulus of a contest with the waves before they do their wonderful best. In quiet times, they are more likely to sleep.

Usually, when one is seen, it is sleeping; otherwise they are too alert to be easily discovered. And even at this time, none but the keenest, best-trained eyes can find them, for the body is sunk below the surface; only the head and flippers show. They have no look of a furry animal form, but pass for a black drift log with snags.

As they float their merman way, they are frequently seen with hind flippers raised and spread, as if catching the breeze to sail or drift before it.

On sunny days, so rare in their wild, stormy ocean home, their big brown eyes might suffer from the unwonted glare; but as they float and dream, they sometimes shade their eyes with one idle paw; just as one of us might do, if we could swim that way, and faced the light.

They do not have those big, fawn-like eyes for nothing. Their vision is all it should be, for such perfect organs. Their ears, too, are keen; but of all their senses, smell is their best, the safest sentinel that guards their life.

Best equipped is he, in this respect, of all the wild things in the wild and windy North. But also hardest pressed. For that matchless robe, worth its weight in gold, has the force of a blood-price on his head. Many a brave and valiant man has been hunted down for less reward than the fetch of this wondrous pelt.

As one reads the full account by Steller, one gets the impression of humanization almost too much, and might question the accuracy of some details. But they have been endorsed by Elliott, Scammon, Stejneger, Snow, and many others since.

And as one reads of its mild and human face, its fish-like hinder parts, its human arms, in which the mother, two-breasted, carries her whimpering babe, and croons it to its slumber, or plays with it, as she sports in the rolling surf, can one not readily believe that in this we have, perhaps, the original of the mermaids and the mermen of the ancient tales?

LOUIS ROULE (1861–1942)

The Tomb of the Pearl

From *Les Poissons et le Monde Vivant des Eaux.*
Vol. II, *La Vie et l'Action.* 1927

[Louis Roule was, in some ways, the most notable of recent French ichthyologists, and his versatility was equalled only by the great amount of his output and the clarity and accuracy of his studies. As professor of the National Museum of Natural History in Paris his work dealt with the more technical aspects of strange and little-known types of fish, such as those from the antarctic regions and from the icy, black abysses of the ocean.

What gives him an honored place as a naturalist are the ten volumes entitléd *Les Poissons et le Monde Vivant des Eaux.* Into these he put the summation of his extensive knowledge, and presented it in most excellent language, which in some ways possesses the charm of Fabre. If his personal philosophy comes to the fore now and then with somewhat too great emphasis, we must judge it indulgently, as the right of one who is providing us with a superb survey of this piscine field of natural history. The life of a man such as Roule must be considered as unusually well balanced — a man who can crown many long years of technical investigation with a final set of books successfully addressed to the unscientific but intelligent reader.

Several of these volumes have been translated into English and from one of these I confidently choose a chapter.]

BEFORE ME, on the table at which I work in my laboratory, lie four pearl shells. The name is commonly used to denote the large, thick, almost flat valves of the shells of pearl oysters which are eagerly sought because of the pearls they sometimes contain, and for the mother-of-pearl of which they themselves are composed. Although the shells of other molluscs are also made of mother-of-pearl, and therefore deserve the same name, especially since they are sometimes used in the same way, in this particular case the shell has a beauty, a compactness and size which give it a genuine superiority. The more so, because the pearl oysters, which live in the warm

345

seas, often gather in large beds, like the oysters we eat, settle beside one another, and thereby offer fishermen an opportunity which they are not slow to take.

These four shells, which have come from the Caribbean Sea, where the pearl oyster beds are keenly exploited, do not differ in any essential particular from those we see in seaside shops with a picture painted inside them. They have two faces, one bulging slightly, which is the outer part when the oyster is alive, greyish in colour and rather dull; the other, inside, hollowed out, with the naked mother-of-pearl, white and sparkling, glittering exquisitely. But, on this pearly side, there is a special thickening not found elsewhere, a conspicuous, rounded, winding band, also made of mother-of-pearl. These four shells, selected because of that band from thousands of others, all have it, but it differs in size and shape. The rarity and the lack of resemblance indicate that here we have an accidental case, and we are curious to know what it means.

I take up one of the shells. With the point of a scalpel, I break off in little pieces the surface of the pearly band. This, made of very thin layers, breaks up into small flakes, and I soon discover that it forms a sort of shell, enwrapping something, upon which it applies itself and models itself exactly. Patiently and carefully, for it is fragile, I uncover the object thus surrounded, and I find that it is a small fish, dead, of course, and petrified, with a longish body, caught in the mother-of-pearl and so preserved. I can discern its head, the eyes sunken in their orbits, the snout through which the bones protrude, and I can also see the fins, all the rays of which are preserved, and the whole body, in which the vertebrae may be seen one behind the other. This object is the mummy of a fish, preserving its normal form. The thick band which protrudes in the inside of the shell is a tomb of mother-of-pearl in which the mummy is buried.

It is not difficult to reconstruct the phases of the extraordinary event which has turned this oyster shell into a sort of mausoleum, wherein is a tomb in which a corpse is buried. This corpse was first impregnated with calcareous matter, as we may easily see if we treat it with weak acid. Thus embalmed, it was covered with one layer of mother-of-pearl after another, until it was completely surrounded. Nature had built for it a tomb of rare and precious stone, a magnificent coffin; she had done for this little fish what the Egyptians used to do for the remains of their kings.

This shell, in which the tomb is placed, once belonged to a pearl oyster. When it was alive, it had the structure of its fellows. Its body was sheltered in the large shell with two thick valves, and it constructed the shell itself by continually forming calcareous matter which spread in thin layers of mother-of-pearl one upon another. It was able not only to build its own house, but to keep enlarging it all the time, making it continually thicker, adding new layers to those which were already there. And, as the valves enlarged, the oyster grew.

For making the shell, the body makes use of two large symmetrical expansions which envelop it completely like a cloak, and, in fact, the name "mantle" is actually given to them because of their nature. These two parts of the cloak form and lay down, on their outer side, the calcareous matter of the mother-of-pearl of which the shell is made. If we examine a whole oyster when it is alive, we first separate the two halves of the shell and see that the calcareous mass is lined inside by a thin plate of living tissue forming the two expansions which make the cloak. In the spacious cavity which they bound, but occupying only a small space in it, is the body itself with its organs, the most notable of which are those of respiration, branchial plates lodged in the angle between the cloak and the sides. This type of organization is also that of other molluscs of the same class, though of different appearance, the "Lamellibranchs," the ordinary oysters of commerce, mussels, clams, etc., which also have a two-lobed cloak and a shell with two valves surrounding the whole body.

When alive, in their native waters, the pearl oysters, on the bottom where they live and where the fishermen gather them, leave their shell half open, and their valves allow the water from outside to penetrate freely into the large interior cavity. Two currents are set up, one going in and the other coming out, so that the water is continually renewed, bringing dissolved oxygen to the gills, and particles of food to the digestive tube. The thick, pleated edges of the mantle, which line those of the shell, have fluted contours so arranged as to direct these currents into channels and take them where they have to go. Although they are fixed to one place and cannot move from it, the oysters thus receive what they need, and live in due conformity with the requirements of their particular structure.

They live so well that they swarm, however little circumstances are in their favour. Side by side, and even one on top of another, they form extensive banks, with little ones beside the big ones, and all together form large heaps spread out over the sea bottom. These banks, made of compact and strong shells, themselves serve as supports and shelters to other creatures. On them, under them, between them, flourishes a rich vegetation of seaweed, mingled with the rocky arborescences of coral. Bright-hued sea-anemones, different kinds of molluscs and other creatures, settle wherever they can find a suitable place. A regular population, varied and highly coloured, swarms and develops to an extraordinary degree upon this solid support. Fishes are there too, profiting by the number of likely victims, especially since they are so easy to secure. Some of them are particularly common, among them one whose existence, even more than that of the rest, is bound up with the oysters to such an extent that it deserves the name of "Pearl-fish" which has been given to it. It is this fish which becomes the hero whose end is to be a mummy encased in a coffin of pearl.

This fish belongs to the genus *Fierasfer*. Its specific name is *Fierasfer affinis*. In appearance it is like a very small eel, long and supple, undulating, slightly brownish in colour, with translucent tones and flashes of silver. The resemblance is only in form and appearance, for, in reality, *Fierasfer* belongs to a different group from that of the eels and congers, and has neither their structure nor their mode of life. In its ordinary state, it sometimes lodges, folded up on itself, in the interior of the pearl oyster, and makes its abode in the roomy cavity surrounded by the mantle. The oysters live well, and the pearl-fish, settling down with them, finds shelter, if not a livelihood. Situated beside the oyster's gills, taking advantage of the entering current which flows towards them, protected by the thick envelope of the shell which covers the mantle, it can enjoy, in peace, and remote from every risk, the charms of a life of complete repose. It takes cover in this shelter which it has chosen. It enters through the space left between the open valves, and goes out the same way when the need of food compels it to do so; it returns again when it has fed. The fish asks the oyster to allow it to share the shelter of its shell and the oyster agrees; both live together in complete harmony.

But things sometimes go wrong, and the agreement is occasionally broken. It may happen that the visitor, when young and not very big, mistakes the door. Instead of passing through the opening in the edges of the mantle and so into the cavity to which it is allowed access, it tries to find its way between the mantle and the shell, to enter thereby into the private quarters of its host. Then the oyster defends itself and punishes the intruder. When the fish has forced its way into this private ground, the mantle contracts, tightens round it, and imprisons it against the shell. The intruder is unable to move, cannot free itself, breathes with difficulty and finally dies, where it is in the narrow gaol in which it has imprisoned itself. Caught in a trap, it suffers there the punishment of death. Then the oyster, not being able to get rid of the corpse, sets to work to embalm it, to mummify and to immure it. Its secretion of calcareous matter begins by infiltrating the tissues of the fish, and then incrusts them. The secretion continues and the new deposits spread in little blades of pearl which surround the mummy and gradually totally enclose it. Thus the tomb of pearl is built. Its presence, its position, its contents, tell a tragic story which, transposed into human terms, would be that of a young rake who, as a result of the error of his ways, comes first to prison, then to death. If Nature uses a precious substance to make a coffin for him, if she adorns what she herself has done, she gives, none the less, an example and a lesson.

This story is that of an accident which happens comparatively rarely. It is brought about by a chance circumstance, the coming of a little fish into a place where it ought not to be. The consequences throw light upon the essential point, the continual production of mother-of-pearl, of which it is only an episode. The pearl-fish is not the only one to try to introduce itself between the shell and the mantle; other animals, even smaller, do likewise. The result is the same. Just like it, the intruders are immobilized, imprisoned, stifled, then mummified and covered with a pearly shroud. More often still, some impurity, a tiny fragment of shell, a grain of sand, which has got there by accident, also finds its way into the forbidden area. Its presence ends in the same way; it too is swallowed up by the pearl and put aside as it were, isolated. This is the usual origin of those protuberances often implanted in the inner face of the valves. From their form and composition they are called "nacreous pearls."

Their size and appearance are subject to variation, but their origin is similar in every case. They are concretions formed around a foreign body, the first centre of a local irritation which has brought about and directed the secretion of the nacreous matter. It is in this way that their fine enveloping layers are produced and deposited, as in the case of the fish's tomb. The story has been generalized and has given up its real secret.

We know the shells which have inside them figurines — tiny figures of chinamen — encrusted in the pearl, usually little magots, for these pieces come from the Far East. They are pearls formed in a freshwater mussel of the genus *Dipsas*, as a consequence of an artificial stimulation. The producer has chosen a certain number of fine specimens; introduced, between their shell and the mantle, porcelain figurines prepared beforehand; then he has gone away and left to the natural function the job of completing the work. It has not failed to do its task. When he comes back, he fishes out the shells, empties them, and finds the figurines as he wished to find them, fixed to the valve, immured in mother-of-pearl so fine that all the details of the figurine appear through it. The fish's tomb is here a simple covering, for Nature, not making a distinction, acts in the same way in all cases.

But the story continues beyond the nacreous pearls until we come to fine pearls, to the real pearls. These are much rarer, and instead of being attached to the inside of the valves, occur free in the living tissues like concretions or cysts, not attached in any way. Hence their regular shape, most often spherical, and the pure smoothness of their surface. They are made of fine concentric layers, and the light, playing in them with more splendour, gives them that peculiar lustre which is the source of their charm and value. But in origin they are not the least different from the others. There is a little kernel at the centre of each, sometimes microscopic, whose presence has set the stimulus at work. After the secretion has been set going, successive enveloping, concentric layers have been deposited and the size of the concretion has increased, until we have a round, regular pearl, with all its charm.

This kernel, the initial cause, is any little thing which has passed through the superficial layer of the tegument and penetrated the living tissues, carrying along with it some elements of the layer, and the pearl is the result. The living tissues have secreted calcareous matter as they always do; then the local irritation has facilitated the formation of the initial cyst; finally, the remainder has been deposited, layer by layer, thickening the envelope thus brought into existence.

This object from which the first impulse comes, varies in different cases. It may originate in the creature itself, some tiny fragment which has accidentally become detached from the outer part of the body, and pushed inside it. It may have been introduced by the hand of man, as in the case of artificial pearls. Most often, it is a tiny creature, a larva of a parasitic worm, which, after attaching itself in the first place to the tegument, perforates it, and passes into the nearby tissues. Then the mollusc reacts in its peculiar way, struggling against the intruder. Most animals, in such circumstances, and when they have the power, make use of their phagocytes, their lymphatic cells, which attack the intruder and gradually destroy it. The mollusc uses a different method. Instead of breaking up and annihilating its enemy piecemeal, it surrounds and immures it. On a smaller scale, and in another way, it recapitulates the episode of the mummified fish and its tomb, or that of the petrified figurines. In its very flesh, it surrounds the parasite with a layer which cuts it off, thickens it, increases it, continually depositing new layers, and ending by turning into a gem what was originally a tiny coffin.

A Persian poet, attributing a celestial origin to pearls, declared that they were born of dew-drops which fell from the air and were solidified. In his mind, the resemblance, the similitude of flashings and sparkle, justified the attribution of such an origin. In Nature, the reality is deeper and even more poetical. The iridescent pearl, the burial place of an invisible body, brings together in itself the principal aspects of the problems of life and death. It begins by destruction, is followed by an intense degree of vital activity, and ends by creating beauty.

The great collar of pearls, an ornament of high value, with its ropes of glorious gems, is a collection of tiny sepulchres, woven into a splendid set of jewels. Falling over a silken dress, or over beautiful shoulders, its exquisite gleam of pure lustre, at once translucid and brilliant, illuminating things both in appearance and in depth, gives each single gem its true value. Death is a passing accident, a personal affair. Life goes beyond it, utilizes it, employs it for its own purpose, that it may itself continue. The pearl is built around a particle of flesh about to decompose; many a gay insect comes from a chrysalis that developed in foulness and decay; the most glorious flowers bloom beside a tomb; the richest harvest ripens on the battlefield. Life uses everything, even death.

The story of the pearl reminds us of the tomb of pearl, and of the *Fierasfer* which is buried in it. This genus, which contains a fairly large number of species, is found in warm and temperate seas. All its members have the same appearance, the same customs. They avoid the light, try to find dark corners, bury themselves in the sand or hide in the crevices of rocks, but especially they prefer the shelter of a living body, provided always that they are assured of easy access. There they find the shade of which they are so fond, the protection which defends them, and, in addition, to excite their own vitality, contact with another vitality functioning at full strength. One species found in the Mediterranean exhibits this characteristic in an astonishing degree.

This is the Needle Fierasfer, *Fierasfer acus*. It is small, seldom exceeding five or six inches in length, and, in breadth, only a fraction of an inch. Its long, almost threadlike body, slightly flattened, ends in a short, round head and a pointed tail. Almost all along its length, its belly carries a longitudinal median fin, like a thin, soft blade. Its skin has no scales, and is smooth and occasionally bright. It justifies its generic name Fierasfer which Cuvier adapted from

a Greek word given to objects with a polished surface. It is so transparent that all its main internal organs can be seen. Everything goes
to make the creature seem delicate, and this especially in the young,
which are practically translucent. They have above them a long nod-
ular filament, spread out like a crest, which later falls off and is not
found in the adult.

The American *Fierasfer* sometimes shares the cavity of the mantle
of the pearl oyster, but the European species, still more strangely,
lives after the same manner in the intestine of the sea cucumber,
thus becoming what is practically a parasite. It is, in fact, an intestinal
fish, as the tapeworm is an intestinal worm. But there is an important
difference; the worm supports itself at the expense of its host, it lives
and has its abode in this way; it feeds upon food which it takes from
its host, whereas the fish only accepts shelter, for it comes out at
intervals to seek food for itself, and then returns to its accustomed
place. Circumstances, that is to say the arrangement of the organs in
the sea cucumber, allow of such strange comings and goings, which
could not happen anywhere else.

The sea cucumbers are long-bodied Echinoderms, with thick skins,
roughened by the possession in their substance of a number of calcareous concretions. They have not carried this to the extent of other
Echinoderms, notably the sea-urchin, which has a regular suit of
armour surrounding its body. Their genera and species, in considerable numbers, populate the waters of the Mediterranean, where,
because of their weight, they live upon the bottom. Some of the sea
cucumbers, belonging to the genera *Holothuria* and *Stichopus,* are
frequent near the shore, and the fishermen often bring them up,
Holothuria like large black-puddings, *Stichopus* like broad, thick,
pink tongues. Sometimes some of them contain Fierasfers, especially
Holothuria tublosa.

In the sea cucumbers, the digestive tube has two openings, one
for each of the extremities of the body, both spacious in the well-
developed animal; the mouth, surrounded by a crown of tentacles,
the anus, which is bare. The mouth gives access to a long convoluted
intestine, which, before it ends in the anus, broadens out into a large
rectal swelling. Appended to this enlargement, and opening into it,
are two spacious tubular organs which look like hollow, tree-like
growths with very thin walls. Because of this appearance they are
sometimes called the Holothurian's "lungs," for they are used for
breathing. It would be better to call them "gills," since the Holothurians live in water and have to breathe in it. Their method of

working can be judged by the mode of their construction. The skin of the Holothurians is not of much use for breathing because it is so thick. So is the intestine, which is filled with sand and little pebbles which the creature swallows on the bottom, leaving to its digestive wall the function of picking out and absorbing the little fragments of food mingled with this debris. The only parts left free are the rectal swelling, with the organs appended to it, and the anus, which is fully open, allowing the outside water free access to the interior. In this way respiration is effected, and the anal orifice adds this new role — of breathing and life-giving — to the office it ordinarily performs.

Fierasfer takes advantage of this. Like its fellow of the pearly oysters, it seeks, besides shelter, protection and contact with living flesh, with the currents of fresh water which facilitate its breathing. If it cannot find anything else, it will live in a hole in the rock, a depression in the sand, an empty shell; but here, in this body full of life, it exalts its own life, and takes full advantage of all the well-being it is capable of experiencing. It settles down in the spacious rectal cavity, curls up there, and so sheltered, takes the fullest advantage of the dwelling to which it has invited itself, and to which its host has welcomed it.

This is a strange enough affair in itself, but less so than the manner in which the agreement is compounded. These Holothurians, in spite of their apparent torpor and inertia, are sensitive, ticklish, and easily upset, and they show this in an alarming manner, breaking off and ejecting their whole intestine when picked up. The lodger must exercise extreme prudence and make his contact with the utmost care, if he is to avoid such consequences. Yet there is no difficulty in coming to an arrangement. The fish approaches when the anus opens completely and remains wide open. It begins by introducing its small pointed tail into the orifice; then, quickly but softly, slipping in backwards, it takes in the rest of its body, including the head, until, finally, it disappears entirely. It is so delicate in its movements and gets on so well with its host that the latter, apparently without raising any objection, sometimes allows its guest, now become its lodger, to put its head out by the opening and survey the neighbourhood, to go out and draw back as it would from a hole with inert walls. Having given its guest permission to lodge there, having allowed it to take up all the room it needs, it continues to permit it to make itself at home, like a good landlord, with the extraordinary peculiarity that the lodging itself, situated in the very

353

body of the landlord, is in the rectum. Nature often surprises us by its arrangements, but it would be difficult to discover a device more astonishing than this particular one.

We may regard it with astonishment, and indeed so we should, but we ought further to consider it in its true light, and see what it actually implies. It is eminently significant, and as full of suggestion as may be. It marks a definite level from which we may measure others. If it were anything less, we might see in it nothing but an ordinary association, such as we find in those species of fish which frequently shelter themselves beneath a piece of floating wreckage or under the parasol-shaped body of a jelly-fish. If it were anything more, it would take advantage of the situation to feed the guest at the expense of the host, and become a complete parasite, like the intestinal worm. But in its case, we have something betwixt and between, and when we put it in its proper place in the series which begins with the pearl oyster, it helps us to see things as a whole.

Nature, in its immensity and diversity, rearing its individuals in the mass, distributes to all the means to live. In the first place, it procures these means of itself; but, more than this, it brings together as best it can those creatures to which it gives life, and makes them help one another; it compels them to aid one another. If, sometimes, the form this assistance takes surprises us, its significance is the one thing to bear in mind. Nature makes use of everything that it contains. It uses the flesh of one to form that of another; it makes use of death itself to take this maintenance to its ultimate limit; it brings life itself to contribute to the better securing of its action. The whole is one vast symbiosis. Living Nature constrains its creatures to observe that mutual help which it institutes without respite, to adapt themselves to the regulating laws which enable it to establish that mutual aid. Its tireless, never-ceasing labours proceed always according to a governed plan.

GUSTAV ECKSTEIN (1890–   )

Two Lives

From *Lives.* 1932

[DR. ECKSTEIN might well have been one of the great translators of zoology into literature had his inclinations not lain in other directions. I was not able to resist his essay on white rats, although his observations were made indoors, on semi-domesticated animals, with no special technical aim in view. The facts are true, therefore scientific, but you do not think of this when you begin to feel the charm of his writing.]

I HAD GOT TO BE a doctor, a man of science, and took a tiny creature, a thing so small it sat with comfort in the palm of my hand, and cut into its skull and removed a tip of its brain. Science has not got much by that, but possibly a few rats have, for he taught me, that little white rat, and I have changed my mind about many things.

The little white rat survived my cunning. There was no mutilation of any function that is commonly said to lodge in the brain. His thought was clear, his spirit brave, he could guide his body, and his length of life seemed even increased, for he reached what in our terms is a hundred years.

The moon tonight is full and flooding through the window. He runs from his chamber — a box that I have set at one end of my rolltop desk — to his granary, a drawer on the other side and below. Back and forth, back and forth. He has been running that way for a month of nights. In the day he sleeps, only with the darkening opens his amber nervous eyes, casts about him, wonders what he missed while away, then yawns, a mighty yawn, and scratches like a mountebank by the side of his ear, and scrubs his face. Scrubs rather his head, the whole of it, uses both hands and the lengths of his arms. And now he cleans his tail, cleans that particularly, knowing that never a healthy rat but has a clean one. Then back and forth, and back and forth, and back and forth again. Suddenly he stops, just in the middle of the top of the desk, and one would say he was porcelain did he not sway and lean far out. It is the moon. He is bathed in the moon's flood. He is struck. He is queer.

She I bring him is a tiny thing. In the half-dark she trembles like a patch of that very moonlight. She rushes, in these first hours, explores all this new, is pleased, is pleased, but vouchsafes him no solitary glance. He is gone to his corner. He watches steadily from there.

I had always said he might have the top of the desk and the upper of the three drawers, and he had always kept to that, but she now lives everywhere, bites through the back of the upper drawer, lets herself hang and drop, and by that strategy coming thus from the rear, is in possession of all, immediately bringing her belongings — one leap from the top to where I drive my distracted pen, another leap to the drawer, and thereafter subterranean grumblings and thuds and perturbations.

He cannot understand it — this fine slender woman, that she should be so material. What can she think of doing with it all? What does she dream?

I cannot understand either — his bedding, his food, my pencil, my pens, the cork of my ink bottle, the eraser with the chewable rubber at one end and the chewable tuft of brush at the other. Hour after hour diagonally across my work she goes, head held high to keep what she carries above her flying feet. In human distances it must be twenty miles. Certainly more than the tiny burnings of that tiny body drive that machine. Only late does she cease. She looks where he huddles uncovered in his chamber. She seems to think him over. She comes to a conclusion. She waddles toward him, settles into him, drops her head, is ostensibly asleep. He cannot sleep. He does not even close his eyes, squats there motionless, almost breathless, is afraid he may disturb her. I pick up the few gnawed bits of my belongings, turn out the lights.

Three weeks ago was the wedding. So soon as I arrived this morning she made me comprehend it was newspaper she wanted. I brought her a newspaper. She put her foot on it, as if to establish possession, then looked at me hard.

"A single newspaper?"

I brought her an armful.

All day she stuck to the job, did not eat, did not drink. I placed food and drink before her. She ran round them. When I persisted she ran through them, trailed them. The paper she tore into strips, leaped with the strips to the drawer, there continued the tearing, each strip into squares. By noon an inch of squares bedded the bottom of the drawer. By evening, three inches. By midnight, five. And now, shortly after, she is ready to rest. Still she has not eaten. I ask her again. She only turns her head.

Poor husband now and then has tried to tear a little paper too, but it was not in his character, a big bulging character. This new young wife has made great changes in our ways, his and mine. A hush lies over the establishment. I sit before my desk, but do not work. She sits by my side. She is thinking her thoughts. And so is he. And so am I.

Next morning the mother and father are moving about. Mother is thin. They are thirteen, if I count them right, and they wiggle and worm and topple and tumble. She will not eat even now, and he does not find it easy, either. He is bewildered. How can anything like that have happened? I try to explain to him, but I do not rightly understand, myself, and he climbs heavily out of the drawer, and on my arm, and in my pocket hides his confused head. By evening, however, he has talked it over with her, and she has told him something I could not, and in consequence he is licking her, and she is licking them, and he is so interested in what he is doing that he steps all over them, and they squeak and step over one another, the smallest the most stepped on. No one in the heap seems able to take in the whole of the heap. That is somehow sad.

Three out of the litter I intended to leave little mother, but she would have a big family or none. At least she was indifferent to three. Two I found when I came on the fourth morning, dead. The third I never found, though, fearing it might have got into difficulty, I looked under every square of paper. I regret they are gone. They had the color and somewhat the form of the fingers of the new-born baby. The little pink legs were so weak that they dragged, and the little pink tails dragged symmetrically between them. Boneless they seemed, and sightless they were, and they kept up an aimless motion.

Mother appears unconcerned about what has happened, but I am not sure, for mother is hidden deep under her white hair, and perhaps is hidden deeper even than that. At any rate, father, who knows her better than I, is more solicitous today than usual. He picks her vermin with a more insistent care. She lies there very flat, spreads the maximum of surface to the smooth cool table below, and the maximum of surface to great father above. The exertion makes him pant. As to his feelings about the babies, I believe they were mixed. Thirteen was many, and though they were lovely, it must be admitted they were restless.

Both are asleep. I reach into the chamber to pet the back of mother's heaving neck. She starts. She bites me. It is not much of a bite and she is grieved. She probably thought that 'I was coming for one of the brood. She glances nervously about, finds it hard to recollect.

On shipboard I met a man from Guernsey. We talked things easy to talk, things near our hearts. I told of my rats. Then he told how the level of the olive oil dropped day after day in the thin tall bottle on the second shelf from the top in his pantry in the house in Guernsey. Oil does not dry at such a rate. So, being a philosopher, he sat himself quietly before the pantry and stayed the afternoon. With dark she came, a great gray one, scaled the highest shelf, waddled to a point directly above the bottle, studiously inserted her tail, studiously drew it forth, and licked it clean, and left by the road she came. The gentleman from Guernsey attempted to meet this ingenuity with a trap, but her interest was in oil and, that gone, she returned no more.

Whereupon I told of the skimming of the milk. I was only a boy, alone in the kitchen, and my poor mother a woman so clean, so clean, that had she known she would in the dark have slipped to some distant apothecary to purchase the shameful poison. This one also was great and gray, also knew where she was going. One spring to the back of the chair. One spring to the middle of the table. And there, set there every afternoon to cool, stood a flat dish of milk. Carefully she swung her stern, carefully fitted it to the rim, and, a single sweep, the job was done.

These tails, that look so rigid and are so skillful, much could be written about them. But no one would read. "If it were not for their tails." How often have I heard a growing interest cut short with that.

To be sure, the two rats of whom I have just been telling were hungry rats, but a rat will steal when not hungry. Mother rat will steal from my very hand, will leap out of a sound sleep to snatch a bit of chocolate from under father's nose, and rush with it to a place of hiding. One morning, finding a loaf of bread where she commonly finds a slice, she tried at once to steal the loaf, but it was ten times her size and she lacked the strength. Visibly all curiosity left her. Not able to steal it and store it, what interest in bread?

Father, on the contrary, never steals and never stores, eats from my hand as a dog eats. For certain foods he will tussle when mother tries to take them out of his very mouth, but it is always as if he had forgotten himself. Banana, however, it is not easy to yield. He has a great partiality for banana, and there is no fairness, she not having eaten her own, having hid it away, and having immediately and confidently ambled after his. Once I saw him make straight into his straw, bore his way in, let the banana mash about his face and, though she pushed him on one side, then pushed him on the other, and sought to drive her thin nose alongside his broad nose, he was stubborn. Only after a long time did he back out, looking very comical and feeling very ashamed.

I divide a peanut, call them, and they come pressed one against the other, he so big, she so slight, settle on their haunches, take the separate halves in their hands, and drive their tiny jaws with a speed that makes one think of their tiny hearts.

Toward ten every evening the two take turns to bathe. I have fitted a board across the basin under the tap where the water drips one drop at a time. To be wet all over makes them very weak and very unhappy, but to catch one drop, and wash vigorously with that, and then catch another, that is different. I myself look forward to it — to see the way they rise from the board, put out those marvelous hands, and wait for the drop.

Father this morning is lying on his side, his two hands folded just under his nose, as if he had fallen asleep in prayer. Father's sleep is pictured with dreams. I can tell by the way he waves the tip of his tail, and when the dream gets too vivid he turns, settles on his belly, sidles over to where mother is sleeping on her belly. Then he scratches his head. A little later he scratches his head again. This time he wakes sufficiently to realize that though he is scratching his head he feels nothing. Promptly he scratches again, and still feels nothing. It is a condition so peculiar that it breaks through his sleep. He opens one eye, not far, but far enough to make out what has happened, for he is scratching not his head but hers, she having pushed hers under his neck and brought it out just inside his right hand. The discovery does not anger him. It does not even surprise him. Gently he puts her head aside, and scratches his own.

It has come of a sudden, as it does — father's aging. It is all in the last weeks. He is thin and walks cautiously and tries to show in his lettuce a pleasure he no longer feels. He is so shaky it worries me when he leaves his chamber because he cannot sleep, sits at the edge of the desk, the better in that posture to fill his hungry lungs. He has fallen several times in the past, and it might be bad with him if he fell now. The two behave very differently when they fall. Father stays where he lands, knows that if I am out of the room, sooner or later I must return, and when I do he had better be where I shall see him, or where, if I do not see him, he can nudge my ankles to remind me. And when I am reminded he is as pleased as a puppy for the way I sympathize in his misfortune. But when mother falls she hastens at once to the most cornery corner. She wants to be found, there is no question, and when I reach her is most relieved, and yet has let me beg her for hours to come out from next a water-pipe.

Tonight it is father sits in the precarious place. He looks tottering, and he frightens me. I mention his name. He moves his tail that is hanging over the edge of the precipice. I mention his name again. He moves his tail faster. He comes to my side of the desk, stands on his two feet like a diminutive polar bear, his beads of eyes trying ever so helplessly to find where I am.

About two o'clock mother rouses me. Two o'clock in the morning. There is a tempest in her chamber. I must come at once and see, and she mounts high on her toes to be sure that I do see. Her drinking-water, I have put it inside her chamber instead of out. Her drinking-water she does not want inside her chamber. Deliberately she tumbles it over her bedding. And now she is enraged also at the wetness of the bedding. She wants the empty dish out. She wants the wet bedding out. I drudge like a scolded maid.

Father grows older and older. Then one evening he leaves drawers and desk top. He goes to be an eagle. At least he goes to be an eagle if it is true that yearning has its way. His gaze was always at the edges of his universe.

I am filled with the pain of the shortness of everything. That is a common pain. But it is freshened by the shortness of this little life. His great events were a thirtieth the length of my great events, yet they make mine seem not long, but brief. When I saw his death coming, how truly frightful was the feeling that nothing could stop it. And that also is a common feeling. But this life lay right there in my hand and made my helplessness seem so much more helpless. Good care and good food and warmth would save him an uneasy week, perhaps, and were I able to add all the cunning in the world it would save him another week perhaps. How then must I know with a new strong draft of conviction that gentleness and gaiety are the best of life.

He knew that. He knew how to be affectionate to his friends. To mother he yielded not only what she needed but what she wanted, and what she did not want, what the sweet and lavish extravagance of her youth and sex made her wish only to cast to the winds, cast off the precipice into the dark empty spaces of the universe.

Every night the last months he and I used to play a game. He had too heavy a body, and his legs were too short, and where he walked he rubbed the earth, so it was difficult to pretend to flee ahead of my hand, then abruptly in the midst of that flight to rear on his hind legs, give the length of his body an exaggerated shiver, as in some barbaric dance, and then continue to flee. Yet that was the game.

I think of that now. I think as one does of everything, of the night he made it plain he needed a wife, and how she nevertheless confounded him when she came, and of the litters that passed one by one, and the signs of maturity, how they passed, and the signs of age, and all in my hand, he learning every day to be less a rat, excellent though that is, learning to be more and more thoughtful, and then the final sharpness, how he mastered even that, grew gentler and gentler, and one night went to be an eagle.

Poor little mother! Babies gone and father gone. I describe how it is with father this morning, how he is off to the Peruvian mountains, and how, a short distance below the highest peak, where there is a good shelter against the blasts of the south, under great wings, nudging his brothers and sisters, he is beginning again, is waiting, though perhaps he knows it not, for little mother.

I put my hand into the dark of the drawer, and she pretends my fingers are the whilom family. She scrubs them roughly one by one. She crawls under them, crawls over them, goes round the nails and up into the crotches. She scrubs them roughly, and when each is done bites it, bumps it aside. I talk to her. She answers out of the dark. Father never would use his voice. Hers is a kind of cluck, and after she has spoken she is quiet. And I am quiet. Each of us has it in mind to wait on what the other will do. But she never can wait, must at least turn round, shove her little self through a quarter of a circle, then fix me with one great glowing eye. What she sees of me with that eye I have no notion. Nor have I any notion of what she makes of what she sees, more than that it is an embodiment with which in her loneliness she finds it possible to commune.

I have talked of my rats to many a person. Some have entered smilingly into my feeling. Some have bantered. Some have doubted me, have thought I was bantering them. There was one, an oldish fellow, whose seriousness was greater even than my own. But I must hereafter be very careful what I say. I was telling a very clever lady about putting my hand into the drawer, about little mother holding the hand for a quarter of an hour, then something made me stop, something penetrating in the lady's face. The lady was reading me. I felt it — the way one feels when one tells certain people one's dreams.

Kind lady, now that I need not at the same time look into your canny face, let me add a note to your picture. Let me say that all the while it was teaching me I also was teaching my family what I imagine may be useful to eagles. I was teaching it how to be fond. You may think that all that that requires is to throw a piece of cheese, as one throws a dog a bone, or a man gold. No, at least that is not the way I did it. I taught fondness by being fond. And, when I consider, I was able to teach father with very little, a touch as I happened near his box, a touch as he passed me on his walk to the quarters below. Only in the last months it got to be more. His breathing then was not always easy, and that brought fears, and we sometimes must talk long talks to get over the fears. With mother the greater intimacy came after he died. She was in such plain need, every night would sit beside me there in her drawer, immersed in a world that seemed so limited till each time I would remember afresh how limited my own. After all, what thoughts might she not be thinking? Not thoughts like mine, not severe logical incrustations like mine, but the quality I was sure was charming. I was sure because, like father, little mother had ceased in such an extraordinary way to want. Exactly as she received affection she could do without things. Father, in fact, was only an ordinary rat with the interests of an ordinary rat, in cheese and vermin and women, and then I came. And when he died I was sorry for little mother, and being sorry proved, as so often, the prod to the backward affections, and she too gave up being merely a rat, and one day soared away.

GERALD HEARD (1889–   )

The Emergence of the Half-Men

From *The Emergence of Man.* 1932

[MANY ATTEMPTS have been made to visualize realistically some of the primitive activities of the slow-dawning super-ape intellect of our far-distant ancestors. Here is one that seems to me to be conceived as well within the ape man's consciousness, and with numerous small unexpected details which catch our sophisticated modern minds off guard. It is a happy combination of what we know to be chimpanzee and gorilla mental end-products, together with inchoate beginnings of humanity.

Mr. Heard is an English author, most of whose writings have been along the lines of *The Emergence of Man,* the title of the book from which this selection is taken.]

THE CLIMATE is on the whole mild, and when the winter comes it is little more than tonic. Life is happy enough. The parkland stretches for these simple folk to the world's end. The world's centre is this noble Tree. The sun rises on that primal human day, and round the Tree's base the males — who have slept huddled in the lowest crutches of its span — stretch themselves, yawn, clamber down and call. Immediately above in the branches a score of cries reply. All sorts of sounds are bandied. The squealing and shouting takes on the rhythm of a simple chorus. Those on the ground interrogate. Those up aloft answer. It is like an orchestra tuning up. It is indeed. It is the great orchestra of human speech being practised and its technique mastered against those works which, aeons after, are to be composed. Now the pack is shouting, laughing and yodelling from pure virtuosity. It is a delight to hum and howl, running up and down the scale, as the friendly stimulating voices of all the others blend, diverge, glance on an unexpected harmony and break into happy pandemonium. The group's activity increases. Arms and legs swing to the noise. A junior suddenly swings right out of the Tree plumb into the middle of the chanting, swaying elders below. There is a dart of hands, but off he goes round the great bole with squeals ot apprehensive challenge. The Tree rains its living fruit. The pack gambols round its sanctuary. They wrestle, they roll, they chant. Voice and limb are wreathed. The rhythmic game, like a current, tumbles them. They are bathed and refreshed in the group gymnastic. Suddenly from the brake, not further than the great Tree's shadow stretches just before sleep, there is a sound . . . not loud, but none the better for that. Out of the merry commotion of the maypole dance raises himself the bravest male. The rest remain stilled, looking at him and then swinging eyes toward where he stares. His nose is thrust out. His foot rises; he stamps. He is shaken with anger that the fun should be broken in upon. "Ugh!" he cries. "Ugh!" The hatred and disgust in his voice electrifies his hearers. He is brave and defiant, but, through his tone, they realise the horror he is diagnosing. He barks out his defiance again against the mysterious bush, and at that all save his closest pals swarm up into the branches. Warily he watches their retreat, all the while keeping his face to the danger. They did not take to the Tree empty-handed. Each as he swung to safety held a stone in his mouth or fist. The last up, with a snort the hero and his comrades also follow. The Tree becomes as still as though deserted. Then the bush moved and into the dappled sunlight came out, sniffing the air, a full-grown lion. Catlike his nose dipped to the ground as he padded up to the Tree. There about the roots he drew long breaths. Appetite blazed. He lifted his head to the branches, stirred his great loins uneasily, and the sense of bafflement broke from him in a roar..The roar galvanised the Tree. It broke into a frenzy of counter abuse. The lion roared contemptuously back, but suddenly his contemptuous anger was shot with pain. One of the great fangs thrown exposed was struck and broken. Before he could wince, a second stone struck sharply and deep, cutting open his left eyeball. Stones drummed on snout, on ears, on skull, on loins. Half blinded, in keen pain, a new sensation took the beast — Fear. What was in the Tree that struck so swiftly and so hard and yet was not in his claws' reach? With a bound he was clear of the Tree's baleful purlieus. He stood by the brake looking with what vision he still had at the looming object, sniffing with bruised snout that peculiar smell. Food, no doubt, but somehow ill food. He shrank away, licking the blood that trickled in his mouth, blinking the eye that no longer saw.

A shout broke from the Tree. A moment after the maypole ring reformed. Round the great bole they gambolled. A piece of creeper hung around the brave one's neck. As it shifted, hands replaced it. He coiled it himself. They all noticed it. Stones, the good stones, were passed from hand to hand, were licked, mouthed and cuddled. The dear Tree, it, too, was caressed. Tower of strength. How it uplifted and protected. Was it not theirs, was it not greater than everything else? The exultation began to slacken. Why, no one had had anything yet to eat. One of the brave ones stood up, stone in hand, others joined him. Off they went over the open heath — giving wide berth to thickets. The sun was up above now and the Tree's shade pleasant for those that kept home. Every now and then one would run down to the pool that glittered and take his drink and so back to the Tree, there to play and groom and exercise, to wrestle and yodel and wreathe the creepers into knots and plaits. Yes, life was good. The group was healthy. Three nights ago an old one had slipped and fallen out of the Tree. The next morning it was still at the foot dragging a broken leg and whimpering. As they looked down on it, suddenly they felt it had become something disgusting, hostile. Some one ran to the deep cleft where a store of stones was always heaped and, picking one, flung it at the monster. As the stone struck, it yelped. At once half a dozen more struck it. It broke away stumbling toward the bush. The rest howled their excommunication of it. The next day as they hunted they came on scraps of its carcase. A lion had eaten it. But looking at it they no longer felt disgust or even interest. Would the Tree feel love, interest or disgust as it looked down on the leaves it had shed? . . .

So as we watch, there still stands the great Tree, centre, shelter and support of the groups' life, armature around which this the first society is sustained. But, should we slow down our cinematograph projector to a speed which no longer shows man growing as a species, we should see that the tree is in detail not the same tree. Sometimes we see it is an oak, sometimes a sycamore, sometimes a beech. Go still slower, the oak is that same oak that we first saw. Twenty such had risen and fallen since the sub-human group loved its tree. Suns flicker past as sunlight dapples on a forest floor. Seasons, as cloud shadows in April, chase across the downs. One day, as the group sat about its bole, its nostrils full of rich safe scents, a tang struck one nose as today a live wire can sting an unwary hand. The hunting had been good. For the trial men had managed to obey life and eat as widely as might be. They had eaten well of a large gathering of roots and berries and succulent fungus, but from their sharp teeth they were picking fragments of flesh, the children were still gnawing small bones and the mothers were giggling as they placed fragments of pelt sometimes on their heads or on their necks. Aeons before, the rabbit that bolted from under the ape man's foot was snatched up and killed in a moment. The hawk and the jackal ate with relish. The ape man tasted. It was tasty, the warm flesh, and it was invigorating. They that ate animals thrived, so the half-men all sat round the bole in vigorous calm. Then came again that tang. One cried out at the shock of it. At once all, even those who had not been struck, howled with horror. Then the sharp one, he who used both to smell and also to look, he pointed and in a moment they all saw as clearly as he, that face looking at them. It wasn't panic they felt. This was no roarer or render. This was something simply unnatural. People didn't rush into the Tree. Rather they stood looking at it with fascinated disgust. In the silence a young one suddenly vomited. With a bound the leaders were after the thing. It shambled off, but they were too fast for it. The Dryopithecus was surrounded. It showed its teeth and roared its desperate defiance. They closed in on it like inquisitors on an unnatural sinner. Some one flung a stone that, hitting the beast on the neck as it drew itself up, disturbed its balance. At once they were on it. Twice it bit hard but clumsily. They left it the moment it was still, walking slowly off and now and then looking back. Before night and the hyenas came, the men of the Tree heard a whimper and saw reddish forms moping and moving where they had left the body. Every one snarled, but stayed by the Tree. The red forms disappeared soon. After that no one again saw a great ape. The fear of the cunning fury of the Tree-men had spread and the apes gave them wider and wider berth, letting their best feeding-grounds be trespassed and annexed and going lamely to seek others.

The next incident was graver. Again that tang of smell; again the sense of something unnatural, but this time not one shape but several. The Tree group gathered itself to launch against the intrusion but hesitated. As it delayed, out into the full sunlight rolled an unbelievable creature. One of themselves, but, no, it couldn't be. It was horribly close, a forgery on their confidence and sense of right. And the effrontery to come in this way toward their Tree! The confused feelings became organised into steady anger. And then the one became several. They stood looking steadfastly at the Tree. No one made a sound. Then, without a sign, they wheeled and ambled off. The Tree's people followed, at a safe distance. Every now and then the intruders stopped and looked back. Then stopped the Tree's people. Finally when the Tree was getting unpleasantly far away, one of its people flung a stone at the strangers. It went well, striking a laggard on the nape. He howled and ran up to the rest and they, catching his pain and fear, swung off. The Tree's men began running, too, till one of the intruders turned and himself flung clumsily a small stone. As it bounded toward the Tree-men they stopped and looked at it; one bent and smelt its horrid smell with misgiving. The pull of the Tree became stronger. The Tree group recoiled back into it.

Several such passes between the other trial men took place. Every few centuries one type overflowed into the country of another type. The boundaries fluctuated, for already they had boundaries. Their life focused round the Tree, but beyond that spot light or *fovea* there was a circle of country-side — their country, known and lived upon and in which they felt a peculiar concern. But the raids of other species of man were rare and grew rarer. That did not, however, mean that the Tree fell into fat lethargy. The slinking snarlers were less troublesome. More and more they gave way. They recognised this odd being was food, but a more dangerous food than any other. He stung at a distance, further than a spring, and he was pertinacious, a creature like a giant swarm of hornets. Better give him a wide berth. Spring on a laggard if you could, but his Tree — that was dangerous ground. So the young had fewer nightmares. Those green slit eyes suddenly aglare, level with you on the branch — that was a horror less and less often stamped into mind. But the snakes were stupider than the cats and learnt more slowly that the young of the Tree-men were costlier prey than the young of the monkeys. So, still, the young knew, too often to forget, the sudden coming to life of a small branch, its hiss, the stroke and the helpless fall. Fewer and fewer dreamed of cats and shuddered at their smell, but still many in the night howled because of a swift rustle, awoke screaming and clutching their mothers because the safe Tree seemed to have let them drop into the gulf below. Yet fat lethargy would have come had not success brought its own problems. The Tree people's prosperity overloaded the Tree. The pressure made random gambollings become savage squabbles. Sides were taken — the single-celled society was about to divide. One day after a savage brawl as they drew apart three of them lay dead. The stronger group had held near to the Tree. The smaller looked across the dead at those already clambering up into the branches. Suddenly the Tree seemed alien. One or two whimpered, one actually touched a corpse that lay near and moaned. Then one of the stronger who had fought fiercely, shouted, shook his arm at the Tree. The rest gathered round him. He threw a stone at the Tree. It hit the trunk and bounded among the roots. The Tree was silent. It was not their Tree, they became sure. The leader wheeled and they followed. Now the Tree seemed something to get away from. They went on till everything they associated with it was gone. They went out of its country till they saw no more any landmark that they knew. Then as the dusk came on they smelt for water and found a stream. They drank, gathered food, and that night slept on a ledge half-way up a rocky bluff against which the stream bent. The next day they went on their march until about noon they saw a Tree like their Tree, yet other, there, standing up to welcome them. They came home to it with a sense of return and yet renewal. How friendly, how ample, how much for them it was.

So diffusion continued. Sometimes there were throwbacks. One Tree, as it were, fell back into the orbit of another Tree, frontiers were infringed and there was random fighting, but on the whole the urge was centrifugal. It was grand to find a new Tree to claim as one's own when the old Tree had grown unfriendly. So must Archanthropos have pushed out, and before him moved away Pithecanthropus, Eoanthropos, and the others. He did not exterminate them. Nature is no narrow rationalist; she does not push things to conclusions. In chess it is not necessary to take many pieces to win the game. In Natural Selection one species of wild animal forces out another, not by direct action but by indirect pressure. The marsupial wolf, that now hangs on in Tasmania, once was spread over Australia. Then down came from the north the mammal wild dog, the dingo. Two types each suited to the same life could not last side by side. The more efficient took the land; needing the same food, the dog found it more efficiently, reared its young better, and the less efficient passed away. So with the trial men. As Archanthropos spread, the other sub-men gave way before him, left' him their choice lands in which they had thriven, withdrew to poor districts and misery spots where they pined and finally died out.

Then in the summer climate a cloud appeared that did not go away. The slow exposure camera (which we are using on this tract of history so as to record only the permanent changes) begins to show the Tree less creeper hung, the Tree itself more gaunt and the pool has a glaze on it, only a hint of ripple. The brown-green of the ground is greyer as into the web of the days more and more snow-white ones are blended. Then, as grey becomes the dominant colour, the fun goes out of life. The Earth is hard, vegetation spare and stringy, animals savage and scarce. The apes and the last of the ape men lumber off. They find gradually that things are still easy in one direction. They gravitate south. Eden is re-won. There the trees are more bountiful than ever. True, the ground is even more dangerous, and so they hang more and more up in the branches. The hand that was opening to grasp the world, and to untie its close-wound knots, shrinks again. Only one thing is required of it — not every possible thing. For a happy quiet life it can best serve by becoming a hook, and so the hands of the apes have become increasingly more hook-like than they were. The ape has retreated geographically, and that retreat was confirmed by a physical decline. Then it was that Archanthropos stood his ground. The apes fell from being of Life's blood royal. The men-like alone were left. This bleak world was ceded to them and they chose to take it. That bluff where the river turned — in its rocks were deep clefts. It looked south. At midday a little sun made the shelf a place pleasant to lie in and at night when the stream underneath was silent in frost, the first cave-men crowded into the clefts with pieces of pelt over the shoulders and round the feet. All packed together, it was possible to doze in some comfort and safety. The snakes had gone with the ape. Only large animals now remained to threaten and they could not get into these crannies. The stone of the cleft was soft; as you scratched at it it came away. Sometimes you scooped out a bit which had pushed into your back as you were crowded up against the wall. Yet the wall was queer, for suddenly from being soft and white it became just for a spot, darker and very hard. You could wriggle your finger round the hard bit and then suddenly it would plop down on to the ground near you, sitting there, looking at you like a toad. You couldn't help poking at it with your foot, but it didn't stir. You had to push it harder before it would move. So it was picked up and hit against the wall. How it liked that. It bit into the soft wall far better than any finger-nail. Great mouthfuls came away. It was fun. And then it met another toad also buried in the wall. Crack it went against it. It would dig out its companion. Bump, bump it went and the chalk flew. Then crack, it hit its fellow again. It was dark in the cleft. Even outside it was a dreary thing to call a day. But surely the other stone had spoken to his fellow. Another blow and a spark flew. The wielder dropped his nodule. That was not right; that was bad. Had the others seen? No, they were huddled by the entrance, some gnawing, some grooming, all looking out at the valley. The scooper lolloped off and sat with them. But every now and then he looked back at the nodule lying quiet on the floor. Yes, it knew something. Then he forgot. Till one night he was pushed to the same corner. The half-freed nodule pushed him on the back. He began to pull at it. It came away in his hand. The others grunted and jostled him to keep quiet. He fell to sleep with it in his lap. In the morning it was warm and familiar, and how well it came to his hand. He hit the cave side.

The nodule cut it like mud. He took it to the entrance. He noticed other nodules just sticking out round the cave's rim. He had never noticed them before. He hit one and it rolled out and down the bluff. In the light he saw no spark. But it had been angry. When he looked at his nodule's end a corner had been bitten off. It was black inside. It had a faint smell. He licked the smooth black piece, then dropped it angrily. It had bitten him. His tongue was cut. He drew away and the others stopped watching, but he still kept on coming back to it — every time he felt his sore tongue. Finally he again touched the nodule gingerly. It did not bite him. He raised it. It fitted his hand as well as ever. He swung it and the end came down on a bit of root that here thrust across the cave's threshold. He thought the stone would bump back as sticks often did when you drummed on that bit of root. It didn't, and when he looked there was a deep clean bite in the tough root. The nodule had bitten it. So, with many lapses of forgetfulness they learned to make two nodules bite each other and then they would bite roots and sinews and bones. Biting stones became things the tribe tended to make. They sang as they struck the flints together. But mostly you struck them in the open. If you struck them in the dark of the caves they looked at you for a moment as though they were going to do something worse even than bite you. At times they did bite, they bounded up and struck you. Then you left that one alone and worked with kind ones. When they grew tired of biting, you could make them get back their bite by setting them again to bite each other. Some people had pet stones — they fitted their hands so well that they would make them recover their bite again and again. But other people felt this wasn't right. Certainly if you went on trying to make a nodule bite better and better sooner or later it gleamed at you, and that of course was wrong. Every one felt sure of that, and one day it was proved. One who loved his nodule very much — it was a very fine one — was making it get back its bite. He was very cunning. He could make the other nodule only take away a tiny flake and yet after that there was the bite sharper than ever. He was doing this in the cave's mouth at the end of a fine hot day in autumn. The cave mouth was deep with whittlings from sticks other people had sharpened with their biting stones, feathers and down plucked from birds' bodies, tufts of hair from pelts and leaves and bracken which had been dragged here, to lie on in the sun. Several were watching, but no one saw the stone get angry. It must have sparked, but the first thing people knew was that the hot horror was actually in the middle of them. They saw his signal, the blue thread that can't be handled and which smells so rank. Then some one saw himself very small and red moving among the bracken bedding. With a crackle he sprang up. In panic the group rushed into the cave. The flame stood up in front of the opening. They could feel its hot breath. Once it bowed and seemed as though it was going to come in after them. They barked with terror as its choking horror swept in round them. But, a moment after, it flew back, standing again straight up. For a lifetime they seemed to be cowering in front of the fury, until at last the strain was too great. They sat still in coma. Then one at the back began to groom the shoulder nearest him. The terror lifted and they saw the flame had sunk. The sun also had set. So the burnt bedding glowed in the dusk. Then with a shift, a cataract of sparks, part of the emberage fell down the bluff. In a few moments it was possible to steal out of the cave warily, keeping face to the remaining glow. With the sunset had come the frost. It was too cold to stay out. They crept back into the cave, and there it was as though the sun was still up. They sat in the glow first silently, then chattering with pleasure. In the cool part of the ashes some one picking about found roast acorns, another a big bone split with heat. Both had an added flavour. As the glow died they crept toward it. They did not want this kindly warm thing to disappear. When they woke it was, however, gone. No one could call it back, but also no one remembered to be vexed with the sharpener. No one knew he had made fire. They had forgotten how the fire came, because from horror it had turned to comfort. Neither he nor a long line of succeeding flakers struck fire again, but the sparks no longer frightened men. Then again tinder caught. Fire was found and lost many times. At last it stayed. The flint had brought first the power to cut and shape and pierce, and then fire.

And not too soon, for still the cold grew, still the great glaciers creaked and groaned forward, gouging out the valleys, heaping their moraines on either side. The beasts that stayed grew more savage. The cave bear standing twice a man's height no more feared flint-armed man than a coal-heaver fears an urchin armed with a chestnut. But a pack of humans armed with hot brands was a serious matter. He wanted their cave for his lair, their bodies for meat, but when they sat with that sentinel at the door he slunk off. Even when he was ensconced, fire would make it possible for them to drive him out. Its smoke blinded and choked him, and when he rushed out sharp poles with glowing points struck him in eyes and nostrils, till panic-stricken, he broke from his victors a beaten monster. So their courage grew. They learned how to drive him till he fell over the bluff, his great bulk striking the rocks below, and as he lay maimed, they hurled rocks at him till one split his huge skull and he lay inert while flint choppers and knives dug him to pieces. The bear conquered, it was but a step to conquer the mammoth. When this earth-shaking beast fell down the slope men feasted till the valley stank. Then they thought of driving some laggard of the herd into a swamp. One had been found bogged and there beaten to death. The little morass was cleared and lightly covered. It became the first elephant trap.

Meanwhile, the last of the trial men had fallen out. The Neanderthal had come so far. He had found how to use fire. He had taken to caves and endured part of the awful Ice Age. He had carried on the craft of stone weapon-making until at one stroke the flint yielded straight away an edged knife which only needed finishing. The flint was now part of his life and the dead, when put away, had their flints put with them, for were these not as their hands and teeth? Was it wholly safe to use what had been their members? But Neanderthal was too conservative. He clung to his vegetarian diet. Fruit and roots were scarce. Man in an arctic climate must be a carnivore or disappear. So Life chose *Homo sapiens,* subtle, adaptable, fierce and curious, patient and clubable, a creature of mercurial energy, capable of passionate attachment and resentment. He stuck out that icy ordeal. It left its mark on him. Such a hammering made him harder. He had beaten into him virtues so hard that they had in them a high alloy of vice. He was a fiercer creature than before, more resourceful too, and yet the group had been so hammered that in spite of the units' greater cunning, they were no more differentiated from it than were the sub-men from their pack in the far distant Genial Age. But these later packs showed their advance by an intenser aggressiveness against all that came from without. Yet the energy he generated to stand against the cold did not discharge wholly in violence. Those long dark hours in the cave began to be lit by imagination. As the firelight flickered on the walls he saw, in the irregularities of the surface, glimmer out at him his brooding fancies. The most fertile-minded saw most clearly. "Don't you see the bison's snout just there? I see it clearly. Look, I'll point out where it is." As he pointed they saw, but when he ceased to show where the shadows had helped out his fancy, for them the bison vanished. "Make it again," they cried. He took a piece of charcoal-tipped stick. In the morning there was the bison's head. The next day a bison was drawn to the trap and killed. The picture had power. Who could doubt it? The best hunters before setting out asked the draughtsman, "Have you seen bison today?" "Yes, in my mind I have them." Then he would draw over again the head. It was alive under his hand. He had power to make it obey. Some one drew a spear in its side. "Good, good," they shouted. "Good hunting tomorrow." Soon they saw bison and deer in every jut and contour of the cave, and large felines and the boar. Soon such skill had the draughtsman that he no longer needed the jut of the cave wall to give him his theme. He passed from that dream stage when we take the bed curtain for a visitant and then, waking, cannot even see why its fold should have suggested a figure. He only asked a fair surface and on it he could set down each flow of the animal's outline. On bone and ivory he etched them. And as his interest grew he noticed and recorded fishes and birds. So by his game he was drawn into a larger world. As he lay in the cave mouth and drew that great folding line of the buffalo's front, the whirring of a grasshopper tickled his attention. There it sat grinding its legs. He snatched at it and it bounded. More wary, next time, he caught it clean and crunched it. It was pleasant. So he came to sketch this, too, the hors d'oeuvre of his great meat repasts.

And as the artist grew in skill so his authority rose. Must he re-create, every time bison meat was needed, a new bison? No, all he had to do was to draw one again. Make once again the magic pass and the beast was already entangled by his magic line and fore-doomed. Then the hunters could take part in the sacrament of iden-tity and after the magic-artist had made again alive the figure, they might dart at it and so ensure its downfall tomorrow.

The hunting forecast, the foretelling of a fortunate adventure, tends now to become ritualised. Inevitably such day-dreaming gravi-tated toward the true time of the dream, the evening. Then you sat round the hearth at the cave's mouth and chorused your hopes, and, as the light faded and only the flickering embers lit up the depth of the cave, you could see the great magic animals on the walls actually lunge and stagger as to your chaunt the mystery art-men ran over the great beast's outlines and brought them to life and gave them to death. When afterwards you curled up for sleep you saw the very beast fall to earth and you knew that the magic of last night had en-sured the *morte* of tomorrow. But gradually also you noticed that the great beasts drawn about the living part of the cave were not as alive as the fathers had said they were. At night they did move, but if you looked at them in the daylight, especially when the level light of the sun struck in on them, they were not real beasts, and some-times in the morning you needed to have them alive, to know that you could magic into them the dreadful strength of the beast you must meet shortly in the open. But they remained stock still, and you knew that magic was not working, and when the hunt was disastrous and the hunters did not come back you realised that the picture had failed to draw into itself the beast's strength and so inevitably the tribe had been defeated. So the pictures further back, in the deeper parts of the cave where the daylight never came, gained sanctity. Visit them at any time and raise your glimmering light and there they stood lowering at you from the walls. They were never dead or lacking in magic. And when you went to see them the mystery art-man not only could make them alive and dead as he wished, he himself could become like them. He disappeared into the dark, and the moment after there beside the living picture stood a terrible creature, half wild beast, half oneself. The picture moved on the wall but he stood out, came, making horrible sounds, toward the dedicating hunters. They fled in horror, but as they fled they felt a curious kinship had been made between them and their prey. Totemism had arisen. The artist had taken a step toward the vestmented priest. Yet it was no leap. The apes hang themselves with garlands. The ape men must have done the same with pelts and tails and crests. It was but a step for the artist to order such irrelevant additions into a masquerade and for the rest to realise that the dressed-up one was partaking of a double nature, had become a bridge between them and their prey, and was both their victim and their sustainer. So, while still in the cave mouth, humanity has arrived at the threshold of art, ritual and religion.

STEPHEN A. IONIDES (1880-)
MARGARET L. IONIDES

Looking Forward

From *Stars and Men*. 1933

[GIVEN A MAN whose life work is as a consulting mining and chemical engineer and whose favorite book is *Alice*, and the resultant *Stars and Men* is more understandable. Also the possession of a daughter with sufficient interest and ability to be co-author is another unusual ingredient in its production.

When I reached the present point in this Anthology I felt the vital need of something more than natural history and with at least four dimensions. So I picked up *Stars and Men* and found a chapter which was a counter-irritant or at least complementary to some of my introductory material, and so I have included *Looking Forward* in full.

This may be straining our definition of natural history to the breaking-point, but why not? I have always held that every rule and law in the world are made to be broken by the right man at the right time and place, and this is an excellent opportunity. Every word of the Ionides' is worth reading and rereading, and I have memorized the limericks and the quotation from *Punch*. If any reader resents the intrusion I feel like promising that "money will be refunded," as they say in the cheap clothing advertisements. Anyway, here it is.]

WE STAND TODAY in the middle of a bewildering world. Only a few years ago the progress of science seemed safe and sure. Not every problem had been solved, but every problem was on the way.to solution. Even the lay mind could understand the nature of the universe without too much effort. "A quarter of a century ago," writes E. T. Bell in his *Men of Mathematics*, "those who were unable to see the great light which the prophets assured them was blazing overhead like the noonday sun in a midnight sky, were called merely stupid. Today for every competent expert on the side of the prophets there is an equally competent expert against them. If there is stupidity anywhere it is so evenly distributed that it has ceased to be a mark of distinction. We are entering a new era, one of doubt and decent humility."

The progress of humility is not a bad way to mark off mankind's scientific growth. In the beginning we thought that the universe was closely related to us. It might be friendly or hostile but it was neither indifferent nor aloof. Every rock was a friend or foe. Every star was a personal guide. The gods of nature might be ruthless, but they were thoroughly human, capable of sudden beneficent impulses and at times controlled by the same laws of fate which hung over man. And these laws, too, were personal.

Today we regard the universe as utterly indifferent, a vast expanse into which we have wandered, perhaps by accident, perhaps by divine providence. We are not at all necessary to it, but we know that it is necessary to us, and we must learn its rules if we are to survive.

Astronomy, more than any other science, has brought about the change.

We can mark the difference, if we like, by the jolts which have come to man's ego. In the first jolt he discovered that the world was spherical. The directions east and west ceased to have absolute meaning. Nor was his tribe divinely situated in the middle of the world. Then for the first time he had to broaden his viewpoint. But his race at least was safe. The Earth was standing in the middle of space. The solar system was created to go around the Earth, the stars to go around the solar system. Everything still worked for or against man. He prayed to the powers of good and propitiated the powers of evil. He was in the hands of nature, and he was quite certain that nature had hands.

For a little while Greek scientists attained an impersonal view. In those few centuries they gained more knowledge than has ever been gained except during modern times. But on the slower minds of their contemporaries these scientists had little effect. With the advent of the Middle Ages man grasped at a new theory, less personal than the old, but still thoroughly egocentric. Sticks and stones might be inanimate, but the universe as a whole was still centered around man. Humanity and the cosmos were reflections on each other. Man was in small what the universe was in large.

Look at the structure of Dante's cosmos: the rings of heavenly bodies are filled with angels and devils. Man is the intermediate being and everything is pulling him one way or the other — upward toward heaven or downward toward hell.

It is difficult for us to realize the magnitude of the Copernican revolution. We are far enough removed by now, and we have to a certain extent adjusted our minds. The arguments for a stationary Earth look vain and silly; the dogmatic opposition of the Church seems incredible. But we forget that far more than an astronomical theory was involved.

Literally no man knew where he stood. The evidence of the senses was confounded. Eminent churchmen were heard insisting that there was no place for heaven or hell left in the new cosmology. And what was Christianity to do without those two locations on the map?

Very few people were willing to hear anything so absurd. At first Copernicus had only a few disciples, men like the turbulent philosopher Bruno who visited Oxford in 1583, and complained that "Masters and Bachelors who did not follow Aristotle faithfully were liable to a fine of five shillings for every divergence." Statutes like that were very hard on an impecunious Copernican.

Christianity came to a positive decision. The monks of the eleventh century had been absolute in authority — they could afford to argue the pros and cons of any new theory without prejudice. The sixteenth-century church had its back against the wall and was fighting for its life; in such circumstances no institution can afford to be liberal. Books on the Copernican theory were put on the papal index; Copernicus was denounced as a heretic. The Inquisition arose in all its fury, Bruno lost not only his five shillings in Oxford, but his life at a fiery stake in Rome.

Perhaps all the excitement only served as publicity for the book. Certainly the new theory survived, slowly made its way into the scientific mind, and thence into the popular. In a great eastern university the geocentric and heliocentric theories were taught side by side until well into the nineteenth century; but eventually one had to be dropped. About the same time the Catholic Church capitulated at last; and after nearly three hundred years the volumes of Copernicus were quietly and unostentatiously dropped from the proscription list.

We have accepted the inevitable now. Coldly and dispassionately we admit that we are not the appointed darlings of the universe. We have accepted, but it is doubtful whether we have even yet adjusted ourselves.

The ramifications of the Copernican revolution are incalculable, but three or four at least stand out. Philosophers, deprived of the evidence of their senses, began to doubt the existence of the world. Perhaps the world was only a figment of the imagination. The great Bishop Berkeley rose in a fire of divine skepticism and proclaimed that nothing existed outside of mind but that all things were contained in the mind of God. Or, as in the words of another reverend, and equally philosophic gentleman:

> "There was a young man who said, 'God,
> But it certainly strikes me as odd
>> That that sycamore tree
>> Should continue to be
> When there's no one about in the quad.' "

To which the answer was made:

> "Dear Sir, it may strike you as odd,
> But I'm always about in the quad
>> And that sycamore tree
>> Shall continue to be
> In the sight of,
>> Yours faithfully,
>>> God."

There has always been an opposition between the world which we see, and the world of which science tells, but never before had the question risen to such extraordinary heights.

In another way too, man became more skeptical. While he thought that nature was working either for or against him, he was secure in his own importance. Now it began to look as if he were not merely secondary, but actually negligible in the scheme of things. His ego had been shattered for the second time.

When nations are conquered they begin to extol themselves. When Greece fell, Aristotle stopped writing metaphysics and turned to ethics. The symptom of an inferiority complex is sometimes boastfulness and sometimes introspection. With the Copernican revolution, man began to look at himself. "Well, anyhow," he said, "I'm a pretty good thing." "I am certain of nothing," said Descartes, "except my own existence. I think, therefore I am." Psychology, which in its present form is a new science, came into existence. "Whether I'm important to anything else or not," man seemed to say, "at least I am important to myself." Such an assertion had never been necessary before.

Shock after shock followed the initial jolt of the Copernican revolution. The rapid perfection of instruments and the increased visibility through space resulted in theory after theory, and the minds of men were led through a merry dance. No sooner had you believed that the Earth went around the Sun than you were asked to believe that the stars were likewise suns, possibly with planets going around them. All of them were arranged in a formation which was itself an island, and beyond it were other islands — and beyond them — ? Well, beyond them, said the weary philosophers, we cannot think. Anything may exist beyond the stars. That may be the land of all lost theories, of unicorns and Ptolemaic systems and small, safe, sane worlds. "We don't *know* about that," replied the physicists, "but we can make a pretty good guess about what happens in this cosmos at least." They proceeded to guess with wonderful results.

In the midst of the terrific excitement of the seventeenth century up rose Sir Isaac Newton. He was only twenty-four at the time and before he got up he had been sitting under an apple tree. Possibly he had been studying Euclid. At a Stourbridge Fair some years before he had bought a book on astrology, but to his dismay he found that the diagrams of the heavens were beyond his understanding. Someone around Cambridge told him to buy a volume of Euclid; possibly Euclid would help him to understand astrology. The young man bought the book, but the propositions in it seemed to him so self-evident that he tossed Euclid aside as "trifling." Only later when he failed in an examination on the subject was he forced to admit that there might be more substance to the Greek geometry than he had at first noticed.

All this happened before the historic summer of 1666. That was the year when the Royal Society was founded. It was the year when the great fire burned the City of London. It was the year of the plague, and young Isaac Newton together with all the other healthy students at Cambridge was forced to flee into the country. It was the year when he went down to Woolsthorpe and sat upon the grass. It was the year when an apple fell upon his head.

That apple, like another of its kind in legend, came straight from the tree of knowledge. Upon Newton's laws most of the scientific researches of the next two centuries depended. Using Galileo's telescope to look through, and Newton's laws to explain, the astronomers began to explore the universe anew. All scientists explain what is strange in terms of what is familiar to them. Now descriptions of the universe began to sound like the descriptions of a huge machine.

But whatever the analogy, the hypothesis proved extremely serviceable. Under its guidance knowledge of the universe increased by leaps and bounds. Halley tracked down comets, Kepler pursued planets, Herschel found that there was a pattern in the stars. Outwardly and inwardly the world was opening before the advances of research. A safe, sane little world indeed! It was bigger and far more impersonal than anyone had ever dreamt.

Where then was mankind? We still ask the question. In whirling matter are there no other inhabitants? Do we constitute the whole population of interstellar space?

The stars are out of the question. The temperature of the coolest is far too great for the existence of organic matter. At first thought it seems likely that a number of stars would be surrounded by planets, and at least a few of them would permit the existence of some kind of life. If there were numerous animated spots we should feel better somehow about our own existence. Even the presence of vegetables on some planet out in space would make us feel less conceited and less lonely. A single cabbage does not seem much to ask.

Yet, as a matter of fact, the chances for even a cabbage are extraordinarily small. If there is one, then shoes and ships and sealing wax will probably follow as a matter of course, for the existence of a cabbage is an extremely complicated thing. How unsatisfactory — to put it mildly — conditions seem to be for life on the other planets in our own solar system. Only one of the nine major planets has any appearance of life; and that one is our own Earth.

As far as any telescopes have been able to show us there are very few stars — if any — like our Sun. Double stars are common, but in most cases both parts are luminous and too hot. Rare cases, such as Algol, exist where a dark body comes at regular intervals between the bright star and ourselves and noticeably diminishes the light. There may be dark bodies in abundance, but we have no way of knowing of their existence unless like Algol, and like the comets, they shut off light or reflect it in such a way that their return can be forecast even when they themselves cannot be seen.

Scientists of the nineteenth century worried a great deal over the possibility of human existence elsewhere. If this huge galaxy must be impersonal, they seemed to say, at least let it be populated. There was something wistful in their request. More than any other men, they had recognized the complete inhumanity of the cosmos. They were trying to be logical and hard-headed as no men had ever tried before.

There is nothing personal about a machine, and by the middle of the last century the mechanistic theory had reached its culmination. Physicists assumed that everything was simple and easy. After all we are living in a huge machine, they said; it isn't wonderful, or even particularly beautiful. Humanity is nothing but a fortuitous concourse of atoms. Fairly soon we shall be able to manufacture it too. Of course there are some problems still to be solved, but if we look at the universe long enough we shall see it simply as a system of "jellies, spinning tops, thrust bars and cog-wheels."

In an effort to realize the impersonality of the universe they had gone the whole hog and denied everything which seemed personal at all. Free will was swept into the machine and ground away. You cannot act as you please if all your actions are predetermined by the mechanical laws which govern atoms of which you are entirely composed. Hundreds of years before, the Roman philosopher-poet Lucretius had set the same problem and come to the same conclusion. Deny free will and you deny all human attributes. Such words as justice and virtue, right and wrong become meaningless.

That was too bad, said the scientists, but if free will failed to fit the scheme, then free will must go. It was a delusion of grandeur, and a nice idea while it lasted, but it belongs with the unicorns now. Science is certain at last. We have the Ariadne thread of knowledge in our hands.

Alas for certainty! The physicists cast into outer darkness not only free will but everything else for which they could find no place in the machine. They neglected the little clues, the tiny inconspicuous items which seemed incoherent with the world at large. There was for instance an unexplained movement of the perihelion point of the planet Mercury. Le Verrier had noticed it in 1847, and it was observed several other times. All of Newtonian physics would not account for it. "But that will be explained in time," answered the physicists. "Either it will be explained, or your instruments are in error and it does not exist at all."

There is always danger in such neglect. Beside a mountain road just outside of Denver there are the tracks where a dinosaur once walked through the mesozoic clay. The clay is tilted now; it forms a steep slope beside the road, and the three-toed footprints seem to be coming downhill. Just at the bottom, almost at the road, the front claws have pierced deeper into the rock as if the animal had stopped just there. It is our private theory that as Dinah walked through the clay, a little unspecialized animal, who was the ancestor of the mammal, ran scuttling with fright across the monster's path. Dinah stopped, arched her long neck and looked at the curiosity. Then she decided that no such timid creature as that could possibly survive in this hard world. It was not worth her majesty's attention and she continued on her way.

The little mammal whom Dinah scorned was the ancestor of the present lords of the Earth. The little details which the nineteenth-century physics refused to recognize were the very things which shattered their mechanistic theory of the universe. And the shattering of the mechanistic theory brought the third jolt to the ego of mankind.

Long before, when scientists first began to explain the world in terms of a machine, they not only neglected relevant details, but they also began to invent substances to support their theory. If the universe behaved in a mechanical fashion, then, they thought, the mechanical action must be transmitted by some substance. If light was composed of waves, then there must be something for it to wave in. There was no substance ready at hand, but they assumed one anyway and called it "ether." Ether was all pervasive, and ether could undulate. It was a sort of jelly, or rather it was many sorts of jelly. As Clerk Maxwell said, "Ethers were invented for the planets to swim in, to constitute electric atmospheres and magnetic effluvia, to convey sensations from one part of our body to another, till all space was filled several times over with ether." Whatever science failed to explain in any other way was at once attributed to this marvelous substance.

In addition, ether was supposed to have the great merit of furnishing an absolute gauge by which the rate of motion of all other bodies could be reckoned. Whether the ether itself moved rapidly or slowly was beside the point. A man sitting on the observation platform of a train, with his feet up on a rail and his eyes pointed skyward, may notice that the streamlined train which passes him is moving faster than he. He may look at the birds circling, and watch the changing aspects of the clouds, but he sees these movements only as they relate to the speed of his train. He can make no absolute reckoning until he takes his feet off the rail and regards the solid ground. The fact that the "solid" ground is all the time rotating on its axis and revolving around the Sun, is for his purpose a matter of absolutely no importance, since it is basic to all these other motions. In the same way the movement of the ether was unimportant, since all the planets and solar systems and galaxies moved relative to it. If only science could find ether, there would be a base from which to judge. It was the philosopher's stone, quite literally the *Deus ex machina* which would put right all the tangled problems of the universe.

Of course an unknown and unproved substance of this sort may act as a valuable hypothesis, but in itself it explains nothing; and the temptation to explain everything by an addition which explains nothing has always been irresistible.

In vain did the scientists try to prove their pet belief, in vain did they call the mechanical universe to their aid. Whatever the nature of their substance, it stoutly refused to help them out. No pagan god had ever been so elusive. It seemed to have charmed all nature into a secret bond against betrayal. The scientists set their traps with the utmost care, but the ethereal substance remained true to its name. And worse was to come. By the turn of the last century the weary scientists were forced to admit that even if ether did exist it would not serve their purpose. Instead of making everything simple the existence of ether would make everything far more complicated!

But ether was absolutely necessary for the mechanical theory of the universe.

Centuries before, when no one had dreamed that anything was mechanical, a little-known medieval philosopher had uttered a maxim which came to be known as "Ocham's razor." "We must not," said William Ocham, "assume the existence of any entity until we are compelled to do so." The nineteenth-century physicists stood convicted of this fallacy. The compulsion which had forced them into assuming the existence of ether was the mechanical universe. But suppose the universe was not mechanical after all!

What could the universe be if it were not a machine? In June of 1905, a young German scientist, Albert Einstein, published a short paper in which he said, "Nature is such that it is impossible to determine absolute motion by any experiment whatsoever." Except for the Copernican revolution, no scientific announcement has ever created so much commotion. The primary reason to assume the existence of ether had been as a gauge for absolute motion. If no such gauge were possible, then there was no need for the assumption.

Ten years later, in 1915, Einstein published a second account of his theory, and this took hold upon the popular imagination like wildfire. True, it was said that no unscientific person could understand either the first paper or the second, and of the scientists not more than a dozen knew what Einstein really meant; nevertheless the public grasped at this change of ideas as if it had been designed to fill a popular need. Einstein's photograph is today as well known as that of a movie actor. His most casual statement is reported by the newspapers. Even *Punch* has commented upon his "family connections":

PRECIOUS STEINS

What with GERTRUDE, EP and EIN
When I hear the name of STEIN
I go creepy down the spine.

EIN has caught the ether bending,
GERT has sentences unending,
EP is really most art-rending.

EIN's made straight lines parabolic
EPPIE's "Night" is alcoholic,
GERTIE's grammar has the colic.

EIN and Space are down to tin-tacks,
EP hews boulders with a flint-axe,
GERT has no respect for syntax.

What with GERTRUDE, EIN and EP,
Life and Art are out of step.
Are we then down-hearted? Yep!

To fill the gap left by the failure of ether, Einstein invented no fabulous substance. He merely took into account the obvious factors of the universe, and he included one to which the physicists had previously paid no attention — the factor of time. You have failed to explain the universe by your three mechanical dimensions, he said in effect, now try this fourth and see the result.

The fourth dimension. In geometry a point has no dimensions whatsoever. A line has one, a body with mobility in one dimension can escape from a body with none. But the disadvantage of a line, as compared with an area, has been sadly and graphically expressed.

"There was a young man who said 'Damn!
I know now just what I am,
   A creature who moves
   In determinate grooves,
In fact, not a bus but a tram.' "

That argument was written by a philosopher who had been convinced that there is no such thing as free will, but it serves also to show the difference between one and two dimensions.

Similarly a body that has ability to move in three dimensions will have the advantage over a body limited to an area. Just as a tram can never catch an unwilling bus, so a bus can never catch an airplane if the pilot does not want to be caught. Our small dog is distressingly conscious of her limitations in this respect. In her views, all the small mountain animals play a perfectly fair game and she respects them for it; but timberline birds which run along the ground until she has almost caught them and then suddenly take to the air are unfair and cheats and she despises them. The difference between these three dimensions is clear enough.

But what about the fourth? Einstein says that the fourth dimension is time. There is an illusion in perspective which sometimes occurs when you look at an object and think it is flat until a slight shift of light shows it as three dimensional. In the same way, scientists tell us, we are deluded in thinking that everything in the universe has only three dimensions. As a matter of fact, they say, everything is moving in one direction along a fourth scale.

Therefore, according to Einstein, we are making an abstraction and speaking only partial truth if we give the location of any object. Objects exist within both a spatial *and* a temporal relationship. If we are to give the truth, the whole truth and nothing but the truth, we must mention not only where they exist but when.

With the advent of this theory, understanding of the universe passed from the engineer mechanic to the mathematician, and we are assured that the fourth dimension cannot be explained except by the most abstruse of mathematical formulae. "What we need," writes Clement Wood, in *An Outline of Man's Knowledge,* "is a solid of four dimensions, each of whose sides is a cube. . . . Altogether our four dimensional solid will contain 32 such cubes. We cannot construct this out of wood or any other solid; for our eyesight, as a rule, stops at three dimensions. Mathematicians, who deal in this fourth form, admit that it cannot be constructed in its solid form; but they can symbolize it, and study its qualities. Such a four dimensional figure, with each of its sides a cube, is called a *tesseract.* Drawings of a tesseract are not clear; they resemble seventeen maps of the successive partitions of Poland scrambled upon each other."

The brains of mathematicians must contain some very curious sights; not only do they claim to know what tesseracts look like, but they also tell us the fourth dimension must be multiplied by the square root of minus one before it will fit into equations, and that it then fits perfectly.

But for all its abstruseness there are some aspects of the theory which the layman can grasp. Dr. Boodin, in his book on *Cosmic Evolution,* cites the example of a speedy motorist who is caught by a motorcycle cop. It is common, I believe, for such motorists to plead, "But I didn't know I was going that fast." It is less common for him to plead that he didn't know he was moving at all. A car is small, and presumably the motorist had his eyes on the solid road, which served as a reference. Yet the illusion of motion is common enough for us to imagine. We have all felt, in trains, that we were standing still while everything else moved. Robert Louis Stevenson, in his little poem, "From a Railway Carriage," illustrates this sensation perfectly:

> "All of the sights of the hill and the plain
> Fly as thick as driving rain
> And ever again in the wink of an eye,
> Painted stations whistle by."

Now suppose we had the same illusion while driving. To one side of the road, moving away from us, like everything else, at the speed of seventy miles an hour, is our friend the motor cop. He is gone before we notice him particularly. But in a moment he reappears. He has of course started his motorcycle and decided to join us. But we thought that we were standing still. He was the one who moved at seventy miles an hour, he together with the road and the bushes and the cow in the field. Now, he is coming closer to us. But since he was moving to begin with, while we stood still, it follows that he must appear to us now to be slowing down. The nearer he comes to us the slower he moves until at last he is right opposite our car. He is going seventy miles an hour. We are going seventy miles an hour — and it appears to us as if the rest of the world rushed by, while we two alone were at rest.

Of course, if, in the event of this sad contingency, we try making this explanation to him, we may find that he is not such a good friend after all. And if we take our explanation to a higher court of appeal we will doubtless find ourselves in a place where motion is very limited indeed and the windows are barred.

Nevertheless, if we could call upon Einstein to witness for us in our trial, he would say that we had truth, though perhaps not justice, on our side. All motion is relative; nothing is static. And under these circumstances each mover has the prerogative of thinking that he alone is stationary.

Time itself is not absolute. If we wound up two identical and perfectly adjusted alarm clocks, if we sent one hurtling through space at a terrific speed and kept the other by our side, and if we could keep our eyes trained on both, we should see that our errant timepiece went off considerably later than the timepiece near at hand. If we could hurl it so fast that it went with the speed of light, the alarm would never start. The clock would stop dead in its tick. But according to Einstein the last part of the experiment is purely hypothetical. Nothing, he claims, can ever move with the speed of light.

In the spring of 1938, the experiment of relative speed was tried, not with alarm clocks to be sure, but with oscillating atoms which could be projected at terrific speed. The rate of oscillations determined their color. At the high speed their color showed that the rate of their oscillations had decreased.

In fact the amazing thing about the theory of relativity is that so far every test applied has served only to prove its truth.

In the essence of the theory some strange physical laws are involved. Newton had thought gravity an instantaneous physical force. Einstein said, it is not a force and it is not instantaneous. It travels with the speed of light. Moreover the presence of matter can affect light. By the inclusion of such rules as these, the theory of relativity explains to perfection one of the little details for which all Newtonian physics failed to account. The movement of the perihelion point of the planet Mercury is no longer a mystery. And this was the one experimental verification for Einstein's theory at the time it was first announced.

Other proof, however, was to follow fast. Einstein had said that light passing through a single medium would bend under the influence of matter. It was a most startling announcement to the physicists. Light was the medium by which straight lines were tested. If light could bend nothing was safe. At the Royal Astronomical Society in 1916–1917, Dr. de Sitter presented three papers, corroborating the Einstein theory. Two years later, a British expedition under the direction of Sir Arthur Eddington was studying a total eclipse of the Sun when they noticed that the apparent light of the stars near the Sun was not where it should be. In other words, their observations seemed to show that light passing close to the Sun is deflected by the Sun's mass. This phenomenon cannot be measured except during a total eclipse when the overpowering radiance of the Sun is darkened and the near-by stars become visible. Had the de Sitter papers not been published in advance, the expedition would probably have thought their instruments in error. As it was, their findings were so exactly what de Sitter and Einstein had predicted, that they could not but acknowledge them. And subsequent experiments have proved like results.

Not long after Einstein's theory reached popular ears, the astronomers announced a strange discovery. We have seen that there are galaxies beyond the Milky Way, and island universe beyond island universe. Forty years ago most scientists would have said that these probably went on forever, that the universe was limitless and infinite, but recent discoveries tend to disprove that idea. As we use increasingly larger lenses to make our photographs the number of stars shown on each plate should increase in geometrical proportion; and so they do — up to a point. Beyond that the proportion decreases, as if the stars were becoming rarer and rarer phenomena in the immensity that is space, and we are forced to inquire what space is, or what light is, since that is what our cameras pick out of the heavens.

According to some modern astronomers (Sir James Jeans and Sir Arthur Eddington among them) space is not an infinite extension, but curved and bent around a sphere, so that the bright specks of dust which are stars do not, as Tennyson said, "blindly run" but flow along with the motion of the whole. If we look at a great sluggish river on a quiet day we hardly know in which direction it is flowing except for the leaves and sticks that come floating around the bend into our sight, and so the motion of the stars may be indicative of the direction in which the universe flows, and we must always remember now that the universe is composed of time welded inseparably with space. The stars may be caught in unexpected eddies, wafted around, brought together and dispersed by their own nature; but if we watch them long enough they will show in what direction the river flows.

This is not an ordinary river, and the analogy, like all analogies, must be accepted only in a limited sense. It is not even the Homeric river which girdled the Earth, though that is a closer parallel, because that one came back to the place whence it started; but ours looks as if the surface of a glass ball had suddenly become liquid without losing its form, except that it continually expands, doubling its size (which really seemed quite large enough already) once in every 1,300,000,000 years.

Then if this is true, light cannot be straight either; but it too goes around in a vast circle back to the place from which it started once every 6,700,000,000 years or so. It is just barely possible that the nebula which we see in one part of the sky may be the back view of what we believe to be another nebula opposite it — and (to go one step further) what we see through a telescope is the place where the back of our head was 6,700,000,000 years ago.

From all these problems a multitude of further questions arise, so few of them answered as yet that we can only accept these new theories as the most *probable* explanation of what happens in the world at large. The strangest part of it all is this: when and if we find some of the answers about the most enormous structure, then at the same moment, in the same breath, we may have solved some of the puzzles about the most minute particles of matter. For the atom, on which the mechanistic theory was based, turns out to be no more like a machine than is the universe. Instead it seems exactly like a cosmic solar system, with planets rioting around a sun, as far removed from them in proportion as is the Sun from us, with cosmic particles shooting off at intervals, and other particles dispersing themselves around the vast heavens that go to make up the existence of a single atom.

It may be as well then, that the ancient peoples did not know the implication of their ordinary statement that the Earth reflects in miniature what the heavens show in large. For at this point the stable values, by means of which knowledge progressed, become so relative and inter-related that the human brain reels and begins to wonder if any knowledge is sure,

Is everything disorganized? Is all previous learning disproved? Space is curved and time is not absolute. Light can bend and the atom is split. At first announcement of relativity the simple mechanical explanation of the universe seemed to fall as had the egocentric universe centuries before. The failure of the Ptolemaic theory deprived man of the focal point of his thought. The Einstein theory of relativity seemed to do even worse. Apparently it upset the structure and pattern of logic. In the safe and sane old days Alexander Pope had eulogized Newton:

> "Nature and Nature's law lay hid in night;
> God said, Let Newton be! and all was light."

Now it was no wonder that the lyrical students at Cambridge offered a revised edition:

> "We thought that lines were straight and Euclid true:
> God said, Let Einstein be! and all's askew."

Just as in the days after the Copernican revolution, philosophers insisted on doubting everything, so in the first flush of relativity, mankind darted from certainty into skepticism. This effect came more rapidly perhaps, because the theory became popular immediately after the great war when society was itself unstable, when no one was quite sure which conventions, if any, he believed. Why not make the job complete? Why not throw over physical and logical conventions as well as social? Why believe in anything?

What ensued was almost a chaos. We have seen its results upon the popular mind. It is represented in sentences which start out to mean nothing and accomplish their purpose exceedingly well. It is obvious in an art which links crystal spheres to ziggurats on typewriters, and labels the result "Portrait of the Authors."

The old mechanistic theory was thrown like a fallen angel not into the middle distance but far into the nethermost regions of the inferno, the seventh hell where the Ptolemaic theory undoubtedly sat awaiting its arrival. The psychology of man is such that the greatest efforts, once disproved, always fall the farthest. And with the mechanistic theory men for a time tried to throw out all its bases and postulates. Of course Newtonian physics went hand in hand with mechanism. The geometry of Euclid was scrapped with the rest. Einstein had demonstrated and astronomers had proved that a straight line cannot be measured in a gravitational field. The means of finding a straight line had always been a light ray, but now light itself was bent. The propositions of Euclid depended upon straight lines.

By popular analogy Aristotelian logic went too. It used the same means of proof as Euclid. Moreover it was static in a world bound up with time.

Only one thing was saved from the debacle, and that was a purely philosophic concept which admitted no proof. The doctrine of free will, excluded by the mechanists, came once again to favor. Mathematicians can say about how often particles of the electromagnetic substances will shoot off into space, but they are powerless to say which particles will do the shooting. In the same way statistics show approximately how many people will commit suicide within a given year, but they never have and they never can say who the people will be. Mathematical truth, unlike mechanism, applies only in general. It predicts, as did the old Babylonian astrology, without reference to the individual.

For the philosophic peace of mind, free will was at least given a chance.

But was nothing else left? Were all the forms and shapes of thought to be tossed away? Was Aristotle entirely wrong when he worked out his logic with such care? Were the propositions of Euclid utterly invalid in this curved space? After reading some of the popular accounts which grew out of the first extreme skepticism we are tempted into such heresy. Perhaps it was only relatively speaking that the apple fell upon Newton's unsuspecting head.

But the popular accounts were examples of science misapplied. Their writers had jumped too fast and read their master too hastily. In his 1915 account Einstein had not thrown out Newtonian physics. It remains applicable for all small portions of space. Let us climb once more, and for the last time, upon the observation car of that convenient express train which we have used so many times. We are sitting sideways with our feet up on the rail. With one turn of the head we can see the club car, and with another the tracks behind us, while in front of our eyes are the parallel tracks where presently another train will run, and beyond them are the open plains. In the club car other travelers are chatting and occasionally moving about. One passenger finishes his magazine and puts it on the table. It is a popular number. Two others dash for it immediately. We can say, quite truthfully, that the man with the wig is moving more quickly than the woman with the dyed hair. He reaches the table first and grabs the magazine. Of course both are moving relative to the motion of the train, but for the purposes of getting that magazine the secondary motion is absolutely unimportant.

Now a rush of air makes us look straight ahead. We have caught up to a local on the next track, and as we pass it, we see an identical scene about to be re-enacted in the coach. We have no chance to see who wins in this second race. It looked to us as if the man were running faster, but perhaps that was only because our train was gaining speed at the moment when we passed his end of the car. An entirely new set of factors have entered into our judgment. The simplicity of the first race has completely disappeared. The longer we think of the problem the worse it gets. There is the speed of our train, and the speed of the local and the possible change of speed of either one at any time. We begin to remember that all motion is relative, and that there is nothing absolute.

Yet at any moment, if we wish to stop studying the universe and read instead, we have only to rise and go into the club car on our own train, pick out whatever magazine we desire, or whatever one the other passengers have left us, and sit down again. None of our complicated theories will hinder us. We may behave just as we always have, and the laws which we obey in this simple series of motion will be the laws of Euclidian geometry and of Newtonian physics. As long as we do not try to jump into the other train, grab something from its table and return to our original seat, we need not disturb ourselves at all.

For practical purposes then, within small areas of space and time, the values of Newton and Euclid remain unchanged. As we enlarge the scope of our observations they become increasingly less applicable until for the great areas of space we shall find them not only faulty but actually more complicated than relativity — though such a statement seems at the moment almost impossible.

For small space, we repeat, within the area of our own train, they have a very useful function. But we are still sitting on the "observation" car. We are still theorists, and the desire to know ultimate truth haunts us as of old. Besides, away in the distance stretch the rails over which our train has come, and by an optical illusion they are distressingly reminiscent of the theorem upon which Euclidean geometry first failed. They are parallel lines, but they do look suspiciously non-Euclidean. Just beyond the farthest point of our vision they always appear to meet.

What do the extremists mean when they say in their dogmatic fashion that relativity has disproved Euclid? It has been a very long time since Euclid was thought to have explained the whole world. He demonstrated plane and solid geometry, and he demonstrated it very thoroughly, and these geometrics still admit of his proofs. We have always known that the plane was an abstraction. Now we are told that the solid is not a complete description of reality. Plato had preached that to the students of the Academy long before Euclid was born, and undoubtedly Euclid had read Plato. The suspicion that three dimensions were inadequate to explain reality can hardly be called new. Yet within its limited and strictly defined field the geometry of Euclid was, and is, exactly and perfectly true. Nor is non-Euclidean geometry a result of the theory of relativity. A very long time ago mathematicians had begun to realize Euclid's inadequacy. In 1521, Henry Savile called the theory of parallels a "blemish in the beautiful body of geometry," and throughout the succeeding centuries mathematicians tried so hard to justify this flaw that their efforts became suspiciously like Iago's "Methinks the lady doth protest too much." By 1799, Karl Friedrich Gauss had published a book to show the coherence of non-Euclidean geometry, and later mathematicians, following his lead, deduced the theory of curved space before any astronomical proof was deemed possible.

For the most extraordinary thing about mathematics is this — that its chance findings, the laws which it deduced apparently without reason or application, are found at last to have a place in the universe. We can hardly wonder that one of the greatest astronomers of our day, Sir James Jeans, called the Architect of the universe by the name of "Mathematician," since strange and imaginary numbers, such as the square root of minus one, are found applicable. And the curved space deduced by Mathematics is said to exist in reality.

Yet here we are forced to stop. No modern science has told us what reality is. Einstein claims only that his theory will help us to predict with greater accuracy than we have ever predicted before. Among his disciples, some have gone a step further and thought mathematics to be a picture of reality itself.

That is a mistake which has been made before. The mechanistic theory was thought perfect in its day, the Ptolemaic theory in its. Designs which are thought too perfect have no room to expand, and whatever has no room for expansion is doomed. If we go back to the beginnings of the mechanistic theory, in the days of Isaac Newton we find no such claim of perfection as his later disciples claimed for him. And the tentative rules which he did make are just those rules which Einstein kept and incorporated in a larger whole. The law of gravitation, the law of conservation of angular momentum, all the principles of early physics have served as guides to the more inclusive laws, the more perfect approximations which have succeeded them.

The same holds true for Aristotelian logic and the laws of thought. What Aristotle said was far more inclusive than the precise rules usually ascribed to him in textbooks today, yet perhaps we must enlarge upon Aristotle's logic. Possibly our world has broadened beyond his view. We need telescopes to see around us now. Yet his laws are still the magnifying glass with which we examine the details and pick out the flaws of each new lens. If the theory of relativity completely disproved Aristotle, by whose logic was this greater truth ever reached? For us to abandon the old without a backward glance is as unjustifiable today as it was in the time of Alexandria when the scholars forgot to take the discredited books on the heliocentric theory with them on their long medieval ramblings.

The old is still pertinent to the new. Surveyor after surveyor has been over this land, each one guided by the old charts and correcting them in the light of new knowledge as he went along. Because the old charts were in existence no scientist was ever completely lost; and seldom was a chart so thoroughly wrong that he gained nothing from it. The earliest ones were decorated with pictures and the lines were distorted to give the pictures room. The last but one was detailed and precise but it turned out to be not spacious enough. This present chart of ours is harder for us to follow because more difficult to understand, but as we look back and forth between it and the universe we do seem to find a better correspondence than anyone has ever known before. But it is too soon to be certain of that yet.

The progress of any knowledge is built up bit by bit. If we were to throw away the old charts now we would be like disillusioned children crying out, "The stories I read were false; therefore I will deny *all* that they taught me." We would be forgetting, in our anger, that what we really gained from these stories was our ability to read.

With day unto day uttering speech, and night unto night showing knowledge we have made some progress toward a reasonable explanation of the universe. What we have now has come about only by generations of men passing their learning on to future generations, only by criticism of the old and hypothesis for the new. Neither Copernicus nor Einstein pulled his theory from the blue air, but from observation combined with stories which were themselves resolved from stories older yet, bit by bit they built up their work. In the last few centuries the universe has opened vistas so wide that sometimes we are frightened by the mere sight of the stars on a clear night. Yet this fright is not the same as that which led our ancestors to throw themselves down upon their faces and pray the gods of the planets to have mercy upon them. We do not throw ourselves down upon our faces now. Oddly enough with each step in the shattering of our egos, something has been gained. For a personal world is not one which men will attempt to conquer even in their minds. If the planets are controlled by gods we should bow to them, but we should not analyze their movements nor try to formulate general rules. It would be impertinent somehow to speculate about their private lives. The intellectual mastery which we have gained is the direct result of our shattered egos. As we became more humble, at the same time we became more aware.

In this book the authors have tried to show how some of the developments in theory took form and grew, how the most theoretical scientists benefited the world with the most practical discoveries. We are too close to the astronomy of our own time to see its implications; but just as Tycho Brahe's star tables made possible explorations of which Tycho Brahe never dreamed, so these new discoveries may have results which no man living today can possibly imagine.

. The vastness of the universe is a challenge; and all that has gone before us serves at once as example and as guide. We know now that we are little people, imprisoned on a tiny planet which is not even central in the gigantic whirlings of suns and galaxies. Yet we have learned much of how these huge bodies whirl. We can predict something of their future courses. We can even make use of their emanations and turn them to our own benefit. Through the ever increasing arch of our experience, the margin of the untraveled world still fades before us as we move. We are human, we have minds and books and records. What we have learned so far is no mean achievement, and our purpose, like the purpose of Tennyson's "Ulysses," holds,.

> "To follow knowledge like a sinking star
> Beyond the utmost bounds of human thought."

JULIAN S. HUXLEY (1887–   )

The Uniqueness of Man

First published in the *Yale Review*, 1937. Reprinted in *The Uniqueness of Man,* 1941

[THE GREATEST COMPLIMENT that could come to Julian when the name Huxley is mentioned is the question: "Which Huxley?" The grandson is a worthy successor to his grandfather Thomas H. Huxley, and his active brain has produced outstanding results in observational, comparative, and creative fields of science. From among his semi-popular essays I have chosen one that is somewhat apart from our usual concept of natural history, one which treats of *The Uniqueness of Man.*]

MAN'S OPINION of his own position in relation to the rest of the animals has swung pendulum-wise between too great or too little a conceit of himself, fixing now too large a gap between himself and the animals, now too small. The gap, of course, can be diminished or increased at either the animal or the human end. One can, like Descartes, make animals too mechanical, or, like most unsophisticated people, humanize them too much. Or one can work at the human end of the gap, and then either dehumanize one's own kind into an animal species like any other, or superhumanize it into beings a little lower than the angels.

Primitive and savage man, the world over, not only accepts his obvious kinship with the animals but also projects into them many of his own attributes. So far as we can judge, he has very little pride in his own humanity. With the advent of settled civilization, economic stratification, and the development of an elaborate religion as the ideological mortar of a now class-ridden society, the pendulum began slowly to swing into the other direction. Animal divinities and various physiological functions such as fertility gradually lost their sacred importance. Gods became anthropomorphic and human psychological qualities pre-eminent. Man saw himself as a being set apart, with the rest of the animal kingdom created to serve his needs and pleasure, with no share in salvation, no position in eternity. In Western civilization this wing of the pendulum reached its limit in developed Christian theology and in the philosophy of Descartes: both alike inserted a qualitative and unbridgeable barrier between all men and any animals.

With Darwin, the reverse swing was started. Man was once again regarded as an animal, but now in the light of science rather than of unsophisticated sensibility. At the outset, the consequences of the changed outlook were not fully explored. The unconscious prejudices and attitudes of an earlier age survived, disguising many of the moral and philosophical implications of the new outlook. But gradually the pendulum reached the furthest point of its swing. What seemed the logical consequences of the Darwinian postulates were faced: man is an animal like any other; accordingly, his views as to the special meaning of human life and human ideals need merit no more consideration in the light of eternity (or of evolution) than those of a bacillus or a tapeworm. Survival is the only criterion of evolutionary success: therefore, all existing organisms are of equal value. The idea of progress is a mere anthropomorphism. Man happens to be the dominant type at the moment, but he might be replaced by the ant or the rat. And so on.

The gap between man and animal was here reduced not by exaggerating the human qualities of animals, but by minimizing the human qualities of men. Of late years, however, a new tendency has become apparent. It may be that this is due mainly to the mere increase of knowledge and the extension of scientific analysis. It may be that it has been determined by social and psychological causes. Disillusionment with *laisser faire* in the human economic sphere may well have spread to the planetary system of *laisser faire* that we call natural selection. With the crash of old religious, ethical, and political systems, man's desperate need for some scheme of values and ideals may have prompted a more critical re-examination of his biological position. Whether this be so is a point that I must leave to the social historians. The fact remains that the pendulum is again on the swing, the man-animal gap again broadening. After Darwin, man could no longer avoid considering himself as an animal; but he is beginning to see himself as a very peculiar and in many ways a unique animal. The analysis of man's biological uniqueness is as yet incomplete. This essay is an attempt to view its present position.

The first and most obviously unique characteristic of man is his capacity for conceptual thought; if you prefer objective terms, you will say his employment of true speech, but that is only another way of saying the same thing. True speech involves the use of verbal signs for objects, not merely for feelings. Plenty of animals can express the fact that they are hungry; but none except man can ask for an egg or a banana. And to have words for objects at once implies conceptual thought, since an object is always one of a class. No doubt, children and savages are as unaware of using conceptual thought as Monsieur Jourdain was unaware of speaking in prose; but they cannot avoid it. Words are tools which automatically carve concepts out of experience. The faculty of recognizing objects as members of a class provides the potential basis for the concept: the use of words at once actualizes the potentiality.

This basic human property has had many consequences. The most important was the development of a cumulative tradition. The beginnings of tradition, by which experience is transmitted from one generation to the next, are to be seen in many higher animals. But in no case is the tradition cumulative. Offspring learn from parents, but they learn the same kind and quantity of lessons as they, in turn, impart: the transmission of experience never bridges more than one generation. In man, however, tradition is an independent and potentially permanent activity, capable of indefinite improvement in quality and increase in quantity. It constitutes a new accessory process of heredity in evolution, running side by side with the biological process, a heredity of experience to supplement the universal heredity of living substance.

The existence of a cumulative tradition has as its chief consequence — or if you prefer, its chief objective manifestation — the progressive improvement of human tools and machinery. Many animals employ tools; but they are always crude tools employed in a crude way. Elaborate tools and skilled technique can develop only with the aid of speech and tradition.

In the perspective of evolution, tradition and tools are the characters which have given man his dominant position among organisms. This biological dominance is, at present, another of man's unique properties. In each geological epoch of which we have knowledge, there have been types which must be styled biologically dominant: they multiply, they extinguish or reduce competing types, they extend their range, they radiate into new modes of life. Usually at any one time there is one such type — the placental mammals, for instance, in the Cenozoic Epoch; but sometimes there is more than one. The Mesozoic is usually called the Age of Reptiles, but in reality the reptiles were then competing for dominance with the insects: in earlier periods we should be hard put to it to decide whether trilobites, nautiloids, or early fish were *the* dominant type. To-day, however, there is general agreement that man is the sole type meriting the title. Since the early Pleistocene, widespread extinction has diminished the previously dominant group of placental mammals, and man has not merely multiplied, but has evolved, extended his range, and increased the variety of his modes of life.

Biology thus reinstates man in a position analogous to that conferred on him as Lord of Creation by theology. There are, however, differences, and differences of some importance for our general outlook. In the biological view, the other animals have not been created to serve man's needs, but man has evolved in such a way that he has been able to eliminate some competing types, to enslave others by domestication, and to modify physical and biological conditions over the larger part of the earth's land area. The theological view was not true in detail or in many of its implications; but it had a solid biological basis.

Speech, tradition, and tools have led to many other unique properties of man. These are, for the most part, obvious and well known, and I propose to leave them aside until I have dealt with some less familiar human characteristics. For the human species, considered as a species, is unique in certain purely biological attributes; and these have not received the attention they deserve, either from the zoological or the sociological standpoint.

In the first place, man is by far the most variable wild species known. Domesticated species like dog, horse, or fowl may rival or exceed him in this particular, but their variability has obvious reasons, and is irrelevant to our inquiry.

In correlation with his wide variability, man has a far wider range than any other animal species, with the possible exception of some of his parasites. Man is also unique as a dominant type. All other dominant types have evolved into many hundreds or thousands of separate species, grouped in numerous genera, families, and larger classificatory groups. The human type has maintained its dominance without splitting: man's variety has been achieved within the limits of a single species.

Finally, man is unique among higher animals in the method of his evolution. Whereas, in general, animal evolution is divergent, human evolution is reticulate. By this is meant that in animals, evolution occurs by the isolation of groups which then become progressively more different in their genetic characteristics, so that the course of evolution can be represented as a divergent radiation of separate lines, some of which become extinct, others continue unbranched, and still others divergently branch again. Whereas in man, after incipient divergence, the branches have come together again, and have

398

generated new diversity from their Mendelian recombinations, this process being repeated until the course of human descent is like a network.

All these biological peculiarities are interconnected. They depend on man's migratory propensities, which themselves arise from his fundamental peculiarities, of speech, social life, and relative independence of environment. They depend again on his capacity, when choosing mates, for neglecting large differences of colour and appearance which would almost certainly be more than enough to deter more instinctive and less plastic animals. Thus divergence, though it appears to have gone quite a long way in early human evolution, generating the very distinct white, black, and yellow subspecies and perhaps others, was never permitted to attain its normal culmination. Mutually infertile groups were never produced; man remained a single species. Furthermore, crossing between distinct types, which is a rare and extraordinary phenomenon in other animals, in him became normal and of major importance. According to Mendelian laws, such crosses generate much excess variability by producing new recombinations. Man is thus more variable than other species for two reasons. First, because migration has recaptured for the single interbreeding group divergences of a magnitude that in animals would escape into the isolation of separate species; and secondly, because the resultant crossing has generated recombinations which both quantitatively and qualitatively are on a far bigger scale than is supplied by the internal variability of even the numerically most abundant animal species.

We may contrast this with the state of affairs among ants, the dominant insect group. The ant type is more varied than the human type; but it has achieved this variability by intense divergent evolution. Several thousand species of ants are known, and the number is being added to each year with the increase of biological exploration. Ways of life among ants are divided among different subtypes, each rigidly confined to its own methods. Thus even if ants were capable of accumulating experience, there could exist no single world-wide ant tradition. The fact that the human type comprises but one biological species is a consequence of his capacity for tradition, and also permits his exploitation of that unique capacity to the utmost.

Let us remind ourselves that superposed upon this purely biological or genetic variability is the even greater amount of variability due to differences of upbringing, profession, and personal tastes. The final result is a degree of variation that would be staggering if it were not so familiar. It would be fair to say that, in respect to mind and outlook, individual human beings are separated by differences as profound as those which distinguish the major groups of the animal kingdom. The difference between a somewhat subnormal member of a savage tribe and a Beethoven or a Newton is assuredly comparable in extent with that between a sponge and a higher mammal. Leaving aside such vertical differences, the lateral difference between the mind of, say, a distinguished general or engineer of extrovert type and of an introvert genius in mathematics or religious mysticism is no less than that between an insect and a vertebrate. This enormous range of individual variation in human minds often leads to misunderstanding and even mutual incomprehensibility, but it also provides the necessary basis for fruitful division of labour in human society.

Another biological peculiarity of man is the uniqueness of his evolutionary history. Writers have indulged the speculative fancy by imagining other organisms endowed with speech and conceptual thought — talking rats, rational ants, philosophic dogs, and the like. But closer analysis shows that these fantasies are impossible. A brain capable of conceptual thought could not have been developed elsewhere than in a human body.

The course followed by evolution appears to have been broadly as follows. From a generalized early type, various lines radiate out, exploiting the environment in various ways. Some of these comparatively soon reach a limit to their evolution, at least as regards major alteration. Thereafter they are limited to minor changes such as the formation of new genera and species. Others, on the other hand, are so constructed that they can continue their career, generating new types which are successful in the struggle for existence because of their greater control over the environment and their greater independence of it. Such changes are legitimately called "progressive." The new type repeats the process. It radiates out into a number of lines, each specializing in a particular direction. The great majority of these come up against dead ends and can advance no further: specialization is one-sided progress, and after a longer or shorter time, reaches a bio-mechanical limit. The horse stock cannot reduce its digits below one; the elephants are near the limits of size for terrestrial animals; feathered flight cannot become aerodynamically more efficient than in existing birds, and so on.

Sometimes all the branches of a given stock have come up against their limit, and then either have become extinct or have persisted without major change. This happened, for instance, to the echinoderms, which with their sea-urchins, starfish, brittle stars, sea-lilies, sea-cucumbers, and other types now extinct had pushed the life that was in them into a series of blind alleys: they have not advanced for perhaps a hundred million years, nor have they given rise to other major types.

In other cases, all but one or two of the lines suffer this fate, while the rest repeat the process. All reptilian lines were blind alleys save two — one which was transformed into the birds, and another which became the mammals. Of the bird stock all lines came to a dead end; of the mammals, all but one — the one which became man.

Evolution is thus seen as an enormous number of blind alleys, with a very occasional path of progress. It is like a maze in which almost all turnings are wrong turnings. The goal of the evolutionary maze, however, is not a central chamber, but a road which will lead indefinitely onwards.

If now we look back upon the past history of life, we shall see that the avenues of progress have been steadily reduced in number, until by the Pleistocene period, or even earlier, only one was left. Let us remember that we can and must judge early progress in the light of its latest steps. The most recent step has been the acquisition of conceptual thought, which has enabled man to dethrone the non-human mammals from their previous position of dominance. It is biologically obvious that conceptual thought could never have arisen save in an animal, so that all plants, both green and otherwise, are at once eliminated. As regards animals, I need not detail all the early steps in their progressive evolution. Since some degree of bulk helps to confer independence of the forces of nature, it is obvious that the combination of many cells to form a large individual was one necessary step, thus eliminating all single-celled forms from such progress. Similarly, progress is barred to specialized animals with no blood-system, like planarian worms; to internal parasites, like tape-worms; to animals with radial symmetry and consequently no head, like echinoderms.

Of the three highest animal groups — the molluscs, the arthropods, and the vertebrates — the molluscs advanced least far. One condition for the later steps in biological progress was land life. The demands made upon the organism by exposure to air and gravity called forth biological mechanisms, such as limbs, sense-organs, protective skin, and sheltered development, which were necessary foundations for later advance. And the molluscs have never been able to produce efficient terrestrial forms: their culmination is in marine types like squid and octopus.

The arthropods, on the other hand, have scored their greatest successes on land, with the spiders and especially the insects. Yet the fossil record reveals a lack of all advance, even in the most successful types such as ants, for a long time back — certainly during the last thirty million years, probably during the whole of the Tertiary epoch. Even during the shorter of these periods, the mammals were still evolving rapidly, and man's rise is contained in a fraction of this time.

What was it that cut the insects off from progress? The answer appears to lie in their breathing mechanism. The land arthropods have adopted the method of air-tubes or tracheae, branching to microscopic size and conveying gases directly to and from the tissues, instead of using the dual mechanism of lungs and bloodstream. The laws of gaseous diffusion are such that respiration by tracheae is extremely efficient for very small animals, but becomes rapidly less efficient with increase of size, until it ceases to be of use at a bulk below that of a house mouse. It is for this reason that no insect has ever become, by vertebrate standards, even moderately large.

It is for the same reason that no insect has ever become even moderately intelligent. The fixed pathways of instinct, however elaborate, require far fewer nerve-cells than the multiple switchboards that underlie intelligence. It appears to be impossible to build a brain mechanism for flexible behaviour with less than a quite large minimum of neurones; and no insect has reached a size to provide this minimum.

Thus only the land vertebrates are left. The reptiles shared biological dominance with the insects in the Mesozoic. But while the insects had reached the end of their blind alley, the reptiles showed themselves capable of further advance. Temperature regulation is a necessary basis for final progress, since without it the rate of bodily function could never be stabilized, and without such stabilization, higher mental processes could never become accurate and dependable.

Two reptilian lines achieved this next step, in the guise of the birds and the mammals. The birds soon, however, came to a dead end, chiefly because their forelimbs were entirely taken up in the specialization for flight. The subhuman mammals made another fundamental advance, in the shape of internal development, permitting the young animal to arrive at a much more advanced stage before it was called upon to face the world. They also (like the birds) developed true family life.

Most mammalian lines, however, cut themselves off from indefinite progress by one-sided evolution, turning their limbs and jaws into specialized and therefore limited instruments. And, for the most part, they relied mainly on the crude sense of smell, which cannot present as differentiated a pattern of detailed knowledge as can sight. Finally, the majority continued to produce their young several at a time, in litters. As J. B. S. Haldane has pointed out, this gives rise to an acute struggle for existence in the prenatal period, a considerable percentage of embryos being aborted or resorbed. Such intra-uterine selection will put a premium upon rapidity of growth and differentiation, since the devil takes the hindmost; and this rapidity of development will tend automatically to be carried on into postnatal growth.

As everyone knows, man is characterized by a rate of development which is abnormally slow as compared with that of any other mammal. The period from birth to the first onset of sexual maturity comprises nearly a quarter of the normal span of his life, instead of an eighth, a tenth or a twelfth, as in some other animals. This again is in one sense a unique characteristic of man, although from the evolutionary point of view it represents merely the exaggeration of a tendency which is operative in other Primates. In any case, it is a necessary condition for the evolution and proper utilization of rational thought. If men and women were, like mice, confronted with the problems of adult life and parenthood after a few weeks, or even, like whales, after a couple of years, they could never acquire the skills of body and mind that they now absorb from and contribute to the social heritage of the species.

This slowing (or "foetalization," as Bolk has called it, since it prolongs the foetal characteristics of earlier ancestral forms into postnatal development and even into adult life) has had other important by-products for man. Here I will mention but one — his nakedness. The distribution of hair on man is extremely similar to that on a late foetus of a chimpanzee, and there can be little doubt that it represents an extension of this temporary anthropoid phase into permanence. Hairlessness of body is not a unique biological characteristic of man; but it is unique among terrestrial mammals, save for a few desert creatures, and some others which have been compensated for loss of hair by developing a pachydermatous skin. In any case, it has important biological consequences, since it must have encouraged the comparatively defenseless human creatures in their efforts to protect themselves against animal enemies and the elements, and so has been a spur to the improvement of intelligence.

Now, foetalization could never have occurred in a mammal producing many young at a time, since intra-uterine competition would have encouraged the opposing tendency. Thus we may conclude that conceptual thought could develop only in a mammalian stock which normally brings forth but one young at a birth. Such a stock is provided in the Primates — lemurs, monkeys, and apes.

The Primates also have another characteristic which was necessary for the ancestor of a rational animal — they are arboreal. It may seem curious that living in trees is a prerequisite of conceptual thought. But Elliot Smith's analysis has abundantly shown that only in an arboreal mammal could the forelimb become a true hand, and sight become dominant over smell. Hands obtain an elaborate tactile pattern of what they handle, eyes an elaborate visual pattern of what they see. The combination of the two kinds of pattern, with the aid of binocular vision, in the higher centres of the brain allowed the Primate to acquire a wholly new richness of knowledge about objects, a wholly new possibility of manipulating them. Tree life laid the foundation both for the fuller definition of objects by conceptual thought and for the fuller control of them by tools and machines.

Higher Primates have yet another pre-requisite of human intelligence — they are all gregarious. Speech, it is obvious, could never have been evolved in a solitary type. And speech is as much the physical basis of conceptual thought as is protoplasm the physical basis of life.

For the passage, however, of the critical point between subhuman and human, between the biological subordination and the biological primacy of intelligence, between a limited and a potentially unlimited tradition — for this it was necessary for the arboreal animal to descend to the ground again. Only in a terrestrial creature could fully erect posture be acquired; and this was essential for the final conversion of the arms from locomotor limbs into manipulative hands. Furthermore, just as land life, ages previously, had demanded and developed a greater variety of response than had been required in the water, so now it did the same in relation to what had been required in the trees. An arboreal animal could never have evolved the skill of the hunting savage, nor ever have proceeded to the domestication of other animals or to agriculture.

We are now in a position to define the uniqueness of human evolution. The essential character of man as a dominant organism is conceptual thought. And conceptual thought could have arisen only in a multicellular animal, an animal with bilateral symmetry, head and blood system, a vertebrate as against a mollusc or an arthropod, a land vertebrate among vertebrates, a mammal among land vertebrates. Finally, it could have arisen only in a mammalian line which was gregarious, which produced one young at a birth instead of several, and which had recently become terrestrial after a long period of arboreal life.

There is only one group of animals which fulfils these conditions — a terrestrial offshoot of the higher Primates. Thus not merely has conceptual thought been evolved only in man: it could not have been evolved except in man. There is but one path of unlimited progress through the evolutionary maze. The course of human evolution is as unique as its result. It is unique not in the trivial sense of being a different course from that of any other organism, but in the profounder sense of being the only path that could have achieved the essential characters of man. Conceptual thought on this planet is inevitably associated with a particular type of Primate body and Primate brain.

A further property of man in which he is unique among higher animals concerns his sexual life. Man is prepared to mate at any time: animals are not. To start with, most animals have a definite breeding season; only during this period are their productive organs fully developed and functional. In addition to this, higher animals have one or more sexual cycles within their breeding seasons, and only at one phase of the cycle are they prepared to mate. In general, either a sexual season or a sexual cycle, or both, operates to restrict mating.

In man, however, neither of these factors is at work. There appear to be indications of a breeding season in some primitive peoples like the Eskimo, but even there they are but relics. Similarly, while there still exist physiological differences in sexual desire at different phases of the female sexual cycle, these are purely quantitative, and may readily be over-ridden by psychological factors. Man, to put it briefly, is continuously sexed: animals are discontinuously sexed. If we try to imagine what a human society would be like in which the sexes were interested in each other only during the summer, as in songbirds, or, as in female dogs, experienced sexual desire only once every few months, or even only once in a lifetime, as in ants, we can realize what this peculiarity has meant. In this, as in his slow growth and prolonged period of dependence, man is not abruptly marked off from all other animals, but represents the culmination of a process that can be clearly traced among other Primates. What the biological meaning of this evolutionary trend may be is difficult to understand. One suggestion is that it may be associated with the rise of mind to dominance. The bodily functions, in lower mammals rigidly determined by physiological mechanisms, come gradually under the more plastic control of the brain. But this, for what it is worth, is a mere speculation.

Another of the purely biological characters in which man is unique is his reproductive variability. In a given species of animals, the maximum litter-size may, on occasions, reach perhaps double the minimum, according to circumstances of food and temperature, or even perhaps threefold. But during a period of years, these variations will be largely equalized within a range of perhaps fifty percent either way from the average, and the percentage of wholly infertile adults is very low. In man, on the other hand, the range of positive fertility is enormous — from one to over a dozen, and in exceptional cases to over twenty; and the number of wholly infertile adults is considerable. This fact, in addition to providing a great diversity of patterns of family life, has important bearings on evolution. It means that in the human species differential fertility is more important as a basis for selection than is differential mortality; and it provides the possibility of much more rapid selective change than that found in wild animal species. Such rapidity of evolution would, of course, be effectively realized only if the stocks with large families possessed a markedly different hereditary constitution from those with few children; but the high differential fertility of unskilled workers as against the professional classes in England, or of the French Canadians against the rest of the inhabitants of Canada, demonstrates how rapidly populations may change by this means.

Still another point in which man is biologically unique is the length and relative importance of his period of what we may call "post-maturity." If we consider the female sex, in which the transition from reproductive maturity to non-reproductive post-maturity is more sharply defined than in the male, we find, in the first place, that in animals a comparatively small percentage of the population survives beyond the period of reproduction; in the second place, that such individuals rarely survive long, and so far as known never for a period equal to or greater than the period during which reproduction was possible; and thirdly, that such individuals are rarely of importance in the life of the species. The same is true of the male sex, provided we do not take the incapacity to produce fertile gametes as the criterion of post-maturity, but rather the appearance of signs of age, such as the beginnings of loss of vigour and weight, decreased sexual activity, or greying hair.

It is true that in some social mammals, notably among ruminants and Primates, an old male or old female is frequently found as leader of the herd. Such cases, however, provide the only examples of the special biological utility of post-mature individuals among animals; they are confined to a very small proportion of the population, and it is uncertain to what extent such individuals are post-mature in the sense we have defined. In any event, it is improbable that the period of post-maturity is anywhere near so long as that of maturity. But in civilized man the average expectation of life now includes over ten post-mature years, and about a sixth of the population enjoys a longer post-maturity than maturity. What is more, in all advanced human societies a large proportion of the leaders of the community are always post-mature. All the members of the British War Cabinet are post-mature.

This is truly a remarkable phenomenon. Through the new social mechanisms made possible by speech and tradition, man has been able to utilize for the benefit of the species a period of life which in almost all other creatures is a mere superfluity. We know that the dominance of the old can be over-emphasized; but it is equally obvious that society cannot do without the post-mature. To act on the slogan "Too old at forty" — or even at forty-five — would be to rob man of one of his unique characteristics, whereby he utilizes tradition to the best advantage.

We have now dealt in a broad way with the unique properties of man both from the comparative and the evolutionary point of view. Now we can return to the present and the particular and discuss these properties and their consequences a little more in detail. First, let us remind ourselves that the gap between human and animal thought is much greater than is usually supposed. The tendency to project familiar human qualities into animals is very strong, and colours the ideas of nearly all people who have not special familiarity both with animal behaviour and scientific method.

Let us recall a few cases illustrating the unhuman characteristics of animal behaviour. Everyone is familiar with the rigidity of instinct in insects. Worker ants emerge from their pupal case equipped not with the instincts to care for ant grubs in general, but solely with those suitable to ant grubs of their own species. They will attempt to care for the grubs of other species, but appear incapable of learning new methods if their instincts kill their foster children. Or again, a worker wasp, without food for a hungry grub, has been known to bite off its charge's tail and present it to its head. But even in the fine flowers of vertebrate evolution, the birds and mammals, behaviour, though it may be more plastic than in the insects, is as essentially irrational. Birds, for instance, seem incapable of analysing unfamiliar situations. For them some element in the situation may act as its dominant symbol, the only stimulus to which they can react. At other times, it is the organization of the situation as a whole which is the stimulus: if the whole is interfered with, analysis fails to dissect out the essential element. A hen meadow-pipit feeds her young when it gapes and squeaks in the nest. But if it has been ejected by a young cuckoo, gaping and squeaking has no effect, and the rightful offspring is neglected and allowed to die, while the usurper in the nest is fed. The pipit normally cares for its own young, but not because it recognizes them as such.

Mammals are no better. A cow deprived of its calf will be quieted by the provision of a crudely stuffed calfskin. Even the Primates are no exception. Female baboons whose offspring have died will continue carrying the corpses until they have not merely putrefied but mummified. This appears to be due not to any profundity of grief, but to a contact stimulus: the mother will react similarly to any moderately small and furry object.

Birds and especially mammals are, of course, capable of a certain degree of analysis, but this is effected, in the main, by means of trial and error through concrete experience. A brain capable of conceptual thought appears to be the necessary basis for speedy and habitual analysis. Without it, the practice of splitting up situations into their components and assigning real degrees of significance to the various elements remains rudimentary and rare, whereas with man, even when habit and trial and error are prevalent, conceptual thought is of major biological importance. The behaviour of animals is essentially arbitrary, in that it is fixed within narrow limits. In man it has become relatively free — free at the incoming and the outgoing ends alike. His capacity for acquiring knowledge has been largely released from arbitrary symbolism, his capacity for action, from arbitrary canalizations of instinct. He can thus rearrange the patterns of experience and action in a far greater variety, and can escape from the particular into the general.

Thus man is more intelligent than the animals because his brain mechanism is more plastic. This fact also gives him, of course, the opportunity of being more nonsensical and perverse: but its primary effects have been more analytical knowledge and more varied control. The essential fact, from my present standpoint, is that the change has been profound and in an evolutionary sense rapid. Although it has been brought about by the gradual quantitative enlargement of the association areas of the brain, the result has been almost as abrupt as the change (also brought about quantitatively) from solid ice to liquid water. We should remember that the machinery of the change has been an increase in plasticity and potential variety: it is by a natural selection of ideas and actions that the result has been greater rationality instead of greater irrationality.

This increase of flexibility has also had other psychological consequences which rational philosophers are apt to forget: and in some of these, too, man is unique. It has led, for instance, to the fact that man is the only organism normally and inevitably subject to psychological conflict. You can give a dog neurosis, as Pavlov did, by a complicated laboratory experiment: you can find cases of brief emotional conflict in the lives of wild birds and animals. But, for the most part, psychological conflict is shirked by the simple expedient of arranging that now one and now another instinct should dominate the animal's behaviour. I remember in Spitsbergen finding the nest of a Red-throated Diver on the shore of an inland pool. The sitting bird was remarkably bold. After leaving the nest for the water, she stayed very close. She did not, however, remain in a state of conflict between fear of intruders and desire to return to her brooding. She would gradually approach as if to land, but eventually fear became dominant, and when a few feet from the shore she suddenly dived, and emerged a good way farther out — only to repeat the process. Here the external circumstances were such as to encourage conflict, but even so what are the most serious features of human conflict were minimized by the outlet of alternate action.

Those who take up bird-watching as a hobby tend at first to be surprised at the way in which a bird will turn, apparently without transition or hesitation, from one activity to another — from fighting to peaceable feeding, from courtship to uninterested preening, from panic flight to unconcern. However, all experienced naturalists or those habitually concerned with animals recognize such behavior as characteristic of the subhuman level. It represents another aspect of the type of behaviour I have just been describing for the Red-throated Diver. In this case, the internal state of the bird changes, presumably owing to some form of physiological fatigue or to a diminution of intensity of a stimulus with time or distance; the type of behaviour which had been dominant ceases to have command over the machinery of action, and is replaced by another which just before had been subordinate and latent.

As a matter of fact, the prevention of conflict between opposed modes of action is a very general phenomenon, of obvious biological utility, and it is only the peculiarities of the human mind which have forced its partial abandonment on man. It begins on the purely mechanical level with the nervous machinery controlling our mus-

409

cles. The main muscles of a limb, for instance, are arranged in two antagonistic sets, the flexors bending and the extensors straightening it. It would obviously be futile to throw both sets into action at the same time, and economical when one set is in action to reduce to the minimum any resistance offered by the other. This has actually been provided for. The nervous connections in the spinal cord are so arranged that when a given muscle receives an impulse to contract, its antagonist receives an impulse causing it to lose some of its tone and thus, by relaxing below its normal level, to offer the least possible resistance to the action of the active muscle.

Sherrington discovered that the same type of mechanism was operative in regard to the groups of muscles involved in whole reflexes. A dog, for instance, cannot very well walk and scratch itself at the same time. To avoid the waste involved in conflict between the walking and the scratching reflex, the spinal cord is constructed in such a way that throwing one reflex into action automatically inhibits the other. In both these cases, the machinery for preventing conflicts of activity resides in the spinal cord. Although the matter has not yet been analysed physiologically, it would appear that the normal lack of conflict between instincts which we have just been discussing is due to some similar type of nervous mechanism in the brain.

When we reach the human level, there are new complications; for, as we have seen, one of the peculiarities of man is the abandonment of any rigidity of instinct, and the provision of association-mechanisms by which any activity of the mind, whether in the spheres of knowing, feeling, or willing, can be brought into relation with any other. It is through this that man has acquired the possibility of a unified mental life. But, by the same token, the door is opened to the forces of disruption, which may destroy any such unity and even prevent him from enjoying the efficiency of behaviour attained by animals. For, as Sherrington has emphasized, the nervous system is like a funnel, with a much larger space for intake than for outflow. The intake cone of the funnel is represented by the receptor nerves, conveying impulses inward to the central nervous system from the sense-organs: the outflow tube is, then, through the effector nerves, conveying impulses outwards to the muscles, and there are many more of the former than of the latter. If we like to look at the matter from a rather different standpoint, we may say that, since action can be effected only by muscles (strictly speaking, also by the glands, which are disregarded here for simplicity's sake) , and since there are a limited number of muscles in the body, the only way for useful activity to be carried out is for the nervous system to impose a particular pattern of action on them, and for all other competing or opposing patterns to be cut out. Each pattern, when it has seized control of the machinery of action, *should* be in supreme command, like the captain of a ship. Animals are, in many ways, like ships which are commanded by a number of captains in turn, each specializing in one kind of action, and popping up and down between the authority of the bridge and the obscurity of their private cabins according to the business on hand. Man is on the way to achieving .permanent unity of command, but the captain has a disconcerting way of dissolving into a wrangling committee.

Even on the new basis, however, mechanisms exist for minimizing conflict. They are what are known by psychologists as suppression and repression. From our point of view, repression is the more interesting. It implies the forcible imprisonment of one of two conflicting impulses in the dungeons of the unconscious mind. The metaphor is, however, imperfect. For the prisoner in the mental dungeon can continue to influence the tyrant above in the daylight of consciousness. In addition to a general neurosis, compulsive thoughts and acts may be thrust upon the personality. Repression may thus be harmful; but it can also be regarded as a biological necessity for dealing with inevitable conflict in the early years of life, before rational judgment and control are possible. Better to have the capacity for more or less unimpeded action, even at the expense of possible neurosis, than an organism constantly inactivated like the ass between the two bundles of hay, balanced in irresolution.

In repression, not only is the defeated impulse banished to the unconscious, but the very process of banishment is itself unconscious. The inhibitory mechanisms concerned in it must have been evolved to counteract the more obvious possibilities of conflict, especially in early life, which arose as by-products of the human type of mind.

In suppression, the banishment is conscious, so that neurosis is not likely to appear. Finally, in rational judgment, neither of the conflicting impulses is relegated to the unconscious, but they are balanced in the light of reason and experience, and control of action is consciously exercised.

I need not pursue the subject further. Here I am only concerned to show that the great biological advantages conferred on man by the unification of mind have inevitably brought with them certain counterbalancing defects. The freedom of association between all aspects and processes of the mind has provided the basis for conceptual thought and tradition; but it has also provided potential antagonists, which in lower organisms were carefully kept apart, with the opportunity of meeting face to face, and has thus made some degree of conflict unavoidable.

In rather similar fashion, man's upright posture has brought with it certain consequential disadvantages in regard to the functioning of his internal organs and his proneness to rupture. Thus man's unique characteristics are by no means wholly beneficial.

In close correlation with our subjection to conflict is our proneness to laughter. So characteristic of our species is laughter that man has been defined as the laughing animal. It is true that, like so much else of man's uniqueness, it has its roots among the animals, where it reveals itself as an expression of a certain kind of general pleasure — and thus in truth perhaps more of a smile than a laugh. And in a few animals — ravens, for example, — there are traces of a malicious sense of humour. Laughter in man, however, is much more than this. There are many theories of laughter, most of them containing a partial truth. But biologically the important feature of human laughter seems to lie in its providing a release for conflict, a resolution of troublesome situations.

This and other functions of laughter can be exaggerated so that it becomes as the crackling of thorns under the pot, and prevents men from taking anything seriously; but in due proportion its value is very great as a lubricant against troublesome friction and a lightener of the inevitable gravity and horror of life, which would otherwise become portentous and overshadowing. True laughter, like true speech, is a unique possession of man.

Those of man's unique characteristics which may better be called psychological and social than narrowly biological spring from one or other of three characteristics. The first is his capacity for abstract and general thought: the second is the relative unification of his mental processes, as against the much more rigid compartmentalization of animal mind and behaviour: the third is the existence of social units, such as tribe, nation, party, and church, with a continuity of their own, based on organized tradition and culture.

There are various by-products of the change from pre-human to the human type of mind which are, of course, also unique biologically. Let us enumerate a few: pure mathematics; musical gifts; artistic appreciation and creation; religion; romantic love.

Mathematical ability appears, almost inevitably, as something mysterious. Yet the attainment of speech, abstraction, and logical thought, bring it into potential being. It may remain in a very rudimentary state of development; but even the simplest arithmetical calculations are a manifestation of its existence. Like any other human activity, it requires proper tools and machinery. Arabic numerals, algebraic conventions, logarithms, the differential calculus, are such tools: each one unlocks new possibilities of mathematical achievement. But just as there is no essential difference between man's conscious use of a chipped flint as an implement and his design of the most elaborate machine, so there is none between such simple operations as numeration or addition and the comprehensive flights of higher mathematics. Again, some people are by nature more gifted than others in this field; yet no normal human being is unable to perform some mathematical operations. Thus the capacity for mathematics is, as I have said, a by-product of the human type of mind.

We have seen, however, that the human type of mind is distinguished by two somewhat opposed attributes. One is the capacity for abstraction, the other for synthesis. Mathematics is one of the extreme by-products of our capacity for abstraction. Arithmetic abstracts objects of all qualities save their enumerability; the symbol π abstracts in a single Greek letter a complicated relation between the parts of all circles. Art, on the other hand, is an extreme by-product of our capacity for synthesis. In one unique production, the painter can bring together form, colour, arrangement, associations of memory, emotion, and idea. Dim adumbrations of art are to be found in a few creatures such as bower-birds; but nothing is found to which the word can rightly be applied until man's mind gave the possibility of freely mingling observations, emotions, memories, and ideas, and subjecting the mixture to deliberate control.

But it is not enough here to enumerate a few special activities. In point of fact, the great majority of man's activities and characteristics are by-products of his primary distinctive characteristics, and therefore, like them, biologically unique.

On the one hand, conversation, organized games, education, sport, paid work, gardening, the theatre; on the other, conscience, duty, sin, humiliation, vice, penitence — these are all such unique by-products. The trouble, indeed, is to find any human activities which are not unique. Even the fundamental biological attributes such as eating, sleeping, and mating have been tricked out by man with all kinds of unique frills and peculiarities.

There may be other by-products of man's basic uniqueness which have not yet been exploited. For let us remember that such by-products may remain almost wholly latent until demand stimulates invention and invention facilitates development. It is asserted that there exist human tribes who cannot count above two; certainly some savages stop at ten. Here the mathematical faculty is restricted to numeration, and stops short at a very rudimentary stage of this rudimentary process. Similarly, there are human societies in which art has never been developed beyond the stage of personal decoration. It is probable that during the first half of the Pleistocene period, none of the human race had developed either their mathematical or their artistic potentialities beyond such a rudimentary stage.

It is perfectly possible that to-day man's so-called supernormal or extra-sensory faculties are in the same case as were his mathematical faculties during the first or second glaciations of the Ice Age — barely more than a potentiality, with no technique for eliciting and developing them, no tradition behind them to give them continuity and intellectual respectability. Even such simple performances as multiplying two three-figure numbers would have appeared entirely magical to early Stone Age men.

Experiments such as those of Rhine and Tyrrell on extra-sensory guessing, experiences like those of Gilbert Murray on thought transference, and the numerous sporadic records of telepathy and clairvoyance suggest that some people at least possess possibilities of knowledge which are not confined within the ordinary channels of sense-perception. Tyrrell's work is particularly interesting in this connection. As a result of an enormous number of trials with apparatus ingeniously designed to exclude all alternative explanation, he finds that those best endowed with this extra-sensory gift can guess right about once in four times when once in five would be expected on chance alone. The results are definite, and significant in the statistical sense, yet the faculty is rudimentary: it does not permit its possessor to guess right all the time or even most of the time — merely to achieve a small rise in the percentage of right guessing. If, however, we could discover in what this faculty really consists, on what mechanism it depends, and by what conditions and agencies it can be influenced, it should be capable of development like any other human faculty. Man may thus be unique in more ways than he now suspects.

So far we have been considering the fact of human uniqueness. It remains to consider man's attitude to these unique qualities of his. Professor Everett, of the University of California, in an interesting paper bearing the same title as this essay, but dealing with the topic from the standpoint of the philosopher and the humanist rather than that of the biologist, has stressed man's fear of his own uniqueness. Man has often not been able to tolerate the feeling that he inhabits an alien world, whose laws do not make sense in the light of his intelligence, and in which the writ of his human values does not run. Faced with the prospect of such intellectual and moral loneliness, he has projected personality into the cosmic scheme. Here he has found a will, there a purpose; here a creative intelligence, and there a divine compassion. At one time, he has deified animals, or personified natural forces. At others, he has created a superhuman pantheon, a single tyrannical world ruler, a subtle and satisfying Trinity in Unity. Philosophers have postulated an Absolute of the same nature as mind.

It is only exceptionally that men have dared to uphold their uniqueness and to be proud of their human superiority to the impersonality and irrationality of the rest of the universe. It is time now, in the light of our knowledge, to be brave and face the fact and the consequences of our uniqueness. That is Dr. Everett's view, as it was also that of T. H. Huxley in his famous Romanes lecture. I agree with them; but I would suggest that the antinomy between man and the universe is not quite so sharp as they have made out. Man represents the culmination of that process of organic evolution which has been proceeding on this planet for over a thousand million years. The process, however wasteful and cruel it may be, and into however many blind alleys it may have been diverted, is also in one aspect progressive. Man has now become the sole representative of life in that progressive aspect and its sole trustee for any progress in the future.

Meanwhile it is true that the appearance of the human type of mind, the latest step in evolutionary progress, has introduced both new methods and new standards. By means of his conscious reason and its chief offspring, science, man has the power of substituting less dilatory, less wasteful, and less cruel methods of effective progressive change than those of natural selection, which alone are available to lower organisms. And by means of his conscious purpose and his set of values, he has the power of substituting new and higher standards for change than those of mere survival and adaptation to immediate circumstances, which alone are inherent in pre-human evolution. To put the matter in another way, progress has hitherto been a rare and fitful by-product of evolution. Man has the possibility of making it the main feature of his own future evolution, and of guiding its course in relation to a deliberate aim.

But he must not be afraid of his uniqueness. There may be other beings in this vast universe endowed with reason, purpose, and aspiration: but we know nothing of them. So far as our knowledge goes, human mind and personality are unique and constitute the highest product yet achieved by the cosmos. Let us not put off our responsibilities onto the shoulders of mythical gods or philosophical absolutes, but shoulder them in the hopefulness of tempered pride. In the perspective of biology, our business in the world is seen to be the imposition of the best and most enduring of our human standards upon ourselves and our planet. The enjoyment of beauty and interest, the achievement of goodness and efficiency, the enhancement of life and its variety — these are the harvest which our human uniqueness should be called upon to yield.

FRANK M. CHAPMAN (1864–   )

The Big Almendro

From *Life in an Air Castle*. 1938

[FRANK CHAPMAN, dean of American ornithologists, has enjoyed a life full of accomplishment; he has traveled widely, has contributed notably to technical ornithology, and produced popular handbooks of unsurpassed pattern on birds. In addition he has written with clarity and wisdom in volumes of real natural history for the interested layman. His contributions on the subject of birds are so well known that I have chosen a chapter in another field — that of tropical wild life.]

TREES HAVE so many human-like attributes that one who is responsive to their influences inevitably endows them with personality. Their haunt, their size and shape, the appearance of their bark, the form and color of their leaves and blossoms, the nature of their wood, their sap, their fruit, even the movement of their limbs and the sound of the wind in their foliage combine to create the character through which a tree speaks to us — for that trees have voices no tree-lover will deny. It is the endless diversity of their pronounced characteristics and the confusion of their voices that overwhelm one in a tropical forest. The luxuriance of the vegetation sets no limit to their powers of expression.

The relentless cruelty of a strangler fig, the rigid uprightness of a palm, the benevolence of a tree-fern, the hospitality of a ceiba bearing an aerial garden on its huge, wide-stretched arms, the dignity and nobility of an almendro among the more familiar illustrations of tree character in the forests of Barro Colorado. Among them all my closest friendship is with the almendro — and to one almendro in particular do I pay homage. Standing among many great trees of a mature forest, it nevertheless dominates its neighbors. Although fully grown, it shows no signs of age. Rather is it in the prime of virile treehood. Six feet in diameter two yards from the ground, its splendid, smooth-barked trunk ascends column-like with but slight decrease in size seventy-five feet before it branches. Its limbs are open, symmetrical, graceful, and tapering. They stretch upward rather

417

than outward, and their tips are not less than 150 feet above the ground. Few parasites grow on its clean, brown limbs; it suffers no loss of individuality from a draping of vines, and in the calm dignity of its pose stands a prince among its fellows. Beneath its spreading arms grow trees with slender, limbless trunks and rather compact crowns, and beneath these are palms and an open undergrowth of saplings. The forest floor, at this Dry Season, is thickly strewn with dead leaves.

But if the almendro refuses hospitality to orchid, aroid, and fern, it gives it unstintingly to the animals of the forest. It bears annually, but in greater abundance every other year, a flattened elliptical nut about two inches long and half as wide, covered with a thin, fleshy coating and enclosing an almond-shaped kernel, whence (though the tree is not even a member of the almond family) it derives its common name.

The outer covering of the nut has a slightly sweetish taste and is eaten by coatis, kinkajous, and Howling Monkeys. Judged by most human standards it is inedible, but I fully share the liking of squirrels, agoutis, and peccaries for the contained kernel. Dried and roasted, it combines the flavor of a peanut and chestnut, with equally palatable qualities of its own. Some day, doubtless, a place will be given to it on the dinner table, bringing its delectable flavor but no suggestion of the majesty of the tree that bore it or of the romance of its associations.

During January, February, and early March, the period of its fruitage, the Big Almendro supports a large family, and I visit it frequently, not alone for the inspiration of its presence, but also to meet its guests. Chief among them is that raccoon-like animal, the coati. An adult coati weighs as much as thirteen pounds. He is thickset, short-legged, and rather clumsy in appearance. He has a long tail but it is not prehensile, and serves only as a balancing rod when he climbs. Nevertheless, with surprising agility he clambers about the outermost and uppermost limbs of the almendro, picking the fruit direct or drawing in the branches to bring it within reach. Often, in this act, he breaks them, and a bearing almendro much frequented by coatis has, in consequence, many small terminal branches of dead brown leaves.

There is no apparent difference in the external appearance of ripe almendro nuts, and the coati seems to be guided in his selection of food solely by his remarkable sense of smell. He walks slowly along a limb, curling up his elongated snout as he sniffs on this side and that, passing cluster after cluster of pendent nuts without picking one, but when he does help himself his choice is invariably to his liking. Then he stops or seeks a better resting-place, takes the nut in his forepaws and quickly gnaws off the brown skin, leaving an inner bright-green covering which adheres to the shell too tightly to be removed. Then he drops the nut. This is an important part of the food cycle. Not one almendro nut in a hundred falls with its covering intact. The ground beneath a bearing tree may be thickly sprinkled with nuts, but one may hunt in vain for one from which the brown outer coat has not been removed. Vainly I stalked almendro trees to discover the evidently abundant animal that fed on these nuts until one day a green, freshly eaten nut fell on me and I found that it had been dropped by a coati in the branches overhead.

Kinkajous, or so-called "Honey bears," also pick almendros, but they feed only at night, and I am unable to say what share of fallen fruit is theirs. Howling Monkeys feed largely on leaves, but at times add almendro nuts to their fare. It is these animals, therefore, that the agoutis, squirrels and peccaries have to thank for the food that daily and nightly falls to them. The first two are not concerned with the covering of the nut, but with its kernel — and dearly must they love it.

An almendro nut is hard as stone and it takes a sledge-hammer blow to break it. Agoutis and squirrels reach the kernel by gnawing through the hard shell — a well-earned prize — and, with experience one may tell which animal is at work by the key and the rhythm of its gnaw; the note of the squirrel being higher, the time faster. The peccary, on the other hand, cracks the nut along the lateral seam that divides it into halves, a tribute to the hardness of his teeth and the power of his jaws. He also eats it unbroken, doubtless for what remains of its outer covering, since it passes through the alimentary canal entire. These animals, therefore, must play an effective part in the distribution of almendro nuts and hence in the perpetuation of the series.

The *Cativo* (*Prioria copaifera*), a common neighbor of the almendro, bears large nuts which are evidently little, if at all, eaten by animals, and in the Dry Season the ground beneath these trees is densely grown with seedlings — a little forest of them. But one may look in vain beneath an almendro for a nut which has rooted. All, apparently, have been destroyed by the animals which frequent these trees so persistently during the season when its nuts ripen. It is of importance, therefore, that some of these nuts be removed from the area of their greatest abundance and essentially complete destruction. This the peccary does in a manner that seems especially designed to ensure their germination. The observed facts in this case relate to the Collared Peccary, a diurnal species. But it is probable that the White-lipped Peccary, a not uncommon but apparently largely nocturnal species on Barro Colorado, occupies a similar relation to the almendro.

My love of the Big Almendro, and my interest in its guests prompted me to seek a place among them. Obviously I could not hope to enter their circle on the ground floor, but it seemed quite possible that I might join the ranks of the tree-dwellers.

I shall not attempt to explain the significance of my inborn and life-long desire to occupy some kind of seat, perch, or platform in trees. These arboreal habitations were the delight of my boyhood and after fifty years and more the tree-hunting habit is still strong. Various have been my tree abodes. I recall a hollow chestnut large enough to give an uncomfortable but nevertheless enjoyable night's lodging, a "moss"-hung cypress in a colony of egrets and a mangrove roost shared with spoonbills, but none of these equalled in fitness and naturalness, charm and potentiality of environment, the tree-blind beneath the Big Almendro.

It was in a small group of palms that the blind was placed. Four of them formed the corners of a square about two feet across at the base and nearly twice as large at a height of ten feet. To the two trees on one side of this square crosspieces were nailed by way of steps. Ten feet above the ground similar pieces were nailed to the remaining three sides of the square, and small limbs laid on them made the floor of the blind. On this a seat was arranged, and by drawing in and tying the long, pinnate leaves of young palms growing below I secured complete concealment without perceptibly altering the surroundings. It was an ideal hiding place. Perched within it one felt indigenous.

I might now present a composite sketch of what I saw from this lookout, creating the impression of a nicely balanced play with events occurring in controlled succession as though each animal waited in turn to act its part or speak its piece. In truth there was no confusion of life at the Big Almendro and in any event one can speak of only one thing at a time, but I believe I can give a truer, more realistic, if less readable, account of events as I record them by merely presenting, with some comment, my observations as they were written in the blind.

The first day I ascended to my post (February 2, 1930, 7:50 a.m.), I learned how coatis reached the topmost branches of the almendro. Its trunk is obviously too large for them to climb, and it was evident that they must use some other stairway. I was barely settled when I heard footsteps on the dry leaves that cover the forest floor at this season, and a band of eight coatis appeared. When seen without their being aware that they are under observation, coatis impress one as being thoroughly at ease — like a skunk. They make no attempt to conceal their movements, dig here, root there, sniff and snort, and, apparently fearing no foe, exercise no caution and betray no suspicions. So, with long tails erect and waving jauntily, these eight animals came loping through the forest. They made no stops to look for food by the way but, as though traveling a familiar route, went straight toward the base of the almendro. When about ten feet from its trunk, without pause or consultation, the leader went up a sapling about three inches in diameter to a height of ten feet, left it there for a near-growing tree twice as large, ascended that for twenty feet, and then transferred to a pendent, rope-like vine, or liane, not more than an inch and a half thick. Up this he climbed for some thirty feet and then disappeared in the leaves of the lower tree-tops. No sooner was the leader's tail clear of the ground than he was followed by the next member of the band and he, in turn, by a third, so that before number one was lost to view all were climbing and several were on the vine at the same time. The almendro seemed festooned with coatis.

In ascending the small trees that formed the first stages of their journey, the coatis progressed with a galloping motion, that is, the front feet were thrown forward together, the hind feet followed, and the advance was made by jerks. But when they reached the rope-like vine they climbed hand-over-hand, following the movement of each forepaw with an exaggerated right and left wagging of the head which to one who knows the serious nature of coatis was very comical. I sat in my palm-leaf shelter entranced by this performance.

The animals traveled cautiously, with frequent rests, for the coati is terrestrial rather than arboreal and one can almost believe that, in spite of his agility, he has learned to climb at a comparatively recent period in his history. He seems never to feel at home when in trees and, if discovered there, loses no time in reaching the ground. On February 13, 1931, I found a group of coatis in the upper branches of the almendro. So far as I was concerned, they could not have been in a safer place. But as soon as they saw me they left the tree to seek various routes to the ground. On one occasion a half-grown coati, in its haste to follow its elders from the tree in which I had surprised them, missed its footing and fell for about forty feet. It landed in a bush-top, lay motionless for nearly a minute, then, as I advanced, jumped to the ground and scampered away.

From my blind I could observe only what transpired beneath the almendro. Its top was as hidden from me as the roof of a house from its cellar. It soon became evident, however, that the coatis were seeking their breakfast directly above me and I was shortly bombarded by the nuts they dropped. Fortunately the force of the fall of these stone-like fruits was broken by the tops of my palm trees, but thereafter I brought a peak-crowned hat to the blind.

From somewhere in the shadows came the sound of gnawings and crunchings. They were doubtless made by peccaries and agoutis, but I was unable to see them. These animals had finished their morning meal before the sun was high enough to penetrate openings in the forest roof and fleck its floor with golden patches. Then the great Amazona Parrots called *stop it, stop it quick-quick-quick* in a voice so loud and harsh that even as they flew through the tree-tops it was in truth ear-splitting; Short-billed Pigeons uttered with ceaseless fervor their emphatic *Je t'adore,* and often followed this avowal with an unloverlike growl; trogons cooed and cowed, toucans yelped their *Dios té de,* or croaked like frogs, fruit crows cawed, guans, locally known as *Pavos,* piped and drummed, but *Lathria unirufa,* the sentinel, was apparently the only bird to observe me. From a perch almost overhead he challenged with his staccato, explosive *see-you I-see-you,* a long, sweeping silvery whistle which in volume, clearness and commanding quality, I have never heard equaled by a bird. One marvels that so loud a note can be produced by so comparatively small a bird (he is only nine inches long) but like every good vocalist he sings without apparent effort. In color he is uniform brown. I was doubtless trespassing on his territory, possibly he had a home near-by, for invariably he challenged my right to be there.

Like most highly musical mornings it was calm. Not a leaf fluttered and a passing aeroplane shattered the silence with more than usual violence. The Howling Monkeys have not yet become accustomed to this intruder and invariably roar defiance at their only rival in sound producing. A clan not more than one hundred yards away now gave voice and another, distant about two hundred yards, added protest. It is unusual to find groups of these animals so near each other. Doubtless they were close to the boundary line which separates their respective territories.

At 8:30 the coatis, half sliding, half galloping, began to descend. All of them did not return, but at nine o'clock nuts ceased dropping and, as I afterward learned, a number of coatis slept in the almendro. Possibly they passed the day there, coming to earth after their evening meal. A glittering Morpho butterfly, the bluest thing in the world, passed erratically below me; *Lathria*, still suspicious, occasionally questioned; a squirrel near-by, holding an almendro nut in both paws, gnawed persistently. At 10:30 the forest slept to the droning hum of cicadas and I returned to the Laboratory.

It was 7:15 on the morning of February 26th, when again I climbed to my perch in the palm-trees. The sun was just entering the forest and the air was vibrant with the calls of toucans, doves, parrots, and ant-birds. *Lathria*, whistling sharply, soon discovered me. Peccaries were feeding beneath the almendro but left as I approached. No coatis appeared, and in the absence of falling nuts, I assumed that none had arrived. But at 7:45 the bombardment began and, first putting on my hat, I tried in vain to see its authors. A coati at about my level was viciously attacking a large air-plant growing in a neighboring tree. He literally tore it to shreds with his long, powerful claws in a search for insects, their eggs, or larvae. There was an outburst of hoarse squawks from a passing band of White-faced Monkeys, who seem always to be bound elsewhere, and from somewhere in the great canopy overhead I could hear the low conversational tones of howlers.

The nuts ceased falling. There was the sound of animals leaping in the upper branches, and a band of Howling Monkeys, who had apparently been feeding in the almendro, came into view as it passed near my blind. Familiar as I am with these remarkable animals, this was my first encounter with them in their element. Knowing that they rarely leave the upper limbs of the taller trees, a person on the ground listens to their astounding vociferations with composure, but having now assumed the role of an arboreal creature, I found that my point of view both literally and mentally was considerably altered. Moreover, I was now the unquestioned cause of their deafening uproar. An old male seemed particularly threatening. Descending to within fifty feet of me he roared until he choked, then, gasping, roared again. His teeth shone, his lips dripped saliva, his large luminous, protruding eyes set far apart in his broad, ebony face gleamed with savage ferocity to which his appalling bellow gave eloquent expression. Reason assured me that he was harmless, but, fortunately for the range of our mental experiences, reason is not always in control of them, and I was sufficiently moved by this encounter to enjoy it thoroughly.

A mother with a baby clinging to her breast, both its arms and tail encircling her body, struck a lighter note. Hanging from a limb by the tip of her tail, she swung to and fro and half revolved in response to the vigorous motions of her arms and hands as she "batted" insects swarming about her head. Insects, especially bot-flies, are among the howlers' chief enemies. One might imagine that in time they would have developed an immunity to them but in the absence of predatory foes there must be some checks to prevent their undue increase.

There was never as much activity in and about the almendro in the afternoon as there was in the morning. But the early nightfall is the time when pumas seek their evening meal, and it was with a hope that the agoutis usually beneath the almendro might prove a lure to pumas that I climbed to my perch at 2:30 on the afternoon of March 1st.

At this hour the animal world was still wrapped in the silence of its mid-day slumbers. But the trade wind was awake and as its voice rose and fell soothingly in the gently swaying tree-tops the motion of the palms holding my blind seemed to make me a part of the scene. The sky was nearly cloudless, but, except for little flickering patches of sunlight here and there, the ground below me was in shadow.

At three o'clock, in response to some unknown cause, a clan of howlers toward the Wheeler Estero announced their presence. Why the howler howls no man can always say. There are howls of song — as at daybreak — and there are howls of protest — as when an aeroplane passes. But there are also periods of howling not connected with time and for which the event is not apparent. Such outbreaks are led by the old males, and as their raucous roars rise and fall, the long-drawn howls of the females and young carry on the strain until again, with impressive surging rhythm, the voices of the males swell the chorus.

The forest now awoke. *Lathria*, evidently asleep at his post, challenged now sharply and with as startling effect as though I had not been expecting him. From near-by I could hear the fine, even gnaw of a squirrel patiently grinding his way to the kernel of an almendro. A Blue-headed Parrot flew over and a Shrike-vireo began to whistle *one-two-three* with tireless persistence. On the ground below the blind a teetering Northern Water-Thrush, fellow winter visitant, tossed the leaves right and left with as much energy as he might display in Canada. At five o'clock a single peccary fed, crunching on the far side of the almendro. The blind did not give me that intimate view of peccaries which I expected to have from it. It was unusual to approach the almendro at this season without finding a band of these animals beneath it. Nevertheless, few came while I was in the blind and, for peccaries, they seemed to be ill at ease. It is their nose, not their eyes, that puts peccaries on guard, and it is possible that my blind was not high enough to prevent them from getting my scent.

It was ten minutes of six when two agoutis appeared. They advanced with characteristic caution, a step at a time, eternally vigilant and ready to flee. I have seen agoutis numbers of times when it was reasonably certain that they were not aware of my presence and always they seemed frightened to the verge of flight. Their color blends closely with the dead leaves and, when motionless, they are almost invisible; but none of the forest animals is shyer or takes to its heels with less cause for alarm. A similarly colored bird would not fly until it was almost stepped on. But the agouti's lack of faith in the protective value of his coloration may be accepted as proof that it has none. Unlike a bird, he always lays a trail behind him that reveals his hiding place. The bird, when flushed, takes to the air and disappears, but the agouti, no matter how quickly he springs or how rapidly he runs, is earth-tied and leaves his scent behind him.

While the startling *whirr-r* of a flushed grouse or tinamou may be only the unavoidable result of the rapid impact of the bird's stiff wing-quills on the air, it is quite probable that this sound may have a certain protective value as it alarms a foe about to spring! But the agouti's alarm cry, as with astonishing rapidity it bounds off through the forest, seems definitely to be uttered as a means of frightening its enemies. Certainly no more threatening sound ever proceeded from so harmless an animal. It is a loud, explosive, rasping squawk that by mere force of suggestion creates a picture of curved claws tearing flesh. The first time I ever heard it I could almost see an ocelot, and after years of familiarity it invariably stops me in my tracks with a perceptible heart jump.

I wished the two agoutis beneath the almendro no harm but the hope was strong, as the shadows deepened, that a puma was on the track of at least one of them. I had flashlighted pumas on this trail within a hundred yards of both sides of the almendro. I used no bait, while here were two animals that evidently rank high on the puma's bill-of-fare. As if reading my thoughts one of them suddenly fled, but the other sat on his hind legs beneath my blind and, holding an almendro nut in his paws, gnawed industriously and with as much contentment as an agouti ever exhibits.

There was still light in the tree-tops, where toucans yelped and croaked, but although I could hear the agouti I could not see him. As abruptly as though under command the toucans ceased calling and with the silence one received an impression that night had fallen.

Descending to the ground, I was conscious of an equally sudden change in my mental attitude toward pumas. Only a moment before I had searched the undergrowth eagerly for a sleek form; now I felt that, on this occasion at least, we might cancel any engagement we may have had to meet beneath the Big Almendro.

At the same time I found that I possessed a much keener appreciation of the agouti's point of view. Indeed, there were moments during the mile and a half walk through the now darkened tunnel of the trail to the Laboratory when I felt that I *was* an agouti.

CARYL P. HASKINS (1908–   )

The Ties that Bind

From *Of Ants and Men*. 1939

[DR. CARYL HASKINS is ranked as a very able chemist, physicist, and research worker in radiation biophysics. When sixteen years old he became interested, as a hobby, in the social life of ants, and fifteen years later published the book called *Of Ants and Men*. Through ignorance, I am unable to evaluate Dr. Haskins's contributions to the realm of radiation biophysics, but I can state with confidence that the volume on ants would be a dominant achievement in the life of any man. My only regret is that I cannot quote several chapters instead of one.]

WHEN WE COMPARE the motives which bind together the societies of humans and of ants, we are forcibly struck by their similarity. Fundamentally, of course, the purpose of social organization is precisely the same in both creatures — to promote individual welfare and security, to permit the individual to live more peaceably in his immediate environment and to reproduce with greater safety, and to obtain that margin of social security which will provide for his needs in time of famine and uncertainty. Even the attitudes of the individual toward his social group are surprisingly parallel in men and in ants. Individuals of both groups labor under a force which may well be called "social pressure." We, to a far greater degree than we are usually willing to admit, are motivated in our daily actions and in the whole molding of our life-patterns by our desire that others shall think well of us, that we shall have general approval. This force acts in favor of our societies, since it promotes initiative, individual activity, and healthy ambition. But it is a beneficent force only as long as the ends of activity which it dictates are in fact those which coincide with the long-term welfare of the society which the individual serves, and which in turn serves him. This is not always the case, either with the individual or on the social plane. Evidence of the former is abundant. We need but to cite the thousands of lives and careers which have been blighted because their owners insisted upon entering a line of professional activity for which they were constitutionally unfit, merely because that particular line of activity chanced to carry social approval and prestige at the moment. Witness, too, the hundreds of other lives which have been blighted because the social estimate of certain types of activity changed suddenly, leaving all the participants in those lines of work who entered it, not for the joy of the work itself, but under social pressure, quite stranded. Evidence is equally striking on the social level. Witness the atrocities that have been carried out by large masses of people in the names of religion and patriotism since very ancient times, when those two great rallying standards for social pressure, of infinite value to society when the ends which they dictate are really of social advantage, chance to be perverted for the time. Social pressure and social approval in any single act or behavior-pattern is far more often a guide for large masses of society than the intrinsic end value of the behavior-pattern itself.

Ants are strikingly similar in their general social attitude. As with ourselves, social pressure is a force for social security so long as its ends coincide with the real good of the society involved, but it may, not infrequently, become perverted. Witness the worker of the higher type of ant, which, famished after hunting, eventually captures succulent prey. The first move of the starving individual, if the prey be movable, will be to return with it to the nest, and normally she will take no share of it herself until that return has been accomplished. If the food supply she finds be liquid, she will lap up great quantities of it, store it in her crop, and her very first move on her return home will be to regurgitate it to her hungry nest-mates. Such a move is clearly for the benefit of the colony. It is something else, however, when a nurse ant, in order to feed more growing mouths with the same limited food-supply, abandons her own large queen-larvae to rear the much smaller ones of a parasitic ant, or even those of parasitic beetles, as may sometimes happen. A normal ant, slaying intruders in the colony, is unquestionably acting in the interest of colony solidarity. Another worker, with equal fervor destroying her own queen in favor of a newly entered parasite of the type which we have already considered, is acting quite differently. Both insects are acting with equal necessity, under equal social pressure, and they have no sense, any more than we have in the great majority of instances, which will permit them to discern the direction in which that pressure is applied. Social pressure, for both men and ants, is a powerful tool, and one that, while normally acting for the good of

428

the individual and its community, is very readily perverted into channels directly destructive to both.

If, then, the social attitudes of individual men and ants toward their societies are strikingly parallel, the actual details of the agents compelling the actions which we see are so fundamentally different as to deserve a chapter, revealing, as they do, the differences in detail in the way that two ultimately similar societies can be organized.

Human societies, in marked contrast to those of ants, are consociations of small, primitive families, rather than highly developed, single ones. This situation confers advantages on the human group no less than it imposes disadvantages. Since the human state or nation is fundamentally a loose association of an indeterminate number of much more ancient units, it is a very flexible thing, and can quite naturally be expanded or contracted immensely in a short time, simply by adding on or subtracting units. Such additions and subtractions, to be sure, bring their own grave problems. Witness the complications of European affairs after the peace treaties. Among ants, such changes are nearly impossible. The union of autonomous colonies is a very rare thing in nature — their splitting, ordinarily, equally so. Furthermore, the size of most ant communities is strictly limited by the fecundity and life-span of one individual, a fact which is never true of man. The accompanying disadvantages of such a situation for man are quite as great and fundamental. Since the association is both a loose unit and one recently adopted, it cannot be expected that the members will have much instinctive ability in adjusting or resigning themselves to its best good, and this they certainly do not, in very many cases. It is necessary to fall back upon the much older social pressure appeal, under these circumstances, and to impose the force of the approval or disapproval of the society in a measure on all its components, in accordance with the society's intellectual judgment concerning the benefit or harm which their activities may bring it. Such measures, of course, range all the way from presentation of the Nobel awards to confinement on Alcatraz Island. Coercion of this sort is totally unnecessary among ant societies, and, in fact, would have no meaning whatever. The individual in the small ant society is in much the same position as the member of the human family, except that it is endowed with an instinct-pattern sufficient to carry it through whatever specialized duties it needs to perform with far less training than a human would require. The normal human is faithful to his own family, and acts in its best interest even when that may conflict with the newer unit, the state or the

429

nation, which unquestionably, on the average, claims only his second loyalty.

The ant is similarly loyal, but has an additional incentive. The member of the human family is still an independent, unmodified individual, capable of living alone, and moreover, ultimately, of starting a family of his own. He is in the position of the daughters of that first ant-queen, whose remains are not even preserved for us today, so ancient was the time in which she lived, which, winged like herself, ultimately left her to establish communities of their own. The modern ant worker is no longer such an independent unit. She is under physical and physiological compulsion to remain with her own family, for her own best interest. She is rather in the position of the human youngster of five, for whom separation from the family into which he was born constitutes the ultimate in dreaded catastrophes. This condition among worker ants is enormously intensified in the highly polymorphic species. Here the physical, mental, and physiological compulsion extends even to the particular function which she will serve in her family-community. By instinct, by intellect, and by physique, she is literally and unmistakably born into her job, and, with such highly evolved ants as the fungus-growers, there will be no other at which she can serve. Finally, there is no force whatever preventing any worker that may choose to desert her community from so doing. This actually happens, though doubtless unintentionally, in every new worker generation of any growing colony, as numbers of its members become irrevocably lost on their early hunting expeditions. Thus each group of workers is rigidly selected for those with both the inclination and the ability to devote their entire lives to the society which they serve. The penalty is simple elimination from the community, and from life, purely through the action of forces operating outside the community.

Such conditions make it very plain why an ant community is intrinsically a much more stable thing than any colony of humans can possibly be, and why criminal, or antisocial, action on the part of any member of it, in the sense in which we ordinarily understand the term, cannot occur, despite an utter lack of any intellectual regulatory force, such as a police body. The guiding penalties and rewards are there, but they are supplied, either entirely without the control of the ants at all, or quite automatically, without their intelligent direction. Despite this fact, the guiding penalties and rewards are extremely powerful, and, if we understand them, we will have gone far toward comprehending the fundamental structure of ant society.

The young ant community is at bottom a society that has been formed as a front-line defense against starvation. From the time that the half-starved first brood of the young queen emerges in search of provender while their famished and exhausted mother awaits within, the family constitutes a hunger group. The workers are small and weak because they have been only half fed, and they remain but half fed throughout most of their lives; for the greater part of the food they bring in must go to meet the ever-increasing demands of the younger brood. This situation is very little alleviated as the colony grows older. The larvae, to be sure, are much better nourished, and the adults emerging are therefore more robust, but their days of large meals, easily gained, are ended as soon as they reach adulthood. Thenceforth they, too, share in the responsibility of food-getting, and most of what they bring in will go to feed their mother, or those younger than themselves. It is very probable that in any large and flourishing ant colony, every member except the queen is in a nearly perpetual state of hunger, and this condition, of course, passes all bounds in the case of the driver ants. It is only necessary to offer a bit of food to any member of the ant colony, and to notice its immediate acceptance, to realize how far the whole social structure is motivated by its quest for food, and how far the workers represent undernourished images of young queens, unable to lay or to reproduce at all, primarily because they cannot accumulate sufficient reserves within their own tissues.

It is very natural, under these conditions, that every member of an ant community should constantly be in quest of any evidence of edible material, and keenly alive to such evidence. This quest, in fact, will be the dominating passion of its waking hours. It is well equipped by nature for the search, moreover. We have seen that the beautifully mobile antennae of ants in all probability constitute everted "noses" and touch organs of great delicacy of perception. There is much evidence, then, that ants are enthusiastic connoisseurs of odors, forms, and textures, and that certain combinations of things attract them immensely, while others are equally repulsive. The combinations which exert these specific actions certainly vary greatly with the species, but of their strong specific lure there is no doubt.

We seem to have the key to the principal binding and directing forces of ant society in perpetual hunger, and in the delicately specialized lure of various odors, textures, and tastes. The eggs and larvae of most ants have strong appeal for adults of their own, and often of other, species, merely as objects to be handled and licked. Thus ants of one species will usually accept the eggs or larvae of another.

The whole sensory approach is very close to that accompanying simple hunger, as shown by the fact that most ants, after accepting and temporarily showing much attention to the eggs and larvae of another species, will eventually devour them. Ants show much attention to the eggs of their own species, fondling them, licking them, agglutinating them in packets, and rearranging them many times a day. The hunger motive is on the very near horizon even here, for many eggs are eaten, even by their own parents, often as soon as they are laid, and only the surplus above immediate needs or tastes ever hatches. It is presumably the odor or the texture, or a combination of these, which renders the eggs so attractive to their nurses. It is not surprising, therefore, that so many of them should be eaten. The whole situation is very analogous to that of a dog with an especially succulent bone, which licks it avidly for a time before actually gnawing it, and which may not, for some time, gnaw it at all. In which case, among the ants, the bone hatches into a larva!

The larva is better protected against destruction by its nurses, although that protection is by no means invulnerable, since larvae are often eaten in normal colonies, especially in times of famine. Larvae, as soon as they are hatched, exude a quantity of saliva, and apparently also secrete through the skin an oily material apparently containing various esters in suspension. Both the salivary secretions and the exudates of the skin are immensely relished by the ants, which lick them most avidly from the surface of the larvae, thereby incidentally freeing the larvae of molds which would be dangerous to them. Now there is much evidence that the quantity of exudate and salivary secretion which the larva produces is roughly proportional to the amount of food which it receives, and, as they grow older, larvae come to exude these attractive substances principally immediately after feeding. Any young adult nurse soon learns the point, and it is a great help to the larva, both in preserving it from being devoured, and in ensuring prompt and efficient feeding. The average nurse would much prefer to deprive itself of a goodly amount of food, and give this to the larva, in return for the fascinating material which the larva yields in much smaller quantity. Among the higher ants, where the feeding of the larvae is carried on mainly by regurgitation, the nurses "play the game" with their charges very strictly, and ordinarily feed them whenever they themselves desire the exudates. Among the primitive Ponerines, however, where the initiation of many of these colonial ties can be very beautifully seen in its crudest form, the workers often pinch and maltreat the larvae for hours at a time without feeding them, with the hope of extracting a little more exudate material thereby.

The link which prompts the rapid and courageous removal of eggs and larvae from harm's way when an ant colony is disturbed seems fundamentally to be motivated by this concept of the eggs and larvae as a potential food supply. Among the higher ants, the connection has grown much weaker, and the larvae receive care and attention far superior to that which would be given an inert supply, but the author has seen workers of *Stigmatomma pallipes* seize particles of insect food which happened to be lying about the chambers of their nest when it was disturbed and hurry them away into dark corners with exactly the same attention as that which was given to the young.

The pupae of ants, whether enclosed in cocoons or naked, can be attractive only because of their texture or odor, and, among primitive types, they receive far less attention than the larvae, as they are nutritively less interesting. They are often severely neglected among the primitive Ponerines, and frequently hatch to adults quite unaided. The fragments of pupal skin adhering to the young adult, however, are potentially edible, and accordingly, in a Ponerine colony, the newly emerged ant is soon stripped of any adhering fragments. This stripping is for it a great and salutary service and provides its primarily selfish benefactors with a meal. The entire adjustment between individual and social good in such primitive forms is delicate and beautiful.

Among the higher ants, the development has gone much further. Although the pupae are not direct sources of food, unless eaten, they are none the less given great attention, and there is much evidence that they have become attractive in a somewhat derived, "maternal" sense, although the gustatory motive always remains. The desire to extricate the pupal envelope is much intensified; and the pupae are carefully drawn from their cocoons and the pupal coverings removed. Ordinarily, this process saves the life of the young ant, which would perish without it. It is to be noticed, however, that if the young insect be wounded, or for any reason its nurses proceed to devour its actual tissue in addition to the pupal envelope, this procedure will continue, and the unfortunate will be condemned to execution, cut up, and devoured.

These, then, are primarily the bonds uniting the adults with the brood. In these bonds we have essentially surveyed, although of course in oversimplified form, all the immediate motives that we can discern which bind the primitive Ponerine community together. Primitive Ponerine adults stand in a clearly competitive relation to their young, and this the young are not slow in reciprocating. The competitive attitude of the adults toward one another, moreover, is but very thinly veiled, and shows immediately one of their number is aged or incapacitated, as we have already seen. Otherwise, the attitude of adult to adult among the earliest of ants is one of comparative indifference, much as is betrayed between the members of a loosely federated colony of birds. The adults return with their food to the main colony after their foraging expeditions, primarily to feed the larvae, and to obtain the larval exudates that result from the feeding and partly because they are accustomed to the colonial environment, and prefer it as a place to rest. Other adults, which happen to be in the brood-chamber when food is brought in, share the feast, but this is entirely without the intent of the provider. The larvae, indeed, constitute the most effective binding cement of such a community. If the brood is removed, the colony quickly drifts apart. Through years of intimate acquaintanceship with the details of such family behavior of some very primitive Ponerine ants, the author has come to feel more and more strongly the essentially nutritive basis of their social organization.

Among the higher ants, the indifference of worker for worker has been largely discarded, and the colony is infinitely better integrated in consequence. The movement seems to be the result of two circumstances. The first is that of the somewhat "derived" interest which the higher ants become capable of taking in things originally nutritive in their significance. This has led . . . first to much greater care of the relatively unproductive cocoons and pupae, and, secondly, to a real appreciation of the adults for one another. Such appreciation, however, receives tremendous practical reenforcement from the habit of regurgitation, which is totally absent, so far as we know, among the cruder Ponerines. Concurrently with the development of a distensible crop or "social stomach," which can be emptied or filled at will, among the higher ants, has come their habit of filling it to its ultimate capacity when in the vicinity of nourishing liquids on their foraging expeditions, extracting but a very small quantity for their own personal use, and distributing by far the greater portion to their sisters on their return to the formicary. This distribution appears to be a pleasurable act for the giver, and is taken most gratefully by the recipient, which in turn imbibes more than is needed at the moment, and redistributes the excess to other sisters or to the brood. Thus a quantity of liquid food brought in by one foraging worker is shared among a very large number within a surprisingly short time. A whole system of signals, very readily recognizable for any given species of ant, has been developed about this act, whereby a hungry ant solicits a meal from one returning full-fed.

The habit of regurgitation has contributed immensely to colonial solidarity among all the higher ants. The strong bonds which have united the adult to the larva from earliest times, and do so today even among the primitive social wasps, were thus extended in equal intensity, to all adults of the community. The whole act, on a somewhat different plane, is reminiscent of our own social pleasure in regurgitating to a community such informational provender as we may have picked up, exemplified by the telling of stories among primitive peoples, gossiping, lecturing, and the writing of books in modern society. In such procedures, pleasure comes to both the donor and the receivers, and the analogy is probably one of the best which we, in our predominantly intellectual society, can conceive.

The social bond which is set up between adult and adult in the higher ant communities is strongly reminiscent of that set up in human relationships by the social attraction of a magnetic personality, but is carried to a far greater extreme. The individual who has a winning personality at first succeeds in a human social group to which he is unknown, regardless of whether he be a useless or a useful member of that society. Similarly, if he is possessed of unattractive superficial characteristics, he finds initial entry into the group and the attainment of a sympathetic hearing very difficult things, however advantageous it may be for the society to accept him. Much initial effort on his part is required to overcome this handicap with every new group which he enters. Only after a considerable time does it come to evaluate the more fundamental qualities of these newcomers, and ultimately, to be sure, they are judged almost solely in terms of such traits. The case is very similar, but much more extreme, among the ants. Given familiar or pleasing odor-characteristics, the worker of a colony may succeed under all circumstances with its colony sisters, quite regardless of whether it is a benefit to the colony or not. Contrariwise, given the repellent contact-odor of a stranger, the most efficient and energetic of ants will soon be slaughtered in an alien community. In such cases, there is no ultimate recognition of the social value of the individual.

The situation goes very much farther than this, and it is due to its inherent limitations that nearly all the striking and bizarre cases of ant parasitism are enabled to exist. . . . It seems fairly certain that the young parasitic alien queens which so easily gain entrance to

435

healthy colonies and ultimately deprive them of their own fertile females, accomplish this result through the possession of extremely attractive "gustatory personalities," which are never seen through by the individuals which they dupe. *Formica sanguinea* acquires her young colony by sheer force and without guile, but the species of *Bothriomyrmex,* . . . and all their ilk, are much weaker, individually and collectively, than the communities which they invade. They become the objects of much undeserved attention, and eventually bring about conditions of social perversion that can only be associated with extreme sensory satisfaction. There seems very little doubt that this satisfaction is gustatory. Some of the little parasitic queens, indeed, are actually provided with tufts of long hairs or "trichomes," which are apparently saturated with attractive ethereal materials highly prized by the host workers, which spend much of their time licking these trichomes and tending the bearers of them.

Ants are by no means the only insects that know how to exploit these rather rigid devices which make for social solidarity among the independent species. More than three thousand species of insects are harbored in ant colonies, for one purpose or another. Some of these, such as the "cows" of which all of us have heard, the aphids, leaf-hoppers, Coccids, and sometimes the larvae of small butterflies, are kept intentionally, for the sweet secretions which they produce. But a large class is in the formicary purely for its own purposes, and the benefits derived from the association are almost wholly on the side of the parasites. The order of insects having the greatest number of representatives in this situation is that of the beetles, and here we can trace a complete evolutionary series of parasites. It begins at one end with species which are mere marauders of the colony, which lie in wait in dark corners to pull down and devour the aged, infirm individuals and the helpless young, and which are persecuted and hated by the ants, and driven out by them whenever possible. These species, which include particularly the short-winged Staphylinids, are analogous to the wolves and jackals lurking about primitive human settlements. The ants are under no illusions as to their destructive character, but they have much difficulty in detecting or catching them, for these creatures are very alert and move swiftly and stealthily, and the bodies of some of them are so hard that, even if they are caught, they risk little damage and escape very soon, to take up their old occupation.

These beetles clearly possess topochemical characteristics extremely distasteful to the ants. But in a more advanced group of beetle parasites this has very largely changed, and the creatures are regarded with indifference by their hosts. They come to live in the main body of the nest, and slip about without interference, stealing food where they find it, and occasionally devouring the young brood. They are particularly fond of appropriating drops of food in passage from one ant to another during regurgitation, and the cheated hosts, unable to do anything about the matter, can only assume a puzzled and disappointed air.

The most highly specialized of beetle parasites, of which we shall have much to say in another chapter, has adopted exactly the means of ingratiation in the colony assumed by the most perfectly adapted of the parasitic ants, and has carried these devices much further than the Formicid parasites. The beetles of the genera *Lomechusa* and *Atemeles* are large, conspicuous, short-winged creatures, characterized by the presence of long tufts of golden trichomes, which evidently are covered with a film of an attractive substance, since the ants are fascinated with them, and, by extension, with their owners. Lomechusine beetles wander at will about the antnests, begging food from time to time from the ants exactly as an ant larva would, and being fed in the same way, after which its trichomes are licked avidly for secretions, just as the larva is for exudates. The beetle is viciously predatory, however, in more ways than one. Although relatively well fed by the ants at all times, it does not hesitate to devour ant eggs and larvae in quantity to supplement its diet. Its own eggs it leaves scattered irregularly about the nest. Evincing the same mechanical extension of interest in these eggs (which bring them gustatory satisfaction) that has produced such close cooperation among the adults of the higher Formicids, the ants secure the eggs and lavish upon them the greatest attention. The eggs hatch into fat, white grubs, not too unlike ant larvae in their general conformation, and these are promptly adopted by the foster-parents, fed, and given care greatly exceeding that lavished upon their own young. The beetle larvae grow rapidly, lying relatively quiet among the ant-young. They, like their elders, devour the ant larvae among which they are placed from time to time and attain a healthy maturity in short order.

Only one error in the conduct of the life of *Lomechusa* prevents it from immediately overrunning and destroying the host colony of ants whose social instincts it has so completely exploited. The larvae of the beetle normally pupate underground, and the pupae must remain covered with earth until ready to hatch. The necessity of being buried is well understood by those ants whose own larvae are accustomed to spin cocoons, for these are habitually buried while spinning, in order that they may have material to which the delicate silken threads can be attached. However, when spinning is completed, the ant larvae are disinterred and carefully cleaned, and herein the requirements of the ant and the beetle radically differ. True to their traditional methods of larva-culture, the adult ants carefully cover the mature larvae of *Lomechusa* and permit them to transform to pupae. Thereafter, however, they dig up and clean all the beetle pupae which they can find, just as they would their own young. The beetle pupae promptly perish under this treatment, and are thrown away. Were the ants quantitatively accurate at finding all of the larvae which they have buried, they might soon rid themselves of their parasite. Unfortunately, however, they are not, and a new *Lomechusa* population springs from the forgotten young.

Beetles of the genus *Atemeles* have carried this association still further, to the increased disadvantage of their socially blind hosts. They are more agile as adults than *Lomechusa,* pay considerably more attention to the adult ants in the colony, and have actually learned the signals ordinarily used by the adult ants in soliciting regurgitated food from one another. So, while *Lomechusa* is fed and treated like a larva by the ants throughout its own larval and adult life, and is perpetually regarded as a somewhat helpless, passive, and immature creature, socially speaking, the adult *Atemeles* is treated by the ants as one of themselves. It is permitted still further freedom in wandering about the colony, in devouring the young, and in robbing incoming adults of all their hard-won provender, which the ants are only too willing to surrender.

A colony infested with many of these parasitic beetles may go to amazing extremes in its insane hospitality toward them. The beetles and their young are fed and tended at all costs, however much the ant young themselves may suffer from neglect. Heavy parasitism by *Lomechusa* and *Atemeles* ultimately results in the production among the younger generation of adult ants of a highly pathological, perverted form known as the pseudogyne. It appears to result from the last-minute modification of a neglected queen-larva toward a smaller worker, although this derivation is not positive. In any case, it is certainly intermediate between the two castes. It usually possesses three simple eyes, in addition to its two compound ones, like the queen, and its head in general is of the shape of the queen's. Its thorax contains provisions for wing muscles, but there are no wings. Its stature is that of a worker, and it is usually poorly pigmented, and lighter in color. These individuals have very little energy, and are lazy and ineffective in both the work and the defense of the colony.

As colonies become more and more heavily infested with their insidious beetle parasites, these pathological individuals appear in greater and greater abundance, and ultimately they weight the whole social structure so heavily that it breaks down and the colony perishes. Strong communities of such ants as *Formica sanguinea,* which normally would live for many years, can be reduced to pitifully small remnants in a surprisingly short time when occupied by Lomechusine beetles. We shall have occasion to examine these ant parasites in another chapter, and to consider the relationships which they bear to certain social racketeers in our own societies. Here we may be content simply to note the example that they give us of the relatively unseeing, unmodifiable rigidity of the ties that so firmly bind together the members of an ant colony. These ties, under all ordinary circumstances, serve perfectly to unite the community; but the ants are helpless in the face of the proper sort of gustatory exploitation. That the "personality" ties which link us adult-to-adult in our modern societies are not exploited so completely in a similar way is only because of our greater mental plasticity, because we can "see through" our mistakes to a greater degree. The human ties, to be sure, are exploited in the same way, but to a lesser degree, although the error is more quickly rectified. See how far, for example, the charlatan of pleasing personality can go in fooling his public, to their own delectation and disadvantage, and how far the "bluffer" can travel on his bluff in the older sorts of business enterprise. The perception that enables the modern business executive to avoid the employment and support of such unprofitable people has been very hard-won. It is, moreover, far from perfect today, varies immensely with the employer, and is usually achieved only by the combined judgment of many associates. Ants are in the position of being obliged to make "snap" judgments among their associates, and to abide by them, for they have no criteria of subsequent behavior which they can apply within one generation. Were we confined by such limitations, our errors would surely be almost as grave and as frequent as theirs.

The result of the closely-tied association which ants have won for the nselves in normal, advanced colonies, rigid though the methods may be, is a margin of "social security" against a hostile environment that shows itself in many ways. A host of harmless insects comes to dwell within the formicary, taking advantage of that margin while doing little damage. The very parasites which we have considered could not subsist as they do were there not a large surplus of wealth to support them. That ants instinctively appreciate this security which they have won is to be seen in the attitude of workers of the same species in large and in incipient colonies, and when at home and abroad. An ant of an aggressive species, surrounded by her own sisters in a large colony, will risk her life in attacks and assaults which she will studiously avoid, or from which she will openly retreat, if she is alone or if her colony be young and weak. Similarly, if her colony be strong, she will not hesitate to attack and dispatch a stranger from an alien group, even if of her own species. If alone and in unfavorable circumstances, she will be much more tolerant of the same stranger, and will be inclined to temporize. In this way, two stranger ants which have lost their ways and have accidentally met far from home will frequently fraternize temporarily, their social instincts triumphing momentarily over their habit of rigidly maintaining colonial exclusiveness and solidarity.

Ants, to a degree far greater than that of men, are specialized, communal animals. The worker, as we have seen, has been nutritively reduced to the position of a dependent animal, and her actions correspond very exactly to this status. She is, however, not in the position of the specialized worker honeybee, for she can exist alone for long periods if need arise. The bee is incapable of doing this. Once again we see in the ant a greater degree of plasticity than that which we can discover in the bee, although the ant is much less plastic than man.

We have drawn such close attention in this chapter to the analogies between the details by which the social structures of men and ants are constituted that there is a real danger of assuming for the ant a consciousness and a thought-pattern similar to our own. Nothing, of course, would be more erroneous; for we have not the slightest justification for any such assumption. It is hard not to assume for the ant some sort of dynamic mental process, but we have only to think of the fundamental structural, nervous, and sensory differences between the ants and ourselves to realize how incapable we must be of accurately picturing to ourselves any of the particulars of their inner lives. This fact, however, in no way destroys either the reality or the impressiveness of the many social parallels which we can draw between the lives and organizations of two creatures whose fundamental purposes in living are so nearly the same. Once again, we need see and emphasize only the striking parallel in the courses of two rivers, sprung from widely different sources, but flowing to a common sea!

440

DONALD CULROSS PEATTIE (1898–   )

The Seeds of Life *and* The Sleep of the Seed

From *Flowering Earth.* 1939

[PEATTIE is one of the most successful of contemporary translators and paraphrasts of science, and his vocabulary and wealth of simile are phenomenal. For our Anthology I have chosen two from among his more simple, memorable essays.]

THE SEEDS OF LIFE

THIS EARTH, this third planet from the sun, was lifeless once. The rocks tell that much. There is one place in the world where the complete record is written on a single stone tablet. The Grand Canyon of the Colorado River is a cross section of geologic time. Cut by a master hand, the testimony appears to our eyes marvelously magnified. The strata burn with their intense elemental colors; they are defined as sharply as chapters, and the book is flung wide open. A silver thread of river underscores the bottom-most line, the dark Vishnu schist where no life ever was.

Mother-rock, these lowest strata are aboriginal stuff. They are without a fossil, without a trace of the great detritus of living, the shells and shards, the chalky or metallic excreta of harsh, primitive existence. These pre-life eras have been past for a long time — two billion years, perhaps. Perhaps a little more. Astronomical sums of time are so great that they bankrupt the imagination. We listen to the geologists and physicists wrangling over their accounts and compounding vast historical debts with the relish of usurers, but it is all one to us after the first million years.

No matter here how they arrived at their calculations. As plantsmen we are interested in the moment when the first plant began. For there was raised the flag of life.

The first life on earth — I have no doubt of it — was plant life. Any organism that could exist upon a naked planet would have to be completely self-supporting. It would have to be such a being as could absorb raw, elemental materials and, using inorganic sources of energy, making living protoplasm of them. Such describes no ani-

441

mal. But it perfectly describes an autotrophic plant. An autotroph is a self-sustaining vital mechanism.

The geologist's picture of the younger stages of this our agreeable planet home resembles the Apocalyptic doom for the world that I once heard predicted to innocents in a Presbyterian Sunday School. For the geologist sees flaming jets of incandescent gas, bolts and flashes that, condensing as they cooled, became a swarm of planetesimals, fragments comparable to great meteoric masses of stone and metal. These, by all the rules of orthodox astronomy, must rush together whenever their orbits come too close. So, by shocking impacts, the world was slapped together at random. It grew snowball fashion. It probably grew hotter, rather than cooler, from the friction and energy of the collisions, and the increasing pressure on the core must have generated a heat to melt the heart of a stone. So, in a molten state, the heaviest elements sank to the gravitational center, and formed the lithosphere — terra firma itself — while the lightest rose to become the atmosphere.

That atmosphere, it is presumed, was far, far thicker than it is today. It was perhaps hundreds of miles high, and may have had an abundance of now rare gases, like helium and hydrogen, neon and argon, and possibly even very poisonous gases, sulphur-drenched vapors, deadly combinations of carbon with oxygen, of oxygen with nitrogen. Almost certainly there was much less free oxygen and free nitrogen and carbon dioxide, than now, and correspondingly little scope for life as we know it.

But dense mists of water vapor, of steam clouds forever moiling and trailing about the stony little sphere, there must have been. For the oceans were, presumably, all up in the air. Only with cooling they began to condense, to fall in century-long cloudbursts, filling the deeps and hollows. At first, perhaps, striking hot rock, they were immediately turned to hissing steam again. The stabilization of the oceans alone must have been an awesomely long affair. It is doubtful if any sunlight at all got through that veil of primordial cloud, and the earth, viewed from Mars, would have been as unsatisfactory as Venus seen from the earth today, for the clouds of Venus never lift. Darkness then, darkness over the peaks clawed by the fingers of the deluge and dragged into the oceans; darkness over the forming seas that were not salty and full of an abundant and massive life, but fresh water, like that of the present Great Lakes. Fresh, and empty of life, warm, and dark. Darkness, and warmth, and water. Dark and warm as the womb, and awash with an amniotic fluid.

And into this uterine sea fell the seeds of life.

The Seeds of Life

The oldest fossils in the oldest of all fossil-bearing rocks, the Archaeozoic, tell six unmistakable things:

The first organisms of which there is any record on the stone tablets of time were cellular, just like all modern organisms.

They were aquatic, like all the most primitive organisms.

They were plants, unmistakably.

They were microscopic.

And they were bacteria.

Of course these were bacteria of a very special sort. Not in the least like the germs that cause diseases of man or those useful scavengers, saprophytes, that break up dead plant and animal remains and excreta. For these dread parasites and vulturine saprophytes are finicking and highly specialized. The parasites are hothouse species, most of them unable to endure more than a few hours outside very modern and complex bodies; even the saprophytes imply the presence of higher organisms to feed on. Not one is an autotroph. Not one sustains itself.

No, the kind of bacteria that left their marks upon the ineradicable record is a sort never studied by medicine. They are autotrophs, sufficient unto themselves. They invade no living bodies; they are probably not related at all to those which do, and if one kind is bacteria, the other ought really to have a clear name of its own. But there is no other common English name for them; botanists call everything "bacteria" which is so small that very little structure can be discerned.

One at least of these autotrophic bacteria that lived in the dark, hot, fresh-water ocean, was the selfsame plant that is found today in mineral springs heavily charged with iron, in old wells driven through hard-pan, in those rusty or tannic-looking brooks that seep away from stagnant bogs, where bog iron ore is gathering. Its name is Leptothrix. The Archaeozoic rocks are about one billion years old. In all that time the ochre Leptothrix has not changed one atom. As it reproduces simply by fission — the splitting of one bacterial cell into two — it has never died. It is, in body, immortal, and may outlive all other races.

The place to look for Leptothrix is around a mineral spring. On the rocks, in little nubbly reefs, in the brooks running from the springs, there waves a yellowish-rusty slime. This has a greasy feeling to the fingers; it rubs away instantly to nothing. But when you tease a little out in a drop of water, and shove the drop, on its clean glass slide, under the lens, the slime comes to life. For besides a great deal of shapeless rusty blobs and cobwebs, there are imbedded in

443

this mass long unpartitioned filaments or tubes. They look a bit like root hairs under low magnification, and are surrounded by a nimbus of slime.

But the walls of the filament are absolutely definite; they proclaim organization, clear-cut form, something with the shape that only the living take on. And those walls of the filaments are of iron, deposited around the living bacterial cells by accretion.

As for the bacterial cells themselves, they are elliptical bodies, but remarkable for having "tails." So, placed end to end, they look like pollywogs packed into a boy's pea shooter. When overcrowded, some of the bacteria escape. Then by their lashing polar tails they swim free, just like a sperm cell of a seaweed or a mammal. Soon a fresh deposit of iron settles around them. As it lengthens, daughter cells come to fill it, by fission of the mother-cell.

Actual living Leptothrix colonies fully charged with active bacteria are not especially easy to find. Often one hunts for hours on bacterial slides, encountering only empty sheaths. But their fossil imprint is particularly sharp and unmistakable. And the sheaths, being iron, and not living matter subject to decay, have long lasting powers. Thus in the iron-charged waters that overlay some of the most ancient of rocks, Leptothrix flourished for countless dark ages, slowly, slowly dropping the detritus of its outworn shards, building up an ooze that, under the terrific pressure of the water above, became iron ore.

But how, it is right to ask, was Leptothrix able to live without photosynthesis? How was it nurtured in a water that contained few or none of sea water's rich salts of today, but only a bitter diet of iron compounds?

Leptothrix lived then, as it does today, by oxidizing iron. When we oxidize carbon (burn coal) we release enough energy to turn all the mills of the world. When oxygen rushes into the lungs of an asphyxiated man, his anemic blood is refreshed; his eyelids flutter, he comes to life. Life is one vast oxidation, one breathing and burning. Man and his beasts are fueled by the plants; the plants consume the earth stuff they built up by their green sun-power; but Leptothrix, aboriginal, microscopic Leptothrix, taps atomic energy. It literally eats iron.

Such is chemosynthesis, contrasted with photosynthesis. In a darkened world of water, chemosynthesis was then the only possible synthesis — or assembling of materials into life — and how effective it was for how long can be judged from the work of Leptothrix in the waters that once rolled above the Mesabi range, north of Lake Superior. This iron seam, believed to be largely the work of iron bacteria depositing a subterranean reef, is called by engineers simply "The Range," for beside it there is no other comparable. It is the range of all iron ranges, and so great and so heavy is the ore yearly moved out of it, that the locks of the Sault canal, though open only six or seven months of the year, and having a traffic deeply loaded only on the out voyage, transmit more tonnage than any other canal in the world, excepting none.

The iron moves eastward and southward because it is on its way to the coal beds of Pennsylvania and Illinois. And coal is another vegetable resource, the fossilized life of the fern and club-moss forests of the Middle Ages of plant history. So when the primeval bacterial iron meets classic fern coal, there the blast furnaces begin to roar, and the skies to glow. I grew up only two miles from one of the greatest steel mills in the world, and from the cupola on our old mansard roof, I liked to watch the rusty lake boats creep over the horizon toward the belching chimneys by the shore. Turning, I could see the freight trains from the mines crawling at the back of the prairie rim. Boat and engine, each in its pitch, hooted a long deep portent. Iron coming to meet coal, coal to meet iron. Out of that meeting, melted and smelted and tempered and rolled, strides the daemon of our age, steel, clanking thighs and swinging arms.

And still today the fern frond trembles in the forest, Leptothrix rusts the waters of the bubbling spring.

Others of these element-consuming bacteria oxidize carbon or hydrogen or nitrogen or ammonia or marsh gas. When they combust this last, then the will-o'-the-wisp dances over the bogs. Still another has manganese for its staff of life. Manganese, by the way, is an alloy of the steel used to burglar-proof safes. But it is no proof against the microscopic, hard-headed Cladothrix. Variously we are being used or served by these masters of a fundamental and simple way of life, the autotrophic bacteria. Some of them have holdfasts, like a kelp or rooted waterweed, so that instead of floating at random, they can grow forest-wise in the waters they inhabit. These enter water pipes and vegetate there, like some flaccid but indomitable eel-grass in a stream, till the pipes are wholly stopped.

Of the bacterial autotrophs one you may smell on the air, and the odor is very like that of rotten eggs. For this one (and its name, if you like, is Beggiatoa) battens upon brimstone. It lives in the mud of curative baths, and grows in sulfur springs, and by building up a slimy reef it makes a bowl about some geysers, enduring and even luxuriating in a zone of their waters that is hot but just not too hot for it. To look at, this sulfur bacterium is colorless. Under the lens, you may see its strands slither, slipping over each other in a perpetual undulant motion with the indifference of a knot of bored snakes.

Now, this ill-smelling Medusa is important to all of us alive here. Not so much because it is sometimes implanted by engineers in septic tanks as a valuable destroyer, as because of its greed for the sulfur on which it lives. It is after sulfur everywhere, anywhere, that it can get it in Nature.

Abbreviating the chemistry of it, the result of Beggiatoa's use of sulfur is sulfuric acid. This is combined with the limes of the soil, creating a compound of calcium and sulfur that is exactly the fertilizer for which all roots are hungering. They do not use, they cannot absorb, the sulfur and sulfurous compounds around them until Beggiatoa has produced this particular form of it.

And living protoplasm must have sulfur, especially for its nucleus. Just a pinch of this mustard among the elements — but that pinch is indispensable to the cuisine. So Beggiatoa unlocks for all the rest of life this invaluable yellow ingredient.

All these autotrophs, with their strange diets and their labor in the dark, are without color. But there is one more autotrophic group which catch the attention because they are pigmented. And the pigments, although not usually green, are photosynthetic.

The red or purple bacteria must, then, have light for their work. Equally, they must *not* have free oxygen, for it is fatal to them. When we cultured them in the old Agassiz laboratory, we filled the flask to the brim with water, stoppered it against air, and put it in the sunshine at the window. There photosynthesis began.

How, since there was no chlorophyll? The answer refers the imagination to antiquity. The pigment of the reds or purples is called bacterio-purpurin, and I don't think anyone knows very much about it, but this much is plain to any mind: bacterio-purpurin (the red) is the complementary color of chlorophyll (the green). So these two utilize just the opposite parts of the spectrum. Imagine, then, that murky and chaotic age of the world, when sunlight was probably of quite another quality than this upon my desk today, and filtered many of the rays that make so gay the little patio garden beyond the window. What used that strange sunlight, what toiled even then at the beginning of the industry that is the world's greatest, may have been — must have been — the purple bacteria.

Early as these purple laborers were at the mighty business, those pallid brother autotrophs, the iron and sulfur bacteria, were, I think, earlier still. For they required not even the tool of light. They were already active in the day of darkness, in the beginning of things. It is difficult to picture any earlier form of life.

Many a scientist will not think about what he cannot see, so that any sort of speculation embarrasses him. He would not cáre even to imagine some Columbian voyage of spores, propelled through space by the corpuscles of light, perhaps, nor picture that moment when they claimed this new world as the lawful possession of life, to have and to hold, by root and by shoot, from that day forth.

For the origin of life was a definite historical event here upon earth. There was a time when there was no life here, and then suddenly there must have been life.

A thing is either quick or it is dead. Nobody has ever seen any organism, structure, or molecule that was half-alive. No one has ever seen the vital spring from the inanimate. All the experiments and proofs of the last century of biology have dealt the theory of spontaneous generation staggering blows; today it is an axiom that life comes only from life. To fancy that the spark was kindled once and forever out of the rock, as by a magician's staff, quits the evidence of science. If you are letting free your imagination, let it fly farther. Let it conceive of a landfall, a landing. An arrival.

Nor is this just a poetic flight. Men with great names in the discipline of science have boldly believed this. First to dare such a theory were Preyer and Richter in Germany, substantiated by Von Helmholtz himself. Svante Arrhenius, that universal genius, and Lord Kelvin extended the idea. In our times Kostychev the Russian has brilliantly supported it, and Charles Lipman in America searched for tangible witness. With elaborate precautions against earthly contamination, he opened new-fallen meteorites and found at the heart bacteria similar to some of those known in the soil of this our dusty planet home. It seems, to some, only certain that, in spite of the best intentions, earth-born bacteria were carried into the meteoric stone in sectioning. Others accept the evidence of life out in space similar to life on earth.

Harsh and terrible are the conditions of outer space. A breathless darkness stabbed with killing rays of ultra-violet light, islands of flame in a sea of icy emptiness, distances so awful they make us laugh a little with fear, time itself as an ally of death — all these would lie in wait for life once it quit a sheltered corner of the universe.

The organism that would bridge such a shuddering abyss must be so small as to be microscopic, for it must be able to escape by its lightness the gravitational pull of the body that harbors it. It must

447

be so infinitesimal that the corpuscles or bullets of light will propel it through space. It must be able to function in darkness and endure inconceivable cold. It must be able to find atomic sources of energy and batten upon raw elements. It must be an autotroph all but immortal.

Those old iron bacteria, consuming their formidable sustenance back in a darkling beginning, are nearer to this concept than anything alive today, or anything with a fossil record. But it follows the nature of things to presume for them ancestors smaller and even simpler, equally autotrophic, able perhaps to subsist upon the interstellar dusts. Or, at all events, capable of a voyage between one hospitable star and another.

The mind that will not take a flight may instead propose, as a bridge between the living and the non-living, the invisible, non-filterable viruses. They have been of late one of the most spectacular displays of the biological laboratories, and they are fascinating and important. But it is not known yet whether they are living organisms, or only substances derived therefrom. Suppose them alive: they could not live without higher animals and plants to parasitize. Their history can be no longer, then, than the history of the comparatively recent forms which they disturb and destroy. They are obligate parasites. They are not autotrophs.

Then as we search the infinite for the origins of the daily terrestrial miracle, life, we catch a gleam of but one great possibility. We are, perhaps, in the position of the Indians of San Salvador on October twelfth, 1492, asking themselves whether these glittering strangers were of divine origin, or, rather, spontaneous apparitions of the sea foam. Or might they actually have arrived, as they were asserting, from a land across the ocean believed intraversable? Were they then men, like Indians, who in their turn had been begotten by men to and beyond the limits of memory?

If life is indeed pan-cosmic, we shall have to admit that in the fulness of time it would be likely to invade all realms and flourish in every place suited to its requirements. More, we shall have to admit that it is in some measure ultimately unknowable. With this I am content. This I do not dispute.

THE SLEEP OF THE SEED

The meaning of flowering is one and simple and to this purpose: that the speck of protoplasm inside the pollen grain — which is in the anther, which is borne on the stamen, which is enclosed by the petals — shall find its way to the bit of protoplasm which is within the ovule, which is in the ovary, which is inside the pistil at the heart of the flower.

Four hundred miles the pollen grain may travel to its goal, as when a high wind blows through a pine forest where the trees are loosing the dust of their fecundity. Or the consummating act may be immediate within the flower, as in those self-fertilizing blossoms where the anther rests right on the stigma. But whether it travels or breeds at home, the pollen grain, from the moment it quits the anther, is an independent plant.

It is a plant like no other, a plant reduced to a microscopic charge of one half of the life of its species. It is so tiny that it ranges from the giant grain of a pumpkin's pollen, all of a hundred-and-twenty-fifth of an inch through, to that of forget-me-not, which is one six-thousandth of an inch. It is so simplified that it comprises merely a sphere wrapped in one or two coats, within which is sheltered a male pro-embryo. This mere fraction of life, this dot of protoplasm, is crowded with the nuclei which carry the chromosomes, messengers of heredity.

Bearing the illimitable future of its kind, the pollen grain has an individual life span of a few hours, a few days, most exceptionally a year. For it is a plant without chlorophyll, and thus cannot work for a living; it has no reserve of food supply packed away with it, as a seed has. It is not yet even part of a seed; it is a spore, very special, highly evolved, a sex spore encapsuling the male generation of its race.

The ferns and the fungi and the brown seaweeds also confide their futures to the spore, but in these tribes it has the true sperm's power of driving toward its objective. The male pro-embryo of a flowering plant has lost the power of self-motility; it taxies within the golden pollen grain, borne by wind or wing.

At the mercy of such chance, tiny, frail and quick to die, the pollen grain is a spark of life that breathes, thirsts, is capable of growth, and can impregnate. It lives to no other purpose than to reach the female pro-embryo, which is charged with the other half of the race life. This female spore is locked up within the ovule, or unfertilized seed-to-be. One end of it is nutritive, and will form the future food supply of the seed; the other is strictly sexual and conjunctive. It is altogether just a little less nothing at all than the male pro-embryo, but it too carries in its nuclei the inscrutable pattern of chromosomes that spell destiny for its breed.

Observe in passing how the alternation of generations — that marvel and sorrow of the ferns — has changed in proportion. In ferns the sexual stage is found on independent plants, little inconspicuous flat bodies as unlike the bending frond as a caterpillar is unlike its butterfly. But in the seed plants the sexual generation has been foreshortened, cut, telescoped, protected and concealed, into the male pro-embryo within the pollen grain — a passive passenger on a short trip — and the female pro-embryo, a potent prisoner inside the ovule.

Dry and avid of water, without a reserve of nourishment, on the verge of death, a lucky pollen grain arrives at the stigma. This has stickiness to catch the pollen, or a knobbed surface to hold it, or, as in the silk of a maize ear, feathery tendrils to seine the wind for it. And the glistening stigmatic fluid is food and drink to the pollen grain which swells with the moisture absorbed and swiftly thrusts out a tube that carries at its tip, as it grows, the male pro-embryo.

Down through the pistil grows the tube, down through the long style. Or right up it, in a nodding flower, gravity having no influence on the growth of the pollen tube, as it has on roots and stems. Or it prizes its way between the loose pistil cells, or it may secrete solvents that liquidate them as it grows. It lengthens out, and out — and not only from inner propulsions.

For it appears that the stigma and the ovules themselves lead it on, suck it in, by the eager secretion of exciting chemicals. This would seem a secret tightly hidden in the flower, but pollen placed on wet paper will grow straight to fragments of the stigma, to macerated ovules put near it in a film of water.

From the moment of pollination, when the grain arrives at the stigma, to fertilization, when the tube has reached the ovule, the growth of the pollen tube may be swift, as in our common stone fruits, when the act may take as little as nine hours. In pines and oaks, it is a matter of many months; in some of the cycads, of years.

Then when the tube has reached and penetrated the ovule, the male pro-embryo and the female pro-embryo meet. Their nuclei fuse, their chromosomes combine, mingling the dual heredity of all things born of sex. So the new life is conceived. The seed is set. The ovary grows great with it, the petals and stamens withering away from its crescent new estate.

So and so only is life created, by some spark from the burning brand of an old generation just touching and quickening the embryonic next. So and so only arises the wonderful, the fearful individuality of life's children, some forever destined to be less than the blood that made them, some to exceed it, all to struggle with their environment.

Once set, the seed begins its complex self-organization. The coats of the ovule become the coats of the seed, usually a hard outer one and a membranous inner. Within them, what was the structure of the ovule conditions the plan of the seed. At the ovule's nutritive end develops the seed's reserve food supply; the other end, where fusion with the male element took place, becomes the living embryo.

The scheme of storing nutriment is not known to the spores of the lowlier plants, and seeds by its possession have mastered the face of the earth. The food may lie quite outside the embryonic plantlet, visibly distinct, like the meat of a nut or the sweet kernel of a grain. Or it may lie in the seed-leaves, the fat cotyledons, nourishing them until they fill the whole cavity within the coats. Starch and protein, sometimes oil, or sugar or cellulose are packed away by an unthinking wisdom, provision for the days of growth.

For this the squirrels jump in the coloring trees of autumn; for this the chaff blows on the hot August air. But the seed's own vitality lies not in the meat but in its least conspicuous part — the embryo. In the seed of the common rush out on the marshes, it weighs but one sixty-eight thousandth of a grain. In a coconut it is a giant, weighing all of two grains, and still this is only one two-thousandth of the total weight of the coconut without its husk.

The embryo is the least part of a seed, and it is the life of the seed. In it lies the power of response unique to the living. From its beginning, the embryo answers opposite poles in its development. At its blunt end, it feels the pull of the root. At its acute tip, it pushes with the upward thrust of the shoot. Already, in its minutely integrated individuality, it perceives the difference between dark and day, down and up, earth and air. And it obeys the diametrically opposed commands of its nature to follow the cosmic stimuli. For the root obeys gravity, the shoot deliberately disobeys it. The shoot seeks light, the root shuns it. Their destinies are antipodal, and even the immature seed still upon the mother plant seems to know it.

It grows, in root and shoot, by swift repeated cellular division, the cleft passing right through the nucleus, each time shearing and sharing it fairly. At last, by crook, hook and bur, by wind or bird or wave, the seed is parted from the parent plant, and sets forth to seek its fortune.

Many and crafty are the devices by which these adventurers make their way in the world. Once Darwin grew a whole weed garden from seeds taken from the feet of migratory birds. Then there are the creeping fruits that hitch themselves along the ground by hygroscopic contortions of their spines. Dreadful are the travels of the porcupine grass, whose tails screw the grains forward, even into the flesh of prairie cattle and through the clothing of cowboys. Some fruits eject their seeds, as the pods dry and crack, to the distance of a foot, a yard, three yards. I used to hear, when I wrote in a Riviera garden, the whang of acanthus pods as they exploded in the arid summer afternoons. Even the modest violet pops its pods, and the touch-me-not is ever ready to do so. Fruits there are, and seeds, which remain buoyant in sea water for as much as four months, without becoming water-logged or losing their viability. So have the beaches of the Pacific isles been populated with a cosmopolitan strand flora. But·the seeds and fruits with a fitness to travel by wind are beyond all numbering. They range from the minuteness of orchid seeds, fine as pollen, to the shining argosies of milkweed down blowing through our idle summer hours. By its winged seeds the cattail has encircled the earth, one species growing in the marshes of all continents. The seeds of the Spanish moss, which is a flowering plant and neither the moss nor the lichen it appears to be, are equipped with a parachute of sharp bristles that, when they lodge in the crevices of bark, serve to anchor the seed precisely where alone it can thrive.

Howsoever dispersed, whether it makes a happy landing or not, the seed cut off is another thing from the seed on the parent plant. There, in its immaturity, it was soft, watery, plump, vulnerable, and full of active life, its parts dividing and differentiating to a nicety. But the seed on its own seems another plant. And now indeed is the plant reduced to first essentials.

Cramped into a tiny space where it cannot grow, it shrinks and shrinks. It loses water steadily, till it is desiccated to a Mojave dryness; every plant is a desert plant when it is a seed. The coats harden sometimes to stoniness; the tender cotyledons lose their green and become blanched as if in death, and the tiny true leaves of the embryo furl like conquered banners. Unresistant, the spark within allows itself to be immured alive; it is cut off as if in a coffin, from all the brightness of the world. The very breath drops lower, till at last, sometimes, no finest chemical test can prove that the seed does indeed still breathe. Vital chemistry stops like a run-down clock. There is a choked coagulation, and one cannot but remember the fabulous mystics of India who swallow their tongues and so by ceasing to live with any show of life, live on for years as men dead.

Among the longest lived and most impermeable of seeds, the more profound this look of death, the likelier it is that they still live. If they are really dead, they drink water like a dry log, swell up, even seem falsely to germinate, by bursting with the slaking of their thirst. But the living seed seems to remain athirst voluntarily. By the depth of its sleep we know how stubbornly alive it is.

The secret of this life-in-death is the impermeability of the seed coat. In so far as the coat can shut out water and oxygen, and shut in the carbon dioxide evolved by the seed's own low but ever present metabolism, the sealed vitality can resist any summons, can sleep and, sleeping, endure. Without water there is no growth; with excess of carbon dioxide and deficiency of oxygen, there is anesthesia.

But even in the embryo's active youth, even as the petals began to shrivel about it, there was a beginning of this state of dormancy in the fruit. Withering spread like a contagion, like sleepiness when the Beauty pricked her finger on the spindle, over the whole life within the seed castle's walls. The deep sleep of the seed, provident as it appears, is due, then, not to the design of the future, but to the inevitableness of the past. Forever, when we ask, Why? Nature gives no answer. Only when we ask, How? she permits us to discover. That inquiry is the field of science.

Antique organ, the seed was evolved by the seed-ferns in the Paleozoic; it is, historically, the oldest part of any plant that bears it. It is that generation, aspect, phase, in a plant's life that is least specialized, least adapted to the particular conditions of terrestrial life, and most ready for the cosmos.

For you can boil or freeze some seeds without necessarily killing them. Some resist continual boiling for forty-eight hours. Alfalfa, mustard and wheat seeds have been experimentally perforated to make them vulnerable, and then desiccated for six months, placed in a vacuum for a year, frozen for three weeks, next moved to a container where the temperature was like the cold of outer space. And still they germinated. Weed seeds were buried in glass bottles for forty years, and dug up and planted, twenty years after the first experimenter had died, and of them almost half were still alive. Seeds have been taken from herbarium sheets dated fifty years ago, and sown, and of these some still could grow. Lotus seeds four centuries old have been definitely known to retain viability; how much longer they might do so, no man can say. So naturally dormant are many seeds that they have a tendency to die in their sleep, smothered within their own obdurate walls. So that horticulturists, to obtain a satisfactory percentage of germination, have to file the coat, bore into it, break it off, leach it with acids.

If there is any living thing which might explain to us the mystery beyond this life, it should be seeds. We pour them curiously into the

453

palm, dark as mystery, brown or gray as earth, bright sometimes with scarlet of those beads worked into Buddhist rosaries. We shake them there, gazing, but there is no answer to this knocking on the door. They will not tell where their life has gone, or if it is there, any more than the lips of the dead.

EDWARD A. ARMSTRONG

Playboys of the Western World

From *Birds of the Grey Wind*. 1940

[The Reverend Edward A. Armstrong has a parish near Leeds, England, and he is also one of the most versatile of ornithologists. The present essay is taken from a volume which puts the author well in the forefront of natural historian *belles-lettres* and the photographic illustrations are on the same excellent level. He has also written an intensely interesting, documented monograph on avian psychology, *Bird Display,* which in its turn assures him no uncertain position among the company of scientific ornithologists.]

T HE BLACK guillemot is a quaint, jolly, little fellow. The more you know him the queerer and the more likeable you find him to be. You appreciate him, indeed, because of his odd ways. "A friend," says the schoolboy's definition, "is a chap who knows you and yet likes you." If the black guillemot has his little whimsies and foibles that is only to say he is a real personality. And after all, if you are kin to the Great Auk you are entitled to the indulgence which the public are always willing to accord to the eccentricities of the aristocracy.

So far as the British Isles are concerned the black guillemot breeds in the Isle of Man, various parts of Scotland, the Hebrides, Orkneys, and Shetlands. Where suitable rocky coast occurs in Ireland there are colonies, but as the east coast from County Down southwards is not rugged they are comparatively scarce there. In Ulster they are nowhere very numerous but there are so many craggy places, such as the cliffs of Rathlin, in which little communities annually establish themselves that they cannot be considered rare. Here, in clefts or under large rocks the two eggs — sometimes three — are laid about the end of May or beginning of June. Thoroughly domesticated, the black guillemot takes turns with his lady in brooding the eggs.

He is attired in the smart evening dress favoured by his somewhat distant relatives, the common guillemot and the razorbill, but absentmindedly has strolled out-of-doors in his coral-red fireside *pantoufles*. These, being several sizes too large, give him something of the appearance of Little Tich. As for his "'tails," the impression is as if they had been unaccountably slashed away in great shield-shaped segments at either side to show the starched shirt underneath. A strange garb in which to mind the baby! It may be said in his favour, however, that when he is not concerned with family affairs he dons sensible grey tweeds and devotes himself to sea-fishing.

Writers have styled the black guillemot "the dabchick of the sea" and Edmund Selous aptly speaks of "these little dumpling birds." In Ulster, where the dabchick is found on the reservoirs and rivers, I have heard it called the "Tom puddin'" or "dam puddin'." Indeed, on the water it looks like a wee, trim-necked, black pudding bobbing about. The black guillemot is like the little grebe to the extent of being a pert, somewhat dumpy, little diver. We might appropriately give it Shakespeare's apt name for the dabchick — "the dive-dapper." Along the Antrim coast it is called the "sea-pigeon," and in the Shetlands, "tystie."

In April the birds may be seen frequenting their accustomed haunts. I believe they arrive as early as March and appear in pairs. Incubation lasts three to four weeks and usually only one chick is reared. What happens to Number Two? I suspect that one of the eggs often gets chilled owing to the interval between the laying of the second egg and the initiation of brooding. This is why boobies manage to rear only one chick. The birds return by preference to the same nesting site each year and are known to live at least nine years.

The colony with which I am most familiar is in an exceptional situation. Instead of being far from mankind on some lonely, rocky shore, or at the base of a beetling cliff, it is in the centre of a crowded holiday resort in County Down. For some reason or another the builders of the stone pier which juts out into the little harbour left a number of recesses about a foot square at regular intervals in the masonry, and for the last twenty years, or thereabouts, each of these has been tenanted during the summer months by a pair of guillemots. The tystie is never much afraid of man, but these have become exceptionally confiding through constantly seeing people passing to and fro. So far from being "very careful not to enter their nesting-place under observation" as Ussher and Warren describe them in *Birds of Ireland* they fly to the nesting niches and enter them boldly while throngs of tourists and holiday-makers promenade overhead. They couch lazily on the edge of the pier or on top of the sea wall within a few feet of passers-by and conduct their amorous billing and cooing — or rather, sighing — in full view of the interested spectator. It is surprising how few pay any attention to the conspicuous, dapper little birds. Occasionally, when I have been watching them, some unhelpful but kindly-intentioned person has come forward — they usually do so when you are trying to concentrate on the birds' minutest doing — and entered into conversation. I have been asked: "Are they a kind of sea-gull?" and one naive gentleman inquired whether the holes in the pier had been made especially to accommodate the birds. I regretfully informed him that our seaside town councils had not yet reached that height of civilization at which nesting sites are provided for beautiful sea-birds.

Parties of trippers crowd into motor boats beneath the birds' nesting-holes, boys dangle fishing-lines outside them, lubbers in rowing boats flog the water a few feet from the guillemots as they swim in the harbour, but the plump little birds merely cock an inquisitive eye at any person who approaches and take to flight only if the disturber of their peace is exceptionally persistent. Often a few of the birds may be seen resting on the roofs of boats' cabins a few yards from the pier. To a naturalist these confiding tysties are not the least of the town's amenities. Perhaps one day an enterprising poster will announce "Blankshore: The Only Seaside Resort in the World where you can watch the Pierrots and the Guillemots at the Same Time." In such Utopian times the enlightened audience might well desert "The Follies" for the show put on by these flat-footed, avian clowns!

At their nesting haunts they spend a good deal of their time out of the water, especially before incubation has started and when the young have hatched. They love to bask in the sun on the rocks, or in this case, the edge of the pier, and show their sociable disposition by squatting in couples or little parties. They can hardly be said to sit or perch; like many old folk they "are not much good on their feet," and hobble along with a clumsy, nautical roll, to subside very soon with an "Oh, my poor corns!" appearance of exhaustion and relief. On a sloping or unequal surface the tystie looks like a little, decrepit, old man with bent neck and back hunched up.

In spite of their awkwardness on land black guillemots are more at home there than the common guillemot or razorbill. They use their feet — dangled downwards and spread apart — as air-brakes when about to alight, in the same manner as the other members of their family. Flying from a ledge or rock to the sea, tysties often take a hurried step or two on the water to break their descent, though sometimes they alight with such violence that they might almost be described as ricochetting on the surface. When going farther out into the bay or open sea they take off with their spread, red, webbed feet stretched behind as planes or balancers, and after flying a few yards close them with a consequent sudden diminution of the red flash. Occasionally a bird will fly forth from a hole as if to alight on the surface, and, still on the wing, make a few strides on the water and fly on.

In the bay where the guillemots nest — if that term can be used of birds which carry only a few chips or shells into the nesting cavity — there is continual liveliness when Venus stirs the emotions of these her quaint sea-doves. The birds, as much as the waters which they ply, are frequently in commotion, and the patient naturalist will find much interest and not a little amusement in watching them. I shall describe the strange performance which may be seen almost any day during the summer months.

A dozen birds swim in a scattered party on the ruffled sea, most of them in pairs, the others a few yards apart. Watch any of these couples and you will soon see one bird swimming round (or partly round) the other, almost or quite within touching distance. As this kind of manoeuvre is frequently one of the "figures" of the courtship of aquatic birds and the pivotal bird is usually the female, for convenience in description I shall assume that this arrangement also commonly holds in regard to the black guillemot. Mr. Walpole-Bond in *Field Studies of Some Rarer British Birds* states that the male can be distinguished from the female by his brighter plumage, but careful scrutiny at close quarters has not enabled me to detect this difference.

While the cock swims around the hen he utters the call from which he derives the sobriquet "tystie" — "*ist, ist, ist,*" in rapid sequence — almost a twitter; a high-pitched, plaintive, repeated squeak, ending sometimes in a long-drawn, wheezy sigh. All the time he is careful to keep his head turned towards the lady, who has ample opportunity to observe and admire the brilliant interior of his mouth — vermilion, deeply tinged with carmine. Sometimes the ceremony is rather more elaborate; one of the birds will nod its head while calling, in a curious, deliberate, rather undulating way, as if the axis of movement were midway between the back of the head and the tip of the bill. As a rule the pivotal bird makes but little response though she turns about to keep facing her partner and at times reciprocates with a similar squeaking or "cheedling." This dance figure may even take place out of the water. I once saw a bird trying to walk around another on the narrow stern of an anchored boat. It was a ludicrous spectacle! But the observation links the guillemots' dance with the courtship circumambulation of sea-swallows and other birds, performed on land. During their gyrations the tysties both dip their bills constantly as if they were taking little sips of water, though it is highly unlikely that they actually drink. Cormorants, mergansers, and red-throated divers have the same habit. To what extent it has any sexual significance it is difficult to say, as, in the case of tysties and cormorants, I have noticed it practised by solitary birds late in the season and away from the nesting haunts.

While one or two couples in a group of tysties are pivoting, the others occupy themselves in various ways, sometimes bathing or flapping their wings as guillemots and razorbills so often do on their nesting ledges — presumably to air and exercise them, though I have little doubt that it is to some extent a way of showing off and relieving tension. Like "dipping" it is characteristic of many aquatic birds when sexually excited. They keep more or less together in their flotilla but I have noticed two odd tendencies; one to form groups of four, composed apparently of two pairs, so that the dance looks rather like a quadrille; and the other, to form into line. I have seen ten birds floating on the water — not swimming line-ahead nor facing in the same direction — at intervals of about five feet in a perfect line. This behaviour is not merely a local phenomenon, due, as I at first thought, to the birds being influenced by the straight, regular line of the sea-wall, for Dr. Darling has noticed the same peculiarity amongst the tysties about the coast of Wester Ross. Razorbills have similar habits, swimming in single file, then forming a ring in the centre and waltzing in pairs with bills interlocked.

Suddenly a bird dives — usually a female, I believe — and almost simultaneously her coadjutor, who is presumably normally her husband, follows. Then, almost or quite contemporaneously, all or nearly all the neighbouring birds also dive. The effect is peculiar. At one moment there are eight or ten birds manoeuvring or floating placidly, and in the flick of an eyelid the sea is left empty of birds, until, a few seconds later, they begin to bob up again here and there. A sort of submersion contagion seems to affect them. My impression is that the performance is not simultaneous in the sense of the evolutions of starlings or dunlin in flight; one bird sets the example and the others, seeing it dive, follow suit. They are rather like a lot of schoolboys sporting in a pool; Smith, Jones, and Robinson "duck's dive" into the water when they see Brown going down, and chase each other beneath the surface. The likeness is accentuated by the guillemots using their white-medallioned wings in a kind of breast-stroke under water. So far as I have observed, the feet are trailed at such times. When seeking food they swim submerged from rock to rock, prying amongst the weeds — singularly beautiful little submarines, gleaming black and white and rosy-red in the bottle-green depths. But during the nuptial, connubial or recreational chasings the guillemots keep close beneath the surface and flutter along like large exotic water-butterflies.

The dive is commonly the beginning of the second figure of the display — the "pursuit." Often the birds follow a fairly straight course under water; sometimes they change direction in their hurried aquatic flight; and when they reach the surface the chase may continue either along the water, above or below it. We may see one swim after the other in the usual style; but sometimes, apparently to gain speed, a tystie, either the pursuer or the pursued, will flop clumsily along the surface wallowing as if wounded or bereft of the use of its wings. Common guillemots on the sea near their nesting cliffs will flee from an approaching boat in a similar fashion. One might imagine that the bird could not decide whether to dive, swim, or fly and was reduced to something of the condition of the puzzled centipede in the well-known rhyme.

> A centipede was happy quite
> Until a toad in fun
> Said: "Pray which leg moves after which?"
> This raised her doubts to such a pitch,
> She fell exhausted in a ditch
> Not knowing how to run.

The guillemot is accustomed to use its wings both in and out of the water and sprawls rather clumsily when it tries to use them on the fringes of both elements, but this procedure is the outcome of emotional excitement. When it decides to rise from the water it takes off with no great difficulty and its rapid wingstrokes carry it away with considerable speed. The birds cannot readily change course in the air and if they miss their perch, as may happen if another guillemot does not make room on a small ledge, they must fly round in a wide circle to make another attempt.

When the guillemots come up again after their submerged swim they look about them alertly. The leading bird, which I presume to be the female, is ready to dive again or to take to her wings. The pursuing "dive-dapper" who has been "peering through the wave" at the other during their submarine evolutions bobs up all agog for the appearance of his mate — and I am again reminded of boys on their guard lest a playmate grab them underneath the water, and on the *qui vive* to play the trick on some one else. He casts a glance at her as she dives and follows with all possible speed. They evidently realize that they can make better progress beneath the surface than on it. The game is a kind of tag combined with hide-and-go-seek, in and out of the water. How useful the brilliant gleam of the female's coral feet and chequered wings must be to the lady's pursuing admirer as he hurries after her through the swirling water! May it not be that we have a hint here of the biological purpose of the bright-hued feet and legs of some diving birds?

Instead of plunging again the female may elect to take to the air; and, if so, she will be followed by the male in a great arc around the bay, though occasionally they will both fly, two vibrating black and white arrow-heads, out to sea. Sometimes they will dive from the air like the large Asiatic kingfishers which hover like kestrels and plunge like sea-swallows. During nuptial excitement in spring one bird will fly over and drop beside another — or more often where it was — for it immediately dives; and there is greater ardour noticeable in the chasings which take place in April than in those of summer, though plunging parties and underwater pursuits continue with un-abated energy.

The third figure of the guillemots' display is "billing and sighing." One bird alights beside another and they express their pleasure in high-pitched endearments. Being so close to each other their bright red mouths are mutually revealed as they utter their thin twitter, and one cannot resist the inference that, as in the case of many other species, such as the merganser, shag, cormorant, fulmar petrel, kitti-wake, puffin, and razorbill, the coloration of the buccal cavity plays a part in the process of sexual stimulation. Their beaks frequently touch as they converse and the bill of one may enter between the mandibles of the other.

Tysties from all over a bay will rise with one accord from the water and fly around like a flight of giant beetles, their black bodies fringed with a quivering white halo on either side. Sometimes every bird — to the number of twenty — joins in the flight. There is the same unanimity in these manoeuvres as there is in the diving par-ties, and the way in which they arise resembles the sudden unaccount-able flights or "dreads" of nesting colonies of terns and gulls. Such corporate flights as I have seen took place in April and concluded with the birds alighting together. They came in a whirring host and stood, waddled, or squatted on the pier in a long row like penguins. Every here and there a pair stood up and conducted a whining duet, and occasionally one would totter towards another, red beak shin-ing as it called, and force it to retreat.

These revels continue from April until August. Belfast's Easter holiday-makers and Yorkshire's August Bank Holiday trippers might, if they cared for these things, enjoy the performance. "The show is on" alike when the spring squill spreads in silver-blue gossamer drifts over the spindrift-drenched sward along the County Down shore and when the heather empurples the hills. The departing cuckoo looks down upon the display as she did when she arrived from Central Africa. All through the summer the birds twirl and circle with their partners, form "sets" and lines, plunge and splash. The bay is the scene of a long carnival and Harlequin and Punchinello never seem to tire. Yet, as dawn warns the revellers at an Oriental feast that it is time to depart, so the sun sets a limit to the frolics of our parti-coloured little clowns. As he nears the meridian it is the signal to be gone. In the afternoon the bay is a dreary spot, as forlorn as a Cambridge College when the students have "gone down" or the Lido at Venice on a cold autumn evening.

What is the significance of such ardent and prolonged festivities, these pirouettings, waltzings, quadrilles, and sets of lancers danced on the sunlit sea? Our brilliant friend, the red-breasted merganser, may be able to supply us with a clue. These ducks, as we have already noticed, participate in romps which are rather like a crude form of the more elaborate and highly developed water-dances of the black guillemot. They swim, splash, and dive together in little parties, but never with the co-ordination characteristic of the tysties. The whole party will fly up and around, or individuals will fly after each other, as the guillemots do. Pairs, or even several birds, will dive almost simultaneously in a systematic way, though it is not a corporate activity to the extent that it is with the tysties, and it has, apparently, some relationship to feeding activities, whereas with our little black and white divers the diving game or dance is not connected with fishing at all. All this indicates that the black guillemot's frolic is a more highly-evolved affair than the merganser's. There are, however, other ducks which dance some of the figures of the guillemot's display. Shovelers will gather in a large flock on the water, skirmishing, bathing, and revolving, drake and duck together; and paddlings of mallard, shoveler, and teal are sometimes seized by a strange and unaccountable frenzy — splashing, diving, flapping, and quacking in great excitement.

I believe that the tysties' display began as a series of courtship manoeuvres but so fascinated the birds that now it has become a romp carried on throughout the season because of the pleasure which the birds obtain in doing it. The pursuits, which originally had a purely sexual motive, are now cultivated and continued, if not entirely for their own sake, at least partly as a recreation. Mother Carey's hens — to give them the West of Ireland name — being neighbourly in their colonies, and possessing like so many birds an imitative vein, have developed a tendency to make these activities into community frolics.

In these days when we have been taught that the singing of birds is not due to any aesthetic feeling but is merely a way of proclaiming loud and long — "Trespassers will be prosecuted," it is customary to deny the recreational element in birds' activities. The guillemots' dances show that birds can and do enjoy themselves and that their activities may develop into what may accurately be regarded as sport — co-ordinated motion enjoyed in and for itself. It would be easy to multiply instances — the flying somersaults of ravens, the aerial toyings of a pair of buzzards or the cat-and-mouse game which Hudson saw a grebe playing with a fish. In Iceland an ornithologist watched some eider-ducks having a good time where a strong current was running. They would waddle along the shore, launch themselves into the stream and sail into the bay, then land, waddle back and repeat the performance again and again. We may still believe that "God's jocund lyttel fowles" can enjoy a game and that the explanation of their activities does not always and entirely lie within strictly biological categories.

When we reflect on the significance of the guillemots' dances we are forced to conclude that they have come into being in much the same way as our own dances. Primitive man is unable to express his feelings adequately by means of language, so he employs gesture and motion. His dances are seldom or never purely recreational. Their aim is to bring rain to the land, fertility to man (and beast), or destruction to his foes. When we peer into the dark recesses of the past we find that games and athletic meetings, too, arose from magical representations of the struggle between life and death, light and darkness, summer and winter, designed to assist the life-giving powers by sympathetic effort on their behalf. Later, the dance, the drama, and the game, which originally all possessed a religio-magical meaning, were elaborated and enjoyed for their own sakes. Primitive dances, performed in order to stimulate the reproductive urge or to induce the growth of the crops, have become refined and transformed in civilized society into little more than a means of "letting off steam," though the music and behaviour in some of our dance halls retain evident affinities with barbaric antecedents.

Mother Carey's hens, too, have learned how to refine their instincts. They have made a discovery which betokens a high grade of civilization — that of pure fun; something for which the economies of ant and bee have no place. Truly if the refinement of a civilization is indicated by the innocence of its recreations, and an individual's character is best gauged by the way he employs his leisure, the black guillemot should stand high in our esteem.

The young are fed mainly on fish one and a half to two inches in length. So far as my observations go they are almost invariably carried across the beak with the head to the right and the tail hanging to the left side. Now and then a larger fish of about three and a half inches is brought, and fairly often crustaceans of some sort. Once I identified a gurnard in a guillemot's bill. Both birds help to feed the young and when food-fetching is in progress the activity is sometimes intensive. I have notes of supplies being brought at 9:25, 9:37, 9:45, and 9:46 a.m., although the fishing grounds were at least half a mile distant. There was very little feeding of the chicks after midday. The birds, with rare exceptions, went out to sea and apparently stayed there until the following morning, though a few might occasionally return towards sundown to the region of domestic cares. I have seen young being fed as late in the year as the seventeenth of August when the adults were beginning to acquire their grey winter plumage.

I regret to have to record that shortly after my last observations were made a ship illegally discharged oil in the harbour, and some twenty birds, young and old — practically all my jolly little friends — were washed up dead and dirty on the sands and carted away by the street scavengers. How much pleasanter the world would be if men were, not merely kinder, but more given to thoughtful consideration of the needs of others.

GILBERT C. KLINGEL

In Defense of Octopuses

From *Inagua*. 1940

[To WRITE ABOUT octopuses and cuttle-fish seems to be the object of everyone who comes into contact with them. To write sanely about these interesting creatures is the prerogative of very few. Klingel has given us such an account, which should do much to discourage further stories of battles to the death with man-attacking mollusks, tales that have wasted so much good ink and paper.

If the volume *Inagua* is a result of natural science being merely a "pastime and hobby," I regret that the author has not devoted his life to the science of natural history.]

I FEEL about octopuses — as Mark Twain did about the devil — that someone should undertake their rehabilitation. All writers about the sea, from Victor Hugo down to the present, have published volumes against them; they have been the unknowing and unwitting victims of a large and very unfair amount of propaganda, and have long suffered under the stigma of being considered horrible and exceedingly repulsive. No one has ever told the octopuses' side of the story; nor has anyone ever defended them against the mass of calumnies which have been heaped on their peculiar and marvelously shaped heads. We have convicted them without benefit of a hearing, which is a most partial and unjust proceeding. I propose that the octopuses, and their near relatives the squids, are among the most wonderful of all earth's creatures, and as such are deserving of our respect, if not our admiration.

My personal interest in octopi dates back to the moment when I turned to climb out of the drowned ravine at the base of the Inaguan barrier reef. I had reached the lower portion of the final slope and was about to seize on a piece of yellow rock to steady myself when I noticed that from the top of the boulder was peering a cold dark eye that neither blinked nor stirred. In vain I looked for eyelids; the orb apparently belonged to the rock itself.

Then suddenly, I felt a chill wave creep up my spine. Before my

465

gaze the rock started to melt, began to ooze at the sides like a candle that had become too hot. There is no other way to describe the action. I was so startled at the phenomenon that it was a full second or two before I was conscious of what I was watching.

It was my first acquaintance with a live, full-grown octopus. The beast flowed down the remainder of the boulder, so closely did its flesh adhere to the stone, and then slowly, with tentacles spread slightly apart, slithered into a crevasse nearby. The head of the octopus was about as big as a football, but as it reached the fissure, which was not more than four inches in width, it flattened out and wedged itself into the opening. It seemed somewhat irritated at my disturbing it, for it rapidly flushed from pebbled yellow to mottled brown and then back to a livid white. It remained white for about twenty seconds and then altered slowly to a dark gray edged with maroon. I stood stock-still but it made no overt motions and I slowly edged away. Quite possibly it might have been a nasty customer, for the tentacles were about five feet from tip to tip.

This last statement may seem a contradiction to my opening paragraph; and, I must admit, that is the way I felt about the octopus at the time. However, since that hour I have collected and observed a number of these creatures, including the squids. I have found them animals of unusual attainments and they should be ranked among the most remarkable denizens of the sea. They are endowed with considerable intelligence and they have reached a system of living all their own which they have maintained for approximately 500,000,000 years. As far back as the Ordovician period of geology we find their ancestors, and there is good evidence that at one time the forefathers of the present octopi very nearly ruled the world. Had they been able to pass the barrier of the edge of the ocean as the early fish-derived amphibians did there might have been no limit to the amazing forms which would have peopled the earth.

Within the bounds of pure speculation, however, the fact remains that the cephalopods, as the entire octopi-like group of animals is termed, have missed the status of brainy intelligence, of which man is the highest criterion, only by a very narrow margin. There is reason to believe that they are the most keen-witted creatures in the ocean and had they developed an opposable thumb and fingers instead of suckers with which to manipulate various objects the entire course of the earth's existence might have been altered.

There are some very curious similarities between the development of intelligence in man and in the modern cephalopod. Both acquired brains after their individual fashions because the course of organic evolution left them without adequate physical protection against the vicissitudes of nature. Man, the weak and the puny, without claws and rending fangs to battle the beasts and without long legs with which to flee, had to acquire cunning or perish. That marvelous addition, the opposable thumb, made possible holding and using tools and gave a stimulus to cunning that nothing else in the mechanics of evolution could have provided. The thumb is by far the most remarkable portion of man's anatomy. Literature, music, art, philosophy, religion, civilization itself are directly the result of man's possession of this digit.

Like man, the modern cephalopods have been thrown upon the world naked and without the armor protection of their ancestors. For cephalopods are shellfish, blood brothers to the oyster, the clam and the conch; they are mollusks which have been deprived of their shells. The only present day cephalopods which still retain their shells are the Nautiloids which are direct descendants of the ancient types whose fossils are found in the tightly compressed rocks of the Upper Cambrian. Over three thousand fossil nautiloids have been named, an imposing group ranging in size from a tiny seven millimeter creature called *Cyrtoceras* to the immense 14-foot cone of *Endoceras!* Only four closely related species of this mighty shelled host remain, all occurring in the South Pacific.

To compensate for the loss of their shells, which were their bulwarks against fate, these unclothed cephalopods have developed, like man, cunning and intelligence. Alone among the mollusks they have acquired by concentration of their chief nerve ganglia what may be truly considered a brain. With the casting aside of the shell they have also gained their freedom, speed and mobility.

Safety often goes hand in hand with degeneration. It is a curious circumstance that those creatures which live completely guarded lives also have a very dull existence. What, for example, could be safer and more stupid and sedentary than an oyster, clad in its house of lime? The loss of a shell not only rescued the cephalopods from dullness but it probably also saved them from extinction. The most highly ornate shelled cephalopods of all time, the gracefully coiled Ammonoids, which are so named because of their resemblance to the ramlike horns of the deity Jupiter Ammon, and which developed during the Upper Silurian and lasted until the close of the Age of Reptiles, went out of existence because the extent of their external sculpture and complexity of septation rendered them so specialized

467

that they failed to respond to change. Some of these fantastic Ammonoids, of which six thousand species are known, possessed coiled shells more than six feet in diameter!

"Cephalopods," the scientific name of the octopi and squids, immediately characterizes them as something unusual, for it signifies that they walk on their heads. This is precisely what they do, for their tentacles or "feet" are located between their eyes and mouths. No other animals on earth utilize this position or method of progression.

However, it is in their mode of swimming that the motion of these weird beings is most amazing. They are beautifully streamlined when in action, and can dart about at remarkable speed. I recall once being out to sea in a fishing trawler off the Virginia Capes. I was sitting in the dark on deck watching the stars and swaying to the slight roll of the boat when suddenly I heard a rapidly reiterated splashing in the sea. The sound was slightly reminiscent of the pattering noise of flying fish. I knew that I was too far north for any quantity of these volant creatures. I went below and returned on deck with a flashlight. Its beam pierced the dark and glowed on the wave tops. The ship was passing through a school of small surface fish. They were being preyed upon by hundreds of *Loligo* squid. The squid were shuttling back and forth through the water at incredible speed. Most wonderful was the organization with which they seemed to operate. Entire masses of these cephalopods, all swimming in the same direction, would dart at the mass of fish, quickly seize and bite at them, then abruptly wheel as a unit and sweep through the panic-stricken victims which scurried everywhere. Some of these squid were traveling so rapidly that when they approached the top of the water they burst through and went skimming through the air for several yards, falling back with light splashes. In the morning I found several on the deck of the trawler where they had jumped, a vertical distance of at least six feet! There is another record made near the coast of Brazil of a swarm of squids flying out of the water on the deck of a ship which was twelve feet above the surface and which was further protected by a high bulwark, making a minimum jump of fifteen feet! Several score were shoveled off the ship when daylight came.

The cephalopods and particularly the squids might be compared to living fountain pens or animated syringes, for they accomplish their flight-like swimming by pulling liquid into their body cavities and squirting it out again. Their likeness to a living fountain pen is even further heightened when one considers that some of the cephalopods contain ink and a quill. Nor is this all, for nature, not content to offer all these wonders in one creature, has ordained that they may swim, not only forward like all other creatures of the sea, but backward! They can swim forward and sideways, too, but the normal mode is stern foremost.

The quill of these mobile fountain pens is the remnant of the shells of their prehistoric ancestors, and it persists, like our vermiform appendix, as a useless but telltale evidence of former usage. The quill, reduced in the octopi to two chitinous rods, and in the squid to a long narrow fluted pen, remarkably resembling an old fashioned quill, is buried deep within the tissue. In a sense, the octopi and squids are shellfish which have surrounded their shells.

It is in the ink of these cephalopods that we are confronted with a true paradox. This ink, basis of the familiar India ink, is utilized for two diametrically opposite purposes. It is intended to provide concealment and, diversely, to enable the animal to keep in touch with its fellows. When there is fear of an attack by enemies, the ink is expelled into the water to form a "smoke screen" behind which the cephalopod flees to shelter. Thus the modern military technique of employing the smoke screen to conceal retreating movements was conceived by the cephalopods as early as the Jurassic, as is proved by a beautifully preserved fossil of that period which shows the ink bag prominently limned in the highly compressed tissue impression of a squid. However, when night closes down on the water shrouding the blue vastness of the deeps in impenetrable gloom it is by means of this same ink that the members of a school of squids are able to keep in contact with one another. It is believed that the ink is extruded in very small quantities and is picked up by unusually sensitive olfactory organs. The more solitary octopi use it in much the same manner to locate their mates.

I had no idea of the efficiency of this inky fluid until my third or fourth meeting with the octopus of the valley. I had been going down for a half hour or so each day near the same spot in the reef and almost always finished the day's dive with a final excursion to the limit of the hose on the base of the ravine. In these trips I saw a number of octopuses, mostly much smaller than the first. These seemed to live in the crevasses near the base of the reef, and often all that I saw of them was a tentacle or two twitching or writhing languidly from a fissure. Some I discovered by the neat piles of mussel shells and other mollusks near the entrances to their hiding places. Some of these shells were, surprisingly, unopened and, it can be assumed, were being stored against an hour of larger appetite. Also, most interesting, the only locality on Inagua where the mussels were to be found in any abundance was in the area of the surf, a living habit that might be attributed to the ceaseless raids of octopi on colonies in more peaceful localities. The mussels, in self-defense as it were, had established themselves in the only place where they might live undisturbed, which was, in contradiction, the most violent area of all the world of underwater. They were, so to speak, between the devil and the deep blue sea, or to be more exact, between the devil and the hot dry air.

Most of these octopi were exceedingly shy, fleeing into their shelters at my approach, and drawing far back out of reach, a reaction quite at variance with the accepted theories of ferocity and malignancy. I tried to capture some of the smaller ones, but they were too fast for me. The big fellow on the slope of the ravine, however, while it did not seem quite so timid always gave me a wide berth and invariably, the few times I encountered it, withdrew to its fissure where it was never quite hidden, but was revealed by a portion of the body and the restless arms. At first I left it strictly alone, but curiosity about its peculiar color changes prompted me to come closer.

It always seemed irritated at my presence. Its nervousness may have been caused by fear, for it certainly made no pretense of belligerency, and it constantly underwent a series of pigment alterations that were little short of marvelous. Blushing was its specialty. No schoolgirl with her first love was ever subjected to a more rapid or recurring course of excited flushes than this particular octopus. The most common colors were creamy white, mottled Van Dyke brown, maroon, bluish gray, and finally light ultramarine nearly the color of the water. When most agitated it turned livid white, which is I believe the reaction of fear. During some of the changes it became streaked, at times in wide bands of maroon and cream, and once or twice in wavy lines of lavender and deep rose. Even red spots and irregular purplish polka dots were included in its repertoire, though these gaudy variations seldom lasted for long.

I had heard that a light touch on the skin would leave a vivid impression of color and I was anxious to see if this were the case. From the boatman I borrowed a long stick and dropped down to the sea floor again. The octopus was still in place and I walked over to it with the pole in my hand. At first I was hesitant about the experiment. The creature had behaved so nicely that I almost decided to give it up. But the old curiosity prevailed and with my pole I slowly reached out and stroked it along the side of its body.

Then things began to happen. The stick was snatched from my

fingers and went floating to the surface. The octopus flashed out of the fissure and ejected an immense cloud of purplish ink. For a brief moment I saw it swimming away, long and sleek in shape, and then I was surrounded by the haze. The fog was not opaque but imparted much the same quality of nonvision as thick smoke in dry air, except that I did not notice much in the way of wreaths. In fact, I was so confused and startled, that my only thought was to get away. From underneath the helmet there arose a faint odor quite unlike anything else. Fishy musk is the nearest description I can think of. The color was most interesting, as I had always been under the impression that cephalopod ink was black. Rather, it appeared dark purple which later faded to a somber shade of azure. I can also remember, when it thinned considerably, seeing vague shafts of reddish when the rays of sunlight far above caught the substance at oblique angles. The ink spread out in a cloud extending over several yards and in the still depths of the ravine took quite a time to dissipate. Actually it floated away as a hazy smudge before it evaporated.

I was not able to continue my observations on color changes until several days later when I netted a baby octopus from some turtle-weed growing a few yards from shore near the place where the reef reached its final termination in a mass of sandy shoals. I transferred the mite, a youngster of seven or eight inches spread, to the tidepool near my old house where I kept it for several days. It took to its new surroundings very gracefully and made no attempt to escape but made life miserable for the numerous small crabs and fishes that shared the pond. The crabs were its principal prey which it captured by stealth and by lying patiently in wait. Patience was its most evident virtue, and much to my disgust it would sit for hours in one spot without moving, staring endlessly at the moving forms in the water. It used a great deal of intelligence in securing the crabs **and** selected a spot to lurk where it had ready command of an entire corner of the pond.

The rocks of its dwelling were creamy brown, and this was the exact hue it assumed while waiting to make a capture. It had perfect control of its pigmentation. In comparison the renowned chameleons are but rank amateurs. The mechanics of this alteration of hue are very complex but are controlled by the expansion and contraction of a group of cells attached to pigmented sacks, known as chromatophores, residing in the outer layers of skin over the entire surface of the body. In addition there is scattered over the body another great series of cells capable of reflecting light. These are yellow and impart a strange iridescent shimmer, slightly suggestive of the glow of pearls.

The chromatophores, which are of a variety of colors, are opened and shut at will producing any or all colors of the rainbow.

These color cells are manipulated by highly sensitive nerves communicating with the brain and with the eye. The eye principally dictates the choice of color although emotion also seems to have a definite influence. When frightened the octopi usually blanch to a whitish or light tone; irritation will cause them to break out in dark pigments. No other creatures in the world can alter their color as quickly and completely. Emotion will cause a human being to flush with anger or become pale with pain or anxiety; but no one can hold his hand and will it to be green with yellow stripes, or even yellow or plain brown, let alone lavender or ultramarine. An artist may paint a picture; only an octopus can color its skin with the portrait of its emotions, or duplicate exactly the pattern of the soil on which it rests. Only a very highly organized creature, one with a brain and an unusually well co-ordinated nervous system could accomplish the mechanical marvel of operating several thousands of cells at once, rapidly opening and closing them in proper order.

The cephalopods are not limited to color change but are also credited with being able to produce the most brilliant light known in the realm of animals. While this luminescence is limited to a very few deep sea species of decapods, which are the ten-armed squids, their light is so vivid that they outshine the fireflies. These light organs may be found on any portion of the body, including the eyeball itself, and oddly enough, even in the interior of the animal! In these last forms the body tissues are quite transparent, so the light is not necessarily concealed. These light organs are quite varied, some being but mounds of glowing fluid, others complex and carefully constructed lenses with mirrors of reflecting tissue. As yet very little is known of these abyssal octopi and squid, though a few captured specimens have been observed burning with a strong light for several hours. Some day when the means of exploring the vast deeps of the ocean comfortably and safely has been devised we will learn more of these unbelievable cephalopods.

Quite unseen, my octopus would wait until a crab ventured near. Then it would either swoop quickly over the victim smothering it in its diminutive tentacles or suddenly dart out an arm and seize its meal before it had time to flee. It seldom missed but when it did it usually retrieved its dinner by a quick pursuit before it had gone far. Before twenty-four hours were up the entire bottom of the pool was littered with the hollow carapaces of crabs. Peculiarly, the animal almost always devoured its victim bottom side up, biting through the softer lower shell with its small parrot-like beak and rasping out the contents with its filed tongue before casting the empty shell away. The legs and feet were seldom eaten and were usually torn off and discarded. Little of this feeding was done during the day. At high noon I even saw a crab crawl over the relaxed tentacles without being molested or becoming aware of the danger it was courting. In the evening, however, particularly just before sunset the octopus seized everything within reach.

The capture of fish was not nearly so easy, and although I saw it make a number of attempts its only successful capture was a small goby that very injudiciously decided to rest a few inches below the octopus' chosen corner. As in the case of several of the crabs it was blanketed by a mass of writhing tentacles. Once the fish was grasped by the vacuum cups of these tentacles it was finished, for in their method of attacking, the octopi utilize one of the most efficient systems devised, a principle more certain than curving claws or the sharpness of teeth. Only the hand of man with its opposable thumb is superior.

The feel of these vacuum cups on the bare flesh is most unusual. It is not unpleasant, and in a small specimen, gives the sensation of hundreds of tiny wet clammy hands pulling at the skin. The strength of the suckers is amazing. When I tried to lift the youngster off my wrist it clung tenaciously and even when I had dislodged all the tentacles except one I still had to give a strong pull in proportion to its size to release the suckers. There have been cases in which the tentacles have been torn apart before the suckers released their grip. These suckers, which operate on much the same principle as the little rubber cups with which we attach objects to automobile windshields, are actuated by a muscular piston. The rim of the cup is fastened to an object, then the floor of the center is raised and retracted to form a vacuum. The cups, I found, would slip easily from side to side but when pulled directly exercised considerable power. In the octopods the suckers are sessile, or are mounted on low mounds; the squids carry the mechanism a bit further and produce them on stalks. In the giant squid the rims of the suckers are even equipped with fine teeth to render them more efficient. Whalers have recorded capturing whales with dozens of circular scars on their heads, inflicted in gargantuan battles with these monsters of the open ocean. Some of these scars have measured over two inches in diameter so that the creatures that possessed them must have been huge.

How large do the squid and octopi grow? There is an authentic record of a North Atlantic squid which measured fifty-two feet over all! Its tentacles had an abnormal reach of thirty-five feet and the remaining seventeen feet was taken up by the cylindrical body which had a circumference of twelve feet. The eye of this fabulous animal was seven by nine inches, the largest visual organ in the world. The suckers had a diameter of two and a quarter inches, and as some of the scars on captured whales have exceeded this measurement it is not unreasonable to assume that there may exist somewhere in the abyssal depths of the North Atlantic still larger squid of perhaps sixty or seventy feet. Even these amazing squid, however, are preyed upon by the great sperm whales which tear them apart with their long shearing teeth. In that old classic and favorite *The Cruise of the Cachalot*, the author, Mr. Frank Bullen, gives a vivid description of a battle between a large sperm whale and one of these squid.

"At about eleven a.m.," he writes, "I was leaning over the rail, gazing steadily at the bright surface of the sea, when there was a violent commotion in the sea right where the moon's rays were concentrated, so great that, remembering our position, I was at first inclined to alarm all hands, for I had often heard of volcanic islands suddenly lifting their heads from the depths below, or disappearing in a moment, and . . . I felt doubtful indeed of what was now happening. Getting the night glasses out of the cabin scuttle where they were always hung in readiness, I focused them on the troubled spot, perfectly satisfied by a short examination that neither volcano nor earthquake had anything to do with what was going on; yet so vast were the forces engaged that I might well have been excused for my first supposition. A very large whale was locked in deadly conflict with a cuttlefish or squid almost as large as himself, whose interminable tentacles seemed to enclose the whole of his great body. The head of the whole especially seemed a perfect network of writhing arms, naturally, I suppose, for it appeared as if the whale had the tail part of the mollusk in his jaws, and, in a business-like methodical way was sawing through it. By the side of the black columnar head of the whale appeared the head of a great squid, as awful an object as one could well imagine, even in a fevered dream. Judging as carefully as possible, I estimated it to be at least as large as one of our pipes, which contained three hundred and fifty gallons; but it may have been, and probably was, a good deal larger. The eyes were very remarkable for their size and blackness, which, contrasted with the livid whiteness of the head, made their appearance all the more striking. They were at least a foot in diameter, and seen under such conditions looked decidedly eerie and hobgoblin-like. All around the combatants were numerous sharks, like jackals around a lion, ready to share the feast, and apparently assisting in the destruction of the large Cephalopod."

Unfortunately Bullen does not tell the result of the combat but one might assume that the whale was the victor, for the food of sperm whales consists almost exclusively of squid.

If the squid and octopi are accused of being fearsome and savage, it might be argued that they live in an underwater world in which savagery and primitive instincts are the most common passions, and the only way to exist is to conform to the mode. There is no doubt that an enraged large cephalopod could be a formidable antagonist. The authentic instances of octopi or squid attacking human beings or divers, however, are so rare as to be considered non-existent in spite of a large literature to the contrary. Most of their savagery is confined to securing their food, which is a normal and reasonable function.

The tentacles serve still another and more wonderful purpose, for it is by means of their arms that these unorthodox creatures are able to perpetuate their race. The arms that serve in this function are known as hectocotylized arms and this name was derived from an honest and understandable mistake by Cuvier. The name also signifies the arm of a hundred cells, and the mistake was made when the detached portion of one of these many-celled arms was found clinging in the mantle cavity of a female paper nautilus where it was erroneously thought to be some new sort of parasitic worm. The strange worm was named hectocotylus and the error was not discovered until further researches had been undertaken in regard to the animal's breeding habits. It appears that the arm of the male paper nautilus is extended during breeding time until it looks like a long worm-like lash. This lash is charged with the fertilizing spermatophores. When the male and female meet they intertwine their tentacles in a medusa-like embrace, and when they disengage from their fantastic lovemaking, the end of the lash is deposited under the mantle of the female, where it is held for a time, for the female is not yet ready to spawn. When her eggs are eventually extruded, they are fertilized by the waiting sperm. The broken arm is not completely lost, for the male can grow another and still another.

The cephalopods are so delightfully versatile that they have still other systems of reproducing. In some forms the hectocotylized arm is not detached but is specially modified so that it can develop and transfer spermatophores to the females' mantle cavity near the oviduct. The spermatophore is itself the most remarkable creation of all this complex mating. It is a long tubular structure loaded with sperm, an apparatus for extruding it, and most wonderful a cement gland for attaching it to the female. It can be utilized at will; a thoughtful provision considering that the female may then take her good time in depositing her eggs under favorable circumstances. In other species the spermatophore is grasped by the male as it passes from his mantle and is placed in her mantle cavity or attached to the membrane around her mouth where the eggs are sometimes fertilized.

Some of the cephalopods show an amazing amount of mother-love and parental care. The common octopus *vulgaris* has been observed in aquaria guarding its eggs which were attached to the stone walls. It fiercely resented any interference and kept a constant circulation of water flowing over them to insure that no parasites would take hold and that proper oxygenation would occur. The eggs were not even left long enough for the mother to secure food, even though the period of incubation lasted for a considerable time. So intense was this guardianship that another octopus in the same tank which ventured close too frequently was set upon and slain. Mother-love in an octopus seems a strange and outlandish emotion, but no doubt it is actuated by the same flame that causes human parents to sacrifice their pleasures and desires that Junior, or his sister, for example, might go to college.

Cephalopodian care of the egg is responsible for another of the truly paradoxical things about these creatures. In the genus *Argonauta* the female carries about with her a beautifully coiled and graceful shell. This seems a contradiction to an earlier statement that the modern cephalopods are creatures which have cast aside their shells. Actually the shell of the Argonauts is not a true shell but is an egg case formed on the spiral shell pattern, which is mechanically a very strong and structurally efficient shape. The Argonaut is not bound to the shell in any way, for it may leave it whenever it desires, which it has been reported to do under certain conditions. No other mollusk is so equipped. Imagine an oyster, for example, opening its valves and stepping out for an airing! The shell is held in position by two arms which are specially formed for the purpose. Only the female possesses this protection, and she forms the shell, not with the mantle as do all other mollusks, but with her two modified arms with their expanded membranous disks. When the Argonauts are first born they have no shells and they do not begin its construction until they are a week or two old. Unfortunately for the natural history of Aristotle, they do not sail over the surface of the sea like miniature ships with the arms held as sails as that ancient. and inquiring naturalist so quaintly believed, but creep and crawl along the bottom or swim by means of their siphons like any other cephalopod. While the eggs of the Argonauts are well protected and carefully mothered the adult has paid a reverse penalty for its acquisition of a shell, even though that shell is not a true one. The Argonauts have lost some of the intelligence and freedom of other octopods, for they appear to be the most sluggish and stupid of their class.

Inagua from above the sea gives no hint of the host of octopods that must harbor in its reefs, or of the tiny frond-colored squids that shelter in the growths of sargassum weed that float ceaselessly by on the currents, or of the larger and more appalling-looking decapods that move about in small groups in the open water. Nor is there much indication even to the diver of their presence. Unlike the reef fishes they are mostly nocturnal. During the bright hours they lie quiescent, curled up in the crevasses of their coral homes or float suspended and still, in the magic manner of underwater between top and bottom, waiting patiently with staring round eyes for the sun to drop and extend vague shadows over the blue depths. Then they creep from their dens and go slithering over the coral boulders or swim like living arrows through the green waters, pouncing on their prey and doing whatever amazing things fall to the lot of cephalopods.

Whenever I think of the great barrier reef of Inagua I think always of two things: first, of the fairyland of the coral itself and the pastel colors, and second, of the octopus of the drowned ravine with its weird eye and rubbery body. More than any other creature, the octopus is the spirit of the reef; unreal themselves, completely fantastic, unbelievable, weird, they are fitting residents of a world in which all the accepted routines are nullified, in which animals play at being vegetables, where worms are beautiful, where the trees are made of brittle stone, where crabs pretend to be things they are not, where flowers devour fishes, where fishes imitate sand and rocks and where danger lurks in innocent color or harmless shape. That they should, also, be inhabitants of the shadowy night places is the final touch on their characters. The octopi fill a niche of creation claimed by no others and a niche which they occupy to perfection.

RACHEL L. CARSON (1907–   )

Odyssey of the Eel

From *Under the Sea-Wind.* 1941

[MISS CARSON has always had a zoological background in personal interest and thorough training, and for the last eight years has held the position of biologist in the United States Fish and Wildlife Service. I have chosen her account of the life story of the common eel as a dramatic presentation of this almost unbelievable aspect of natural history.]

THERE IS A POND that lies under a hill, where the threading roots of many trees — mountain ash, hickory, chestnut, oak, and hemlock — hold the rains in a deep sponge of humus. The pond is fed by two streams that carry the runoff of higher ground to the west, coming down over rocky beds grooved in the hill. Cattails, bur reeds, spike rushes, and pickerel weeds stand rooted in the soft mud around its shores and, on the side under the hill, wade out halfway into its waters. Willows grow in the wet ground along the eastern shore of the pond, where the overflow seeps down a grass-lined spillway, seeking its passage to the sea.

The smooth surface of the pond is often ringed by spreading ripples made when shiners, dace, or other minnows push against the tough sheet between air and water, and the film is dimpled, too, by the hurrying feet of small water insects that live among the reeds and rushes. The pond is called Bittern Pond, because never a spring passes without a few of these shy herons nesting in its bordering reeds, and the strange, pumping cries of the birds that stand and sway in the cattails, hidden in the blend of lights and shadows, are thought by some who hear them to be the voice of an unseen spirit of the pond.

From Bittern Pond to the sea is two hundred miles as a fish swims. Thirty miles of the way is by narrow hill streams, seventy miles by a sluggish river crawling over the coastal plain, and a hundred miles through the brackish water of a shallow bay where the sea came in, millions of years ago, and drowned the estuary of a river.

Every spring a number of small creatures come up the grassy spill-

478

way and enter Bittern Pond, having made the two-hundred-mile journey from the sea. They are curiously formed, like pieces of slender glass rods shorter than a man's finger. They are young eels, or elvers, that were born in the deep sea. Some of the eels go higher into the hills, but a few remain in the pond, where they live on crayfish and water beetles, and catch frogs and small fishes and grow to adulthood.

Now it was autumn and the end of the year. From the moon's quarter to its half, rains had fallen, and all the hill streams ran in flood. The water of the two feeder streams of the pond was deep and swift and jostled the rocks of the stream beds as it hurried to the sea. The pond was deeply stirred by the inrush of water, which swept through its weed forests and swirled through its crayfish holes and crept up six inches on the trunks of its bordering willows.

The wind had sprung up at dusk. At first it had been a gentle breeze, stroking the surface of the pond to velvet smoothness. At midnight it had grown to a half gale that set all the rushes to swaying wildly and rattled the dead seed heads of the weeds and plowed deep furrows in the surface waters of the pond. The wind roared down from the hills, over forests of oak and beech and hickory and pine. It blew toward the east, toward the sea two hundred miles away.

Anguilla, the eel, nosed into the swift water that raced toward the overflow from the pond. With her keen senses she savored strange tastes and smells in the water. They were the bitter tastes and smells of dead and rainsoaked autumn leaves, the tastes of forest moss and lichen and root-held humus. Such was the water that hurried past the eel, on its way to the sea.

Anguilla had entered Bittern Pond as a finger-long elver ten years before. She had lived in the pond through its summers and autumns and winters and springs, hiding in its weed beds by day and prowling through its waters by night, for like all eels she was a lover of darkness. She knew every crayfish burrow that ran in honeycombing furrows through the mudbank under the hill. She knew her way among the swaying, rubbery stems of spatterdock, where frogs sat on the thick leaves; and she knew where to find the spring peepers clinging to grass blades, bubbling shrilly, where in spring the pond overflowed its grassy northern shore. She could find the banks where the water rats ran and squeaked in play or tusseled in anger, so that sometimes they fell with a splash into the water — easy prey for a lurking eel. She knew the soft mud beds deep in the bottom of the pond, where in winter she could lie buried, secure against the cold — for like all eels she was a lover of warmth.

Now it was autumn again, and the water was chilling to the cold rains shed off the hard backbones of the hills. A strange restiveness was growing in Anguilla the eel. For the first time in her adult life, the food hunger was forgotten. In its place was a strange, new hunger, formless and ill-defined. Its dimly perceived object was a place of warmth and darkness — darker than the blackest night over Bittern Pond. She had known such a place once — in the dim beginnings of life, before memory began. She could not know that the way to it lay beyond the pond outlet over which she had clambered ten years before. But many times at night, as the wind and the rain tore at the surface film of the pond, Anguilla was drawn irresistibly toward the outlet over which the water was spilling on its journey to the sea. When the cocks were crowing in the farmyard over the hill, saluting the third hour of the new day, Anguilla slipped into the channel spilling down to the stream below and followed the moving water.

Even in flood, the hill stream was shallow, and its voice was the noisy voice of a young stream, full of gurglings and tricklings and the sound of water striking stone and of stone rubbing against stone. Anguilla followed the stream, feeling her way by the changing pressure of the swift water currents. She was a creature of night and darkness, and so the black water path neither confused nor frightened her.

In five miles the stream dropped a hundred feet over a rough and boulder-strewn bed. At the end of the fifth mile it slipped between two hills, following along a deep gap made by another and larger stream years before. The hills were clothed with oak and beech and hickory, and the stream ran under their interlacing branches.

At daybreak Anguilla came to a bright, shallow riffle where the stream chattered sharply over gravel and small rubble. The water moved with a sudden acceleration, raining swiftly toward the brink of a ten-foot fall where it spilled over a sheer rock face into a basin below. The rush of water carried Anguilla with it, down the steep, thin slant of white water and into the pool. The basin was deep and still and cool, having been rounded out of the rock by centuries of falling water. Dark water mosses grew on its sides and stoneworts were rooted in its silt, thriving on the lime which they took from the stones and incorporated in their round, brittle stems. Anguilla hid among the stoneworts of the pool, seeking a shelter from light and sun, for now the bright shallows of the stream repelled her.

Before she had lain in the pool for an hour another eel came over the falls and sought the darkness of the deep leaf beds. The second eel had come from higher up in the hills, and her body was lacerated in many places from the rocks of the thin upland streams she had descended. The newcomer was a larger and more powerful eel than Anguilla, for she had spent two more years in fresh water before coming to maturity.

Anguilla, who had been the largest eel in Bittern Pond for more than a year, dived down through the stoneworts at sight of the strange eel. Her passage swayed the stiff, limy stems of the chara and disturbed three water boatmen that were clinging to the chara stems, each holding its position by the grip of a jointed leg, set with rows of bristles. The insects were browsing on the film of desmids and diatoms that coated the stems of the stoneworts. The boatmen were clothed in glistening blankets of air which they had carried down with them when they dived through the surface film, and when the passing of the eel dislodged them from their quiet anchorage they rose like air bubbles, for they were lighter than water.

An insect with a body like a fragment of twig supported by six jointed legs was walking over the floating leaves and skating on the surface of the water, on which it moved as on strong silk. Its feet depressed the film into six dimples, but did not break it, so light was its body. The insect's name meant "a marsh treader," for its kind often lived in the deep sphagnum moss of bogs. The marsh treader was foraging, watching for creatures like mosquito larvae or small crustaceans to move up to the surface from the pool below. When one of the water boatmen suddenly broke through the film at the feet of the marsh treader, the twiglike insect speared it with the sharp stilettos projecting beyond its mouth and sucked the little body dry.

When Anguilla felt the strange eel pushing into the thick mat of dead leaves on the floor of the pool, she moved back into the dark recess behind the waterfall. Above her the steep face of the rock was green with the soft fronds of mosses that grew where their leaves escaped the flow of water, yet were always wet with fine spray from the falls. In spring the midges came there to lay their eggs, spinning them in thin white skeins on the wet rocks. Later when the eggs hatched and the gauzy-winged insects began to emerge from the falls in swarms, they were watched for by bright-eyed little birds who sat on overhanging branches and darted open-mouthed into the clouds of midges. Now the midges were gone, but other small animals lived in the green, water-soaked thickets of the moss. They were the larvae of beetles and soldier flies and crane flies. They were smooth-bodied creatures, lacking the grappling hooks and suckers and the flattened, stream-molded bodies that enabled their relatives to live in the swift currents draining to the brink of the falls over-head or a dozen feet away where the pool spilled its water into the stream bed. Although they lived only a few inches from the veil of water that dropped sheer to the pool, they knew nothing of swift water and its dangers; their peaceful world was of water seeping slow through green forests of moss.

The beginning of the great leaf fall had come with the rains of the past fortnight. Throughout the day, from the roof of the forest to its floor, there was a continuous downdrift of leaves. The leaves fell so silently that the rustle of their settling to the ground was no louder than the thin scratching of the feet of mice and moles mov-ing through their passages in the leaf mold.

All day flights of broad-winged hawks passed down along the ridges of the hills, going south. They moved with scarcely a beat of their outspread wings, for they were riding on the updrafts of air made as the west wind struck the hills and leaped upward to pass over them. The hawks were fall migrants from Canada that had fol-lowed down along the Appalachians for the sake of the air currents that made the flight easier.

At dusk, as the owls began to hoot in the woods, Anguilla left the pool and traveled downstream alone. Soon the stream flowed through rolling farm country. Twice during the night it dropped over small milldams that were white in the thin moonlight. In the stretch be-low the second dam, Anguilla lay for a time under an overhanging bank, where the swift currents were undercutting the heavy, grassy turf. The sharp hiss of the water over the slanting boards of the dam had frightened her. As she lay under the bank the eel that had rested with her in the pool of the waterfall came over the milldam and passed on downstream. Anguilla followed, letting the current take her bumping and jolting over the shallow riffles and gliding swiftly through the deeper stretches. Often she was aware of dark forms mov-ing in the water near her. They were other eels, come from many of the upland feeder creeks of the main stream. Like Anguilla, the other long, slender fishes yielded to the hurrying water and let the currents speed their passage. All of the migrants were roe eels, for only the females ascend far into the fresh-water streams, beyond all reminders of the sea.

The eels were almost the only creatures that were moving in the stream that night. Once, in a copse of beech, the stream made a sharp bend and scoured out a deeper bed. As Anguilla swam into this rounded basin, several frogs dived down from the soft mud bank where they had been sitting half out of water and hid on the bottom close to the bole of a fallen tree. The frogs had been startled by the approach of a furred animal that left prints like those of human feet in the soft mud and whose small black mask and black-ringed tail showed in the faint moonlight. The raccoon lived in a hole high up in one of the beeches near by and often caught frogs and crayfish in the stream. He was not disconcerted by the series of splashes that greeted his approach, for he knew where the foolish frogs would hide. He walked out on the fallen tree and lay down flat on its trunk. He took a firm grip on its bark with the claws of his hind feet and left forepaw. The right paw he dipped into the water, reaching down as far as he could and exploring with busy, sensitive fingers the leaves and mud under the trunk. The frogs tried to burrow deeper into the litter of leaves and sticks and other stream debris. The patient fingers felt into every hole and crevice, pushed away leaves and probed the mud. Soon the coon felt a small, firm body beneath his fingers — felt the sudden movement as the frog tried to escape. The coon's grip tightened and he drew the frog quickly up onto the log. There he killed it, washed it carefully by dipping it into the stream, and ate it. As he was finishing his meal, three small black masks moved into a patch of moonlight at the edge of the stream. They belonged to the coon's mate and their two cubs, who had come down the tree to prowl for their night's food.

From force of habit, the eel thrust her snout inquisitively into the leaf litter under the log, adding to the terror of the frogs, but she did not molest them as she would have done in the pond, for hunger was forgotten in the stronger instinct that made her a part of the moving stream. When Anguilla slipped into the central current of water that swept past the end of the log, the two young coons and their mother had walked out onto the trunk and four black-masked faces were peering into the water, preparing to fish the pool for frogs.

By morning the stream had broadened and deepened. Now it fell silent and mirrored an open woods of sycamore, oak, and dogwood. Passing through the woods, it carried a freight of brightly colored leaves — bright-red, crackling leaves from the oaks, mottled green and yellow leaves from the sycamores, dull-red, leathery leaves from the dogwoods. In the great wind the dogwoods had lost their leaves, but they held their scarlet berries. Yesterday robins had gathered in flocks in the dogwoods, eating the berries; today the robins were gone south and in their place flurries of starlings swept from tree to tree, chattering and rattling and whistling to one another as they stripped the branches of berries. The starlings were in bright new fall plumage, with every breast feather spear-tipped with white.

Anguilla came to a shallow pool formed when an oak had been up-rooted in a great autumn storm ten years before and had fallen across the stream. Oak dam and pool were new in the stream since Anguilla had ascended it as an elver in the spring of that year. Now a great mat of weeds, silt, sticks, dead branches, and other debris was packed around the massive trunk, plastering all the crevices, so that the water was backed up into a pool two feet deep. During the period of the full moon the eels lay in the oak-dam pool, fearing to travel in the moon-white water of the stream almost as much as they feared the sunlight.

In the mud of the pool were many burrowing, wormlike larvae — the young of lamprey eels. They were not true eels, but fishlike crea-tures whose skeleton was gristle instead of bone, with round, tooth-studded mouths that were always open because there were no jaws. Some of the young lampreys had hatched from eggs spawned in the pool as much as four years before and had spent most of their life buried in the mud flats of the shallow stream, blind and toothless. These older larvae, grown nearly twice the length of a man's finger, had this fall been transformed into the adult shape, and for the first time they had eyes to see the water world in which they lived. Now, like the true eels, they felt in the gentle flow of water to the sea something that urged them to follow, to descend to salt water for an interval of sea life. There they would prey semiparasitically on cod, haddock, mackerel, salmon, and many other fishes and in time would return to the river, like their parents, to spawn and die. A few of the young lampreys slipped away over the log dam every day, and on a cloudy night, when rain had fallen and white mist lay in the stream valley, the eels followed.

The next night the eels came to a place where the stream di-verged around an island grown thickly with willows. The eels fol-lowed the south channel around the island, where there were broad mud flats. The island had been formed over centuries of time as the stream had dropped part of its silt load before it joined the main river. Grass seeds had taken root; seeds of trees had been brought by the water and by birds; willow shoots had sprung from broken twigs and branches carried down in flood waters; an island had been born.

The water of the main river was gray with approaching day when the eels entered it. The river channel was twelve feet deep and its water was turbid because of the inpouring of many tributary streams swollen with autumn rains. The eels did not fear the gloomy chan-nel water by day as they had feared the bright shallows of the hill streams, and so this day they did not rest but pushed on downstream. There were many other eels in the river — migrants from other tributaries. With the increase in their numbers the excitement of the eels grew, and as the days passed they rested less often, pressing on downstream with fevered haste.

As the river widened and deepened, a strange taste came into the water. It was a slightly bitter taste, and at certain hours of the day and night it grew stronger in the water that the eels drew into their mouths and passed over their gills. With the bitter taste came unfamiliar movements of the water — a period of pressure against the downflow of the river currents followed by slow release and then swift acceleration of the current. Now groups of slender posts stood at intervals in the river, marking out funnel shapes from which straight rows of posts ran slanting toward the shore. Blackened netting, coated with slimy algae, was run from post to post and showed several feet above the water. Gulls were often sitting on the pound nets, waiting for men to come and fish the nets so that they could pick up any fish that might be thrown away or lost. The posts were coated with barnacles and with small oysters, for now there was enough salt in the water for these shellfish to grow.

Sometimes the sandspits of the river were dotted with small shore birds standing at rest or probing at the water's edge for snails, small shrimps, worms, or other food. The shore birds were of the sea's edge, and their presence in numbers hinted of the nearness of the sea.

The strange, bitter taste grew in the water and the pulse of the tides beat stronger. On one of the ebb tides a group of small eels — none more than two feet long — came out of a brackish-water marsh and joined the migrants from the hill streams. They were males, who had never ascended the rivers but had remained within the zone of tides and brackish water.

In all of the migrants striking changes in appearance were taking place. Gradually the river garb of olive brown was changing to a glistening black, with underparts of silver. These were the colors worn only by mature eels about to undertake a far sea journey. Their bodies were firm and rounded with fat — stored energy that would be needed before the journey's end. Already in many of the migrants the snouts were becoming higher and more compressed, as though from some sharpening of the sense of smell. Their eyes were enlarged to twice their normal size, perhaps in preparation for a descent along darkening sea lanes.

Where the river broadened out to its estuary, it flowed past a high clay cliff on its southern bank. Buried in the cliff were thousands of teeth of ancient sharks, vertebrae of whales, and shells of mollusks that had been dead when the first eels had come in from the sea, eons ago. The teeth, bones, and shells were relics of the time when a warm sea had overlain all the coastal plain and the hard remains of its creatures had settled down into its bottom oozes. Buried millions of years in darkness, they were washed out of the clay by every storm to lie exposed, warmed by sunshine and bathed by rain.

The eels spent a week descending the bay, hurrying through water of increasing saltiness. The currents moved with a rhythm that was of neither river nor sea, being governed by eddies at the mouths of the many rivers that emptied into the bay and by holes in the muddy bottom thirty or forty feet beneath. The ebb tides ran stronger than the floods, because the strong outflow of the rivers resisted the press of water from the sea.

At last Anguilla neared the mouth of the bay. With her were thousands of eels, come down, like the water that brought them, from all the hills and uplands of thousands of square miles, from every stream and river that drained away to the sea by the bay. The eels followed a deep channel that hugged the eastern shore of the bay and came to where the land passed into a great salt marsh. Beyond the marsh, and between it and the sea, was a vast shallow arm of the bay, studded with islands of green marsh grass. The eels gathered in the marsh, waiting for the moment when they should pass to the sea.

The next night a strong southeast wind blew in from the sea, and when the tide began to rise the wind was behind the water, pushing it into the bay and out into the marshes. That night the bitterness of brine was tasted by fish, birds, crabs, shellfish, and all the other water creatures of the marsh. The eels lay deep under water, savoring the salt that grew stronger hour by hour as the wind-driven wall of sea water advanced into the bay. The salt was of the sea. The eels were ready for the sea — for the deep sea and all it held for them. Their years of river life were ended.

The wind was stronger than the forces of moon and sun, and, when the tide turned an hour after midnight, the salt water continued to pile up in the marsh, being blown upstream in a deep surface layer while the underlying water ebbed to the sea.

Soon after the tide turn, the seaward movement of the eels began. In the large and strange rhythms of a great water which each had known in the beginning of life, but each had long since forgotten, the eels at first moved hesitantly in the ebbing tide. The water carried them through an inlet between two islands. It took them under a fleet of oyster boats riding at anchor, waiting for daybreak. When morning came, the eels would be far away. It carried them past leaning spar buoys that marked the inlet channel and past several whistle and bell buoys anchored on shoals of sand or rock. The tide took them close under the lee shore of the larger island, from which a lighthouse flashed a long beam of light toward the sea.

From a sandy spit of the island came the cries of shore birds that were feeding in darkness on the ebb tide. Cry of shore bird and crash of surf were the sounds of the edge of land — the edge of the sea.

The eels struggled through the line of breakers, where foam seething over black water caught the gleam of the lighthouse beacon and frothed whitely. Once beyond the wind-driven breakers they found the sea gentler, and as they followed out over the shelving sand they sank into deep water, unrocked by violence of wind and wave.

As long as the tide ebbed, eels were leaving the marshes and running out to sea. Thousands passed the lighthouse that night, on the first lap of a far sea journey — all the silver eels, in fact, that the marsh contained. And as they passed through the surf and out to sea, so also they passed from human sight and almost from human knowledge.

The record of the eels' journey to their spawning place is hidden in the deep sea. No one can trace the path of the eels that left the salt marsh at the mouth of the bay on that November night when wind and tide brought them the feeling of warm ocean water — how they passed from the bay to the deep Atlantic basin that lies south of Bermuda and east of Florida half a thousand miles. Nor is there a clearer record of the journey of those other eel hordes that in autumn passed to the sea from almost every river and stream of the whole Atlantic Coast from Greenland to Central America.

No one knows how the eels traveled to their common destination. Probably they shunned the pale-green surface waters, chilled by wintry winds and bright as the hill streams they had feared to descend by day. Perhaps they traveled instead at mid-depths or followed the contours of the gently sloping continental shelf, descending the drowned valleys of their native rivers that had cut channels across the coastal plain in sunshine millions of years ago. But somehow they came to the continent's edge, where the muddy slopes of the sea's wall fell away steeply, and so they passed to the deepest abyss of the Atlantic. There the young were to be born of the darkness of the deep sea and the old eels were to die and become sea again.

In early February billions of specks of protoplasm floated in darkness, suspended far below the surface of the sea. They were the newly hatched larvae — the only testament that remained of the parent eels. The young eels first knew life in the transition zone between the surface sea and the abyss. A thousand feet of water lay above them, straining out the rays of the sun. Only the longest and strongest of the rays filtered down to the level where the eels drifted in the sea — a cold and sterile residue of blue and ultraviolet, shorn of all its warmth of reds and yellows and greens. For a twentieth part of the day the blackness was displaced by a strange light of a vivid and unearthly blue that came stealing down from above. But only the straight, long rays of the sun when it passed the zenith had power to dispel the blackness, and the deep sea's hour of dawn light was merged in its hour of twilight. Quickly the blue light faded away, and the eels lived again in the long night that was only less black than the abyss, where the night had no end.

At first the young eels knew little of the strange world into which they had come, but lived passively in its waters. They sought no food, sustaining their flattened, leaf-shaped bodies on the residue of embryonic tissue, and so they were the foes of none of their neighbors. They drifted without effort, buoyed by their leafy form and by the balance between the density of their own tissues and that of the sea water. Their small bodies were colorless as crystal. Even the blood that ran in its channels, pumped by hearts of infinitesimal size, was unpigmented; only the eyes, small as black pinpricks, showed color. By their transparency the young eels were better fitted to live in this twilight zone of the sea, where safety from hungry foragers was to be found only in blending with the surroundings.

Billions of young eels — billions of pairs of black, pinprick eyes peering into the strange sea world that overlay the abyss. Before the eyes of the eels, clouds of copepods vibrated in their ceaseless dance of life, their crystal bodies catching the light like dust motes when the blue gleam came down from above. Clear bells pulsated in the water, fragile jellyfish adjusted to life where five hundred pounds of water pressed on every square inch of surface. Fleeing before the descending light, shoals of pteropods, or winged snails, swept down from above before the eyes of the watching eels, their forms glistening with reflected light like a rain of strangely shaped hailstones — daggers and spirals and cones of glassy clearness. Shrimps loomed up — pale ghosts in the dim light. Sometimes the shrimps were pursued by pale fishes, round of mouth and flabby of flesh, with rows of light organs set like jewels on their gray flanks. Then the shrimps often expelled jets of luminous fluid that turned to a fiery cloud to blind and confuse their enemies. Most of the fishes seen by the eels wore silver armor, for silver is the prevailing color or badge of those waters that lie at the end of the sun's rays. Such were the small dragonfish, long and slender of form, with fangs glistening in their opened mouths as they roamed through the water in an endless pursuit of prey. Strangest of all were the fishes, half as long as a man's finger and clothed in a leathery skin, that shone with turquoise and amethyst lights and gleamed like quicksilver over their flanks. Their bodies were thin from side to side and tapered to sharp edges. When enemies looked down from above, they saw nothing, for the backs of the hatchetfish were a bluish black that was invisible in the black sea. When sea hunters looked up from below, they were confused and could not distinguish their prey with certainty, for the mirrorlike flanks of the hatchetfish reflected the blueness of the water and their outlines were lost in a shimmer of light.

The young eels lived in one layer or tier of a whole series of horizontal communities that lay one below the other, from the nereid worms that spun their strands of silk from frond to frond of the brown sargassum weed floating on the surface to the sea spiders and prawns that crawled precariously over the deep and yielding oozes of the floor of the abyss.

Above the eels was the sunlight world where plants grew, and small fishes shone green and azure in the sun, and blue and crystal jellyfish moved at the surface.

Then came the twilight zone where fishes were opalescent or silver, and red prawns shed eggs of a bright orange color, and round-mouthed fishes were pale, and the first light organs twinkled in the gloom.

Then came the first black layer, where none wore silvery sheen or opalescent luster, but all were as drab as the water in which they lived, wearing monotones of reds and browns and blacks whereby they might fade into the surrounding obscurity and defer the moment of death in the jaws of an enemy. Here the red prawns shed deep-red eggs, and the round-mouthed fishes were black, and many creatures wore luminous torches or a multitude of small lights arranged in rows or patterns that they might recognize friend or enemy.

Below them lay the abyss, the primeval bed of the sea, the deepest of all the Atlantic. The abyss is a place where change comes slow, where the passing of the years has no meaning, nor the swift succession of the seasons. The sun has no power in those depths, and so their blackness is a blackness without end, or beginning, or degree. The wind is unknown there. No pull of moon and sun can move that weight of inert water to surge and lapse in the rhythm of the tides. No beating of tropical sun on the surface miles above can lessen the bleak iciness of those abyssal waters that varies little through summer or winter, through the years that melt into centuries, and the centuries into ages of geologic time.

Down beneath mile after mile of water — more than four miles in all — lay the sea bottom, covered with a soft, deep ooze that had been accumulating there through eons upon eons of time. These greatest depths of the Atlantic are carpeted with red clay, a pumice-like deposit hurled out of the earth from time to time by submarine volcanoes. Mingled with the pumice are spherules of iron and nickel that had their origin on some far-off sun and once rushed millions of miles through interstellar space, to perish in the earth's atmosphere and find their grave in the deep sea. Far up on the sides of the great bowl of the Atlantic the bottom oozes are thick with the skeletal remains of minute sea creatures of the surface waters — the shells of starry Foraminifera and the limy remains of algae and corals, the flintlike skeletons of Radiolaria and the frustules of diatoms. But long before such delicate structures reach this deepest bed of the abyss, they are dissolved and made one with the sea. Almost the only organic remains that have not passed into solution before they reach these cold and silent deeps are the ear bones of whales and the teeth of sharks. Here in the red clay, in the darkness and stillness, lies all that remains of ancient races of sharks that lived, perhaps, before there were whales in the sea; before the giant ferns flourished on the earth or ever the coal measures were laid down. All of the living flesh of these sharks was returned to the sea millions of years before, to be used over and over again in the fashioning of other creatures, but here and there a tooth still lies in the red-clay ooze of the deep sea, coated with a deposit of iron from a distant sun.

The abyss south of Bermuda is a meeting place for the eels of the western and eastern Atlantic. There are other great deeps in the ocean between Europe and America — chasms sunk between the mountain ranges of the sea's floor — but only this one is both deep enough and warm enough to provide the conditions which the eels need for the act of spawning. So once a year the mature eels of Europe set out across the ocean on a journey of three to four thousand miles; and once a year the mature eels of eastern America go out as though to meet them. In the westernmost part of the drifting sea of sargassum weed some of them meet and intermingle — those that travel farthest west from Europe and farthest east from America. So in the central part of the vast spawning grounds of the eels, the eggs and young of two species float side by side in the water. They are so alike in appearance that only by counting with infinite care the vertebrae that make up their backbones and the plates of muscle that flank their spines can they be distinguished. Yet some, toward the end of their period of larval life, seek the coast of America and others the coast of Europe, and none ever stray to the wrong continent.

As the months of the year passed, one by one, the young eels grew, lengthening and broadening. As they grew and the tissues of their bodies changed in density, they drifted into light. Upward passage through space in the sea was like passage through time in the Arctic world in spring, with the hours of sunlight increasing day by day. Little by little the blue haze of midday lengthened and the long nights grew shorter. Soon the eels came to the level where the first green rays, filtering down from above, warmed the blue light. So they passed into the zone of vegetation and found their first food.

The plants that received enough energy for their life processes from the sea-strained residue of sunlight were microscopic, floating spheres. On the cells of ancient brown algae the young eels first nourished their glass-clear bodies — plants of a race that had lived for untold millions of years before the first eel, or the first backboned animal of any kind, moved in the earth's seas. Through all the intervening eons of time, while group after group of living things had risen up and died away, these lime-bearing algae had continued to live in the sea, forming their small protective shields of lime that were unchanged in shape and form from those of their earliest ancestors.

Not only the eels browsed on the algae. In this blue-green zone, the sea was clouded with copepods and other plankton foraging on the drifting plants, and dotted with the swarms of shrimplike animals that fed on the copepods, and lit by the twinkling silver flashes of small fishes that pursued the shrimps. The young eels themselves were preyed upon by hungry crustaceans, squids, jellyfish, and biting worms, and by many fishes who roved open-mouthed through the water, straining food through mouth and gill raker.

By midsummer the young eels were an inch long. They were the shape of willow leaves — a perfect shape for drifters in the currents. Now they had risen to the surface layers of the sea, where the black dots of their eyes could be seen by enemies in the bright-green water. They felt the lift and roll of waves; they knew the dazzling brightness of the midday sun in the pure waters of the open ocean. Sometimes they moved in the midst of floating forests of sargassum weed, per-

491

haps taking shelter beneath the nests of flying fishes or, in the open spaces, hiding in the shadow of the blue sail or float of a Portuguese man-of-war.

In these surface waters were moving currents, and where the currents flowed the young eels were carried. All alike were swept into the moving vortex of the north Atlantic drift — the young of the eels from Europe and the young of the eels from America. Their caravans flowed through the sea like a great river, fed from the waters south of Bermuda and composed of young eels in numbers beyond enumeration. In at least a part of this living river, the two kinds or species of eels traveled side by side, but now they could be distinguished with ease, for the young of the American eels were nearly twice as large as their companions.

The ocean currents swept in their great circle, moving from south through west and north. Summer drew to its end. All the sea's crops had been sown and harvested, one by one — the spring crop of diatoms, the swarms of plankton animals that grew and multiplied on the abundant plants, the young of myriad fishes that fed on the plankton herds. Now the lull of autumn was upon the sea.

The young eels were far from their first home. Gradually the caravan began to diverge into two columns, one swinging to the west, one to the east. Before this time there must have been some subtle change in the responses of the faster-growing group of eels — something that led them more and more to the west of the broad river of moving surface water. As the time approached for them to lose the leaflike form of the larva and become rounded and sinuous like their parents, the impulse to seek fresher, shallowing waters grew. Now they found the latent power of unused muscles, and against the urging of wind and current they moved shoreward. Under the blind but powerful drive of instinct, every activity of their small and glassy bodies was directed unconsciously toward the attainment of a goal unknown in their own experience — something stamped so deeply upon the memory of their race that each of them turned without hesitation toward the coast from which their parents had come.

A few eastern-Atlantic eels still drifted in the midst of the western-Atlantic larvae, but none among them felt the impulse to leave the deep sea. All their body processes of growth and development were geared to a slower rate. Not for two more years would they be ready for the change to the eel-like form and the transition to fresh water. So they drifted passively in the currents.

To the east, midway across the Atlantic, was another little band of leaflike travelers — eels spawned a year before. Farther to the east, in the latitude of the coastal banks of Europe, was still another host of drifting eel larvae, these yet a year older and grown to their full length. And that very season a fourth group of young eels had reached the end of their stupendous journey and was entering the bays and inlets and ascending the rivers of Europe.

For the American eels the journey was shorter. By midwinter their hordes were moving in across the continental shelf, approaching the coast. Although the sea was chilled by the icy winds that moved over it, and by the remoteness of the sun, the migrating eels remained in the surface waters, no longer needing the tropical warmth of the sea in which they had been born.

As the young moved shoreward, there passed beneath them another host of eels, another generation come to maturity and clothed in the black and silver splendor of eels returning to their first home. They must have passed without recognition — these two generations of eels — one on the threshold of a new life; the other about to lose itself in the darkness of the deep sea.

The water grew shallower beneath them as they neared the shore. The young eels took on their new form, in which they would ascend the rivers. Their leafy bodies became more compact by a shrinkage in length as well as in depth, so that the flattened leaf became a thickened cylinder. The large teeth of larval life were shed, and the heads became more-rounded. A scattering of small pigment-carrying cells appeared along the backbone, but for the most part the young eels were still as transparent as glass. In this stage they were called "glass eels," or elvers.

Now they waited in the gray March sea, creatures of the deep sea, ready to invade the land. They waited off the sloughs and bayous and the wild-rice fields of the Gulf Coast, off the South Atlantic inlets, ready to run into the sounds and green marshes that edged the river estuaries. They waited off the ice-choked northern rivers that came down with a surge and a rush of spring floods and thrust long arms of fresh water into the sea, so that the eels tasted the strange water taste and moved in excitement toward it. By the hundreds of thousands they waited off the mouth of the bay from which, little more than a year before, Anguilla and her companions had set out for the deep sea, blindly obeying a racial purpose which was now fulfilled in the return of the young.

The eels were nearing a point of land marked by the slim white shaft of a lighthouse. The sea ducks could see it — the piebald oldsquaw ducks — when they circled high above the sea on their return every afternoon from inshore feeding grounds, coming down at dusk to the dark water with a great rush and a roar of wings. The whistling swans saw it, too, painted by the sunrise on the green sea beneath them as their flocks swept northward in the spring migration. The leader swans blew a triple note at the sight, for the point of land marked the nearness of the first stop on the swans' long flight from the Carolina Sounds to the great barrens of the Arctic.

The tides were running high with the fullness of the moon. On the ebb tides the taste of fresh water came strongly to the fish that lay at sea, off the mouth of the bay, for all the rivers were in flood.

In the moon's light the young eels saw the water fill with many fish, large and full-bellied and silvery of scale. The fish were shad returned from their feeding grounds in the sea, waiting for the ice to come out of the bay that they might ascend its rivers to spawn. Schools of croakers lay on the bottom, and the roll of their drums vibrated in the water. The croakers, with sea trout and spots, had moved in from their offshore wintering place, seeking the feeding grounds of the bay. Other fish came up into the tide flow and lay with heads to the currents, waiting to snap up the small sea animals that the swiftly moving water had dislodged, but these were bass who were of the sea and would not ascend the rivers.

As the moon waned and the surge of the tides grew less, the elvers pressed forward toward the mouth of the bay. Soon a night would come, after most of the snow had melted and run as water to the sea, when the moon's light and the tide's press would be feeble and a warm rain would fall, mist-laden and bittersweet with the scent of opening buds. Then the elvers would pour into the bay and, traveling up its shores, would find its rivers.

Some would linger in the river estuaries, brackish with the taste of the sea. These were the young male eels, who were repelled by the strangeness of fresh water. But the females would press on, swimming up against the currents of the rivers. They would move swiftly and by night as their mothers had come down the rivers. Their columns, miles in length, would wind up along the shallows of river and stream, each elver pressing close to the tail of the next before it, the whole like a serpent of monstrous length. No hardship and no obstacle would deter them. They would be preyed upon by hungry fishes — trout, bass, pickerel, and even by older eels; by rats hunting the edge of the water; and by gulls, herons, kingfishers, crows, grebes, and loons. They would swarm up waterfalls and clamber over moss-grown rocks, wet with spray; they would squirm up the spillways of dams. Some would go on for hundreds of miles — creatures of the deep sea spreading over all the land where the sea itself had lain many times before.

And as the eels lay offshore in the March sea, waiting for the time when they should enter the waters of the land, the sea, too, lay restless, awaiting the time when once more it should encroach upon the coastal plain, and creep up the sides of the foothills, and lap at the bases of the mountain ranges. As the waiting of the eels off the mouth of the bay was only an interlude in a long life filled with constant change, so the relation of sea and coast and mountain ranges was that of a moment in geologic time. For once more the mountains would be worn away by the endless erosion of water and carried in silt to the sea, and once more all the coast would be water again, and the places of its cities and towns would belong to the sea.

SELECTED BIOGRAPHICAL MATERIAL

General

Biology and Its Makers. William A. Locy. Henry Holt and Company, New York. 1908.

The New Calendar of Great Men: Biographies of the 559 Worthies of All Ages and Nations in the Positivist Calendar of Auguste Comte. Edited by Frederic Harrison, S. H. Swinny and F. S. Marvin. Macmillan and Company, London. 1920.

Introduction to the History of Science. George Sarton. Published for the Carnegie Institution of Washington by Williams and Wilkins. 1927–31.

The History of Biology. Erik Nordenskiöld. Translated from the Swedish by Leonard Bucknall Eyre. Alfred A. Knopf, New York and London. 1929.

American Naturists. Henry Chester Tracy. E. P. Dutton and Company, New York. 1930.

The Story of Living Things. Charles Singer. Harper and Brothers, New York and London. 1931.

Green Laurels: The Lives and Achievements of the Great Naturalists. Donald Culross Peattie. Simon and Schuster, New York. 1936.

The Great Naturalists Explore South America. Paul Russell Cutright. The Macmillan Company, New York. 1940.

A Treasury of Science. Edited by Harlow Shapley, Samuel Rapport and Helen Wright, with an introduction by Dr. Shapley. Harper and Brothers, New York and London. 1943.

Individual

Agassiz: *Louis Agassiz; His Life and Correspondence.* Edited by Elizabeth Cary Agassiz. Houghton, Mifflin and Company, Boston and New York. 1900.

Akeley: *Carl Akeley's Africa.* Mary L. Jobe Akeley. Dodd, Mead and Company, New York. 1929.

Aristotle: *Memoir of Aristotle.* Andrew Crichton, in "The Naturalist's Library," conducted by Sir William Jardine; Ornithology, Vol. III, *Gallinaceous Birds.* W. H. Lizars, Edinburgh. 1834.

Audubon: *Audubon the Naturalist: A History of His Life and Time.* Francis Hobart Herrick. D. Appleton and Company, New York and London. 1917.

BARTRAM: *John and William Bartram: Botanists and Explorers.* Ernest Earnest. University of Pennsylvania Press. 1940.

BURROUGHS: *The Life and Letters of John Burroughs.* Clara Barries. Houghton, Mifflin Company, Boston and New York. 1925.

CHAPMAN: *Autobiography of a Bird-Lover.* Frank M. Chapman. D. Appleton-Century Company, New York and London. 1933.

DARWIN: *The Life and Letters of Charles Darwin.* Edited by his son, Frances Darwin. D. Appleton and Company, New York and London, 1887.

FABRE: *The Life of Jean Henri Fabre, the Entomologist.* Abbé Augustin. Translated by Bernard Miall. Dodd, Mead and Company, New York. 1921.

FREDERICK II: Introductory chapters in *The Art of Falconry of Frederick II of Hohenstaufen.* Translated and edited by Casey A. Wood and F. Marjorie Fyfe. Stanford University Press, Stanford, California; Humphrey Milford, Oxford University Press, London. 1943.

GESNER: *Memoir of Gesner.* In "The Naturalist's Library," conducted by Sir William Jardine; Mammalia, Vol. XII, *Horses,* by Charles Hamilton Smith. W. H. Lizars, Edinburgh. 1841.

HUDSON: *The Genius of W. H. Hudson.* Edward Garnett. (The introduction to Hudson's *A Hind in Richmond Park.*) E. P. Dutton and Company, New York. 1923.

HUMBOLDT: *Memoir of Baron Alexander von Humboldt.* In "The Naturalist's Library," conducted by Sir William Jardine; Ichthyology, Vol. VI, *British Fishes,* Vol. II, by Robert Hamilton. W. H. Lizars, Edinburgh. 1843.

HUXLEY, T. H.: *Life and Letters of Thomas Henry Huxley.* Leonard Huxley. D. Appleton and Company, New York. 1901.

LEEUWENHOEK: *Antony van Leeuwenhoek and His "Little Animals."* Clifford Dobell. Harcourt, Brace and Company, New York. 1932.

LINNÆUS: *Linnæus: The Story of His Life.* Benjamin Daydon Jackson. H. F. and G. Witherby, London. 1923.

MUIR: *The Life and Letters of John Muir.* W. F. Badè, editor of Muir's *Writings.* Houghton, Mifflin Company, Boston. 1923–4.

OSBORN: *Biographical Memoir of Henry Fairfield Osborn.* William King Gregory. National Academy of Sciences, Washington. 1938.

PEATTIE: *The Road of a Naturalist.* (An autobiography.) Donald Culross Peattie. Houghton, Mifflin Company, Boston. 1941.

PLINY: *Memoir of Pliny.* Andrew Crichton in "The Naturalist's Library," conducted by Sir William Jardine; Ornithology, Vol. V,

Gallinaceous Birds; Part III, Pigeons, by Prideaux John Selby. W. H. Lizars, Edinburgh. 1835.

RÉAUMUR: *The Life and Work of Réaumur.* William Morton Wheeler in *The Natural History of Ants,* from an unpublished manuscript by René Antoine Ferchault de Réaumur. Translated and Annotated by Dr. Wheeler. Alfred A. Knopf, New York and London, 1926.

ROOSEVELT: *Theodore Roosevelt: An Autobiography.* The Macmillan Company, New York. 1913.

SETON: *Trail of an Artist-Naturalist: An Autobiography.* Ernest Thompson Seton. Charles Scribner's Sons, New York. 1941.

THOREAU: *Thoreau.* Henry Seidel Canby. Houghton, Mifflin Company, Boston. 1939.

WALLACE: *My Life: A Record of Events and Opinions.* Alfred Russel Wallace. Dodd, Mead and Company, New York. 1905.

WATERTON: *Life of Charles Waterton.* Norman Moore. (In *Essays on Natural History* by Charles Waterton, edited, with a Life of the Author, by Norman Moore.) Frederick Warne and Company, London; Scribner, Welford and Company, New York. 1871.

WHEELER: *William Morton Wheeler.* L. J. Henderson, Thomas Barbour, F. M. Carpenter, and Hans Zinsser. In *Essays in Philosophical Biology,* by William Morton Wheeler. Harvard University Press, Cambridge. 1939.

WHITE: *The Life and Letters of Gilbert White of Selbourne.* Rashleigh Holt-White. E. P. Dutton and Company, New York. 1901.